Biologically
Inspired Physics

NATO ASI Series

Advanced Science Institutes Series

A series presenting the results of activities sponsored by the NATO Science Committee, which aims at the dissemination of advanced scientific and technological knowledge, with a view to strengthening links between scientific communities.

The series is published by an international board of publishers in conjunction with the NATO Scientific Affairs Division

A	Life Sciences	Plenum Publishing Corporation
B	Physics	New York and London
C	Mathematical and Physical Sciences	Kluwer Academic Publishers
D	Behavioral and Social Sciences	Dordrecht, Boston, and London
E	Applied Sciences	
F	Computer and Systems Sciences	Springer-Verlag
G	Ecological Sciences	Berlin, Heidelberg, New York, London,
H	Cell Biology	Paris, Tokyo, Hong Kong, and Barcelona
I	Global Environmental Change	

Recent Volumes in this Series

Series B: Physics

Biologically
Inspired Physics

Edited by
L. Peliti
University of Naples
Naples, Italy

Plenum Press
New York and London
Published in cooperation with NATO Scientific Affairs Division

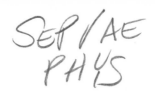

Proceedings of a NATO Advanced Research Workshop on
Biologically Inspired Physics,
held September 3–13, 1990,
in Cargese, France

Library of Congress Cataloging-in-Publication Data

NATO Advanced Research Workshop on Biologically Inspired Physics (1990
  : Cargèse, France)
   Biologically inspired physics / edited by L. Peliti.
       p.   cm. -- (NATO ASI series. Series B, Physics ; v. 263)
    "Proceedings of a NATO Advanced Research Workshop on Biologically
 Inspired Physics, held September 3-13, 1990, in Cargèse, France"-
 -T.p. verso.
    "Published in cooperation with NATO Scientific Affairs Division."
    Includes bibliographical references and index.
    ISBN 0-306-44000-8
    1. Biophysics--Congresses.   I. Peliti, L. (Luca), 1948-   .
 II. North Atlantic Treaty Organization. Scientific Affairs
 Division.   III. Title.   IV. Series.
 QH505.N33 1990
 574.19'1--dc20   91-20198
                                                          CIP

ISBN 0-306-44000-8

© 1991 Plenum Press, New York
A Division of Plenum Publishing Corporation
233 Spring Street, New York, N.Y. 10013

Printed in the United States of America

SPECIAL PROGRAM ON CHAOS, ORDER, AND PATTERNS

This book contains the proceedings of a NATO Advanced Research Workshop held within the program of activities of the NATO Special Program on Chaos, Order, and Patterns.

PREFACE

The workshop "Biologically Inspired Physics" was organized, with the support of the NATO Scientific Affairs Division and the Directorate-General for Science, Research and Development of the Commission of the European Communities, in order to review some subjects of physics of condensed matter which are inspired by biological problems or deal with biological systems, but which address physical questions.

The main topics discussed in the meeting were:

1. Macromolecules: In particular, proteins and nucleic acids. Special emphasis was placed on modelling protein folding, where analogies with disordered systems in condensed matter (glasses, spin glasses) were suggested. It is not clear at this point whether such analogies will help in solving the folding problem. Interesting problems in nucleic acids (in particular DNA) deal with the dynamics of semiflexible chains with torsion and the relationship between topology and local structure. They arise from such biological problems as DNA packing or supercoiling.

2. Membranes: This field has witnessed recent progress in the understanding of the statistical mechanics of fluctuating flexible sheets, such as lipid bilayers. It appears that one is close to understanding shape fluctuations in red blood cells on a molecular basis. Open problems arise from phenomena such as budding or membrane fusion. Experiments on model systems, such as vesicle systems or artificial lipids, have great potential. Phenomena occurring inside the membrane (protein diffusion, ionic pumps) were only discussed briefly.

3. Cellular Structures, other than membranes. Emphasis was placed on cytoskeleton and motile structures of cells. They appear to be amenable to physical treatment, at least in their most basic aspects: elasticity of actin or spectrin cytoskeleton, functioning of the actin/myosine motor system, dynamics of microtubule structures. The relationship with points 1. and 2. are close and instructive.

4. Networks: Although neural network theory was biologically motivated, it appears to have developed into an independent, self-sustaining subject. As a consequence, much emphasis was placed on other interpretations of network concepts, in particular for understanding the immune response. It has also been suggested that network concepts may be helpful in understanding developmental mechanisms at the cellular level.

5. Morphogenesis: There have been important advances in the physical modelling of biological morphogenesis, on the one hand by reaction-diffusion mechanisms, on the other by understanding cellular motion and cytomechanics. It turns out that these advances have been made possible by close interactions between physicists and biologists. It is likely that connections with points 3. and 4. will become more and more relevant in the future.

6. Dynamical Systems: Stress was laid on evolutionary models, considered as dynamical systems of a special kind. The gap between the theory of evolving molecular populations and autocatalytic sets on the one hand, and the experimental realization of evolving molecular systems on the other, is yet to be bridged. In particular, realistic descriptions of the relevant fitness landscapes are still lacking. However, experimental and theoretical investigations are now being conducted in parallel.

In conclusion, we see the following picture emerging:

(i) Connections between physics and biology are good at the level of molecular methods. Physical methods of investigation are successfully applied in many problems, and new applications are constantly emerging.

(ii) Theoretical methods find application with some difficulty, due to the high complexity of biological systems. What can be done is an attempt to introduce new concepts – arising from condensed matter physics – to complex hypermolecular systems.

(iii) However, this task of great potential usefulness is made more difficult by the intrinsic difficulty of the problems concerned, as well as by the present organization of science. Initiatives like the present one do help in the exchange of concepts and ideas and so help to prevent the parcellization of the physics of complex systems in subdomains, often divorced from the consideration of real systems.

It is a pleasure to thank NATO and the Commission of the European Communities for having made this meeting possible, and for the interest manifested in the initiative. We thank Erich Sackmann for having accepted to take part in the organization of the Meeting and for all the help he dispensed. We are also grateful to the staff of the Institut d'Etudes Scientifiques de Cargèse, and in particular to Marie-France Hanseler, for having contributed to making the workshop as pleasant as it has been fruitful. We thank Angela Di Silvestro and Marcella Mastrofini for the invaluable help they provided in editing the resultant proceedings volume.

Luca Peliti
Stanislas Leibler

CONTENTS

MACROMOLECULES

MEMBRANES

CELLULAR STRUCTURES

NETWORKS

MORPHOGENESIS

DYNAMICAL SYSTEMS

PHYSICS FROM PROTEINS

Hans Frauenfelder, Kelvin Chu, and Robert Philipp

Department of Physics, University of Illinois at Urbana-Champaign
1110 W. Green St., Urbana, Illinois 61801

1. THE GOAL

The title of the workshop, "Biologically Inspired Physics" also heralds the goal of this contribution: Can we learn "new" physics by studying biological systems? In *biophysics*, physical tools are used to explore the structure and function of biological systems, ranging from biomolecules to the brain.

In *biological physics*, we hope to find and study new physical concepts and laws, and to obtain new insight into already known laws. Biophysics is already a highly developed field and it is clear that physical tools and physicists have had a very large influence on biology. Without X-ray and neutron diffraction, NMR, Raman spectroscopy, and electron microscopy, biology would not be at the present level. Biological physics is far behind, but interesting physics is even emerging from biomolecules, the lowest rung of the biological ladder.

2. PROTEINS

In the present paper, we are concerned exclusively with proteins [12,17,39]. Proteins are biological machines designed for specific functions, such as the transport and storage of matter and charge. They are long polymer chains, constructed from twenty different building blocks, the amino acids. Amino acids share the same basic design, differing only in the residue group, the side chain attached to the α carbon. The side chain determines the shape, charge, and electronic reactivity of the amino acid. In the proper surrounding, the polypeptide chain folds into the working structure, often with a prosthetic group.

Protein structure is organized into four levels. The primary structure of a protein consists of the linear polypeptide chain of amino acids. The information for the primary structure is coded by and transported to the ribosome for assembly by nucleic acids. The secondary structure of the polypeptide chain is caused by the steric relationships between amino acids in the chain. This gives rise to periodic structures, such as the α helix, the β pleated sheet, and the collagen helix. Tertiary structure describes the three-dimensional structure of the protein. Proteins that are com-

Biologically Inspired Physics, Edited by L. Peliti
Plenum Press, New York, 1991

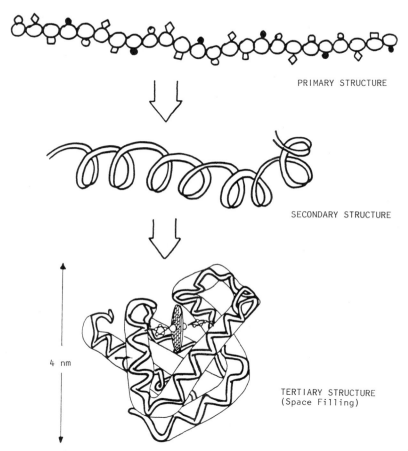

PRIMARY STRUCTURE

SECONDARY STRUCTURE

4 nm

TERTIARY STRUCTURE
(Space Filling)

Fig. 1. Protein Folding. Primary, secondary and tertiary structure.

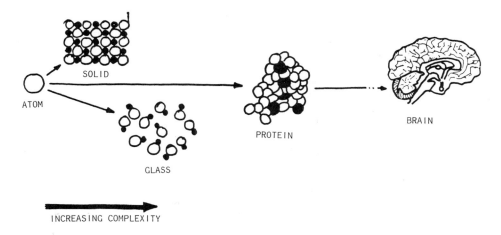

Fig. 2. The Chain of Complexity.

posed of multiple polypeptide subunits display quaternary structure. Schrödinger[36] called biomolecules "aperiodic crystals." Although some properties of proteins are well characterized by this name, there exist profound differences and proteins may be considered a separate state of matter. The crucial difference lies in the bond arrangement. In a crystal, each atom is fixed rather rigidly by "strong" bonds in all directions. Except for very rare jumps, the motion of the atom is restricted to vibrations; crystals are dead. The amino acids in a protein are covalently bonded only along the primary chain. The tertiary structure of proteins is held together mainly by weak hydrogen bonds and disulfide bridges. The overall binding energy of a protein is consequently small (of the order of a few tens of $k_B T$) so that internal motions of even large segments are possible. Dipole moment and charge vary from region to region in a protein, while solids are regular and spatially homogeneous. Investigation of the structure of proteins relies upon techniques such as X-ray diffraction, neutron diffraction and nuclear magnetic resonance (NMR) which determine the average position of every atom in the biomolecule. Monte Carlo and molecular dynamics energy minimization calculations are theoretical tools used to determine the structure and attack the protein folding problem.[29,11,23,8,30]

Perhaps the best characterized protein is myoglobin (Mb), a 153 amino acid globular protein, consisting of eight α helices wrapped around a heme group. Myoglobin serves to store oxygen inside muscle; it is similar to the subunits of hemoglobin, a tetrameric protein responsible for oxygen transport in red blood cells. Mb may thus be considered a model system for understanding globular proteins in general.

3. COMPLEXITY – PHYSICS OF THE FUTURE

Over the past few decades there have been many studies of the future direction of physics. These studies share one common feature: they all miss the most important new discoveries. Despite this dismal record of predictions, we venture to say that the physics of complexity will be increasingly important. Without defining complexity[33],

3

we sketch in Figure 2 a chain of complexity. It is intuitively clear that atoms are simple and that the brain is incredibly complex. It may be hopeless at present to extract new physics from the brain but proteins may be simple enough so that they can be understood; they may yield new laws and concepts. Their properties make them excellent tools for the exploration of complexity: Their mesoscopic scale places them at the border between classical and quantum physics. Four gigayears of research and development have optimized their structure and function. All proteins of a given type have exactly the same primary composition and size. They are sufficiently large so that statistical techniques may be employed, yet they are small enough so that properties of individual groups can be studied and modified. Genetic engineering permits planned site-directed changes at the atomic level.

4. PHYSICS FROM PROTEINS

What physics can we learn from proteins? We take our cues from well established branches of physics. In atomic, condensed matter, and subatomic physics, progress was propelled by experimental discoveries in three areas; structure, energy levels, and dynamics. The explanation of the experiments led to the discovery of many fundamental physical theories. The development of the Einstein and Debye theories in condensed matter physics emerged from studies of the specific heat of crystalline solids. In nuclear and particle physics, the models of Jensen, Goeppert-Mayer, Bohr and Mottelson, and the Gell-Mann - Nishijima relation are based on knowing the energies and quantum numbers of subatomic systems. Can we expect a similar evolution in biological physics? Progress has been encouraging as a few examples will show.

5. THE STRUCTURE OF PROTEINS

We have already sketched the construction of proteins in Section 2 and add here a few remarks concerning the structure and the connection to complexity. X-ray diffraction, neutron diffraction and 2-D NMR tell us where each individual atom in a particular protein sits. In contrast to a crystal, the Debye-Waller factor[19] varies from atom to atom[16,35], implying a close relation between proteins and amorphous substances. There is, however, also a profound difference between glasses and proteins. In glasses, the atoms are randomly arranged. A protein possesses a well-defined coarse structure; looked at with low resolution, a given atom will always be at the same position. This organization is seen clearly in all the structures exhibited in biochemical texts. However, a high-resoution look indicates "disorder." A given atom can occupy slightly different positions, but it will always remain close to the position shown in the average structure. One way to state this result is to say that proteins are ordered at long distances and disordered at short ones.

6. THE CONFORMATIONAL ENERGY LANDSCAPE OF PROTEINS

The energy landscape of a system is the energy of the system as a function of relevant parameters. For a simple system, such as an atom or a nucleus, the conformational energy landscape consists of just one potential well. For a slightly more complicated system, such as the two-level system of NH_3[22], the energy landscape is given by a double minima surface, plotted along a conformational coordinate (cc).

To envision the energy landscape of a complex system we perform a small *gedanken* experiment. If one takes a large number of identical drops of a liquid and freezes them quickly, the drops will all be *macroscopically* identical. *Molecularly*, the configurations of the atoms in each drop will be different. This difference in atom positions will yield different configurational energies, and the properties of the

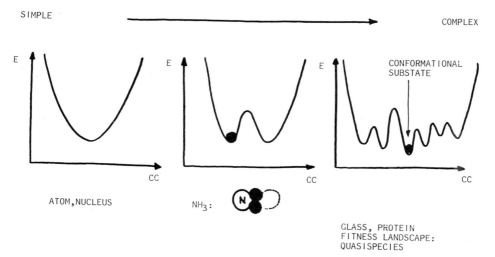

Fig. 3. Energy Landscapes. With increasing complexity, the energy landscape of a system becomes more intricate. Proteins and glasses possess a highly degenerate conformational ground state.

system will vary from droplet to droplet, leading to a plot of internal energy versus conformational coordinate as sketched in Fig 4. Proteins exhibit a similar highly degenerate ground state and we call each well a conformational substate (CS)[18]. If each of the approximately 200 amino acids of a protein can assume roughly 3 different configurations of equal energy, then the ground state of the protein is about $3^{200} \approx 10^{71}$ -fold degenerate. Such an energy landscape, with a highly degenerate ground state, is characteristic of most complex systems, and leads to large differences between the dynamic behavior of proteins and crystals.

Because of the complexity of proteins, one cannot use a single experimental

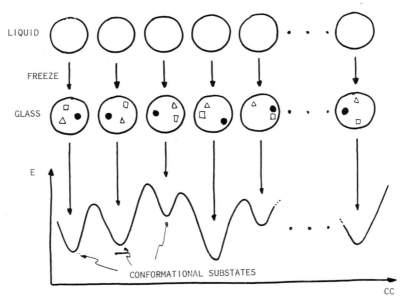

Fig. 4. Conformational Substates. Cooling identical drops of a liquid quickly will freeze the molecules of the drops in different conformations, with slightly different conformational energies.

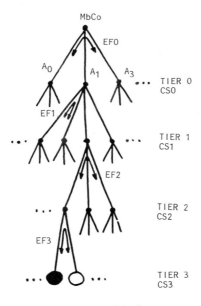

Fig. 5. Hierarchy of Substates. The hierarchical arrangement of the conformational substates.

technique to investigate the energy landscape and must rely upon many different techniques (flash photolysis[4], perturbation techniques[15], rate-window techniques[6], inelastic neutron scattering, Mössbauer[18], fluorescence[27,24], etc.) that explore the reactions, relaxations, and fluctuations of biomolecules. Experiments have just begun to explore the energy landscape of a few proteins; heme proteins represent a significant fraction of those that have been studied in any detail. A rough survey of the conformational energy landscape of proteins has revealed a number of features:

    i The landscape exists and is highly complex.

    ii The landscape is arranged in a hierarchy of conformational substates. We denote the various levels of the hierarchy as tier 0, tier 1, etc., in order of decreasing energy barriers between CS. For example, in Mb, evidence exists for at least four tiers of CS.

    iii Tier 0 consists of a small number of CS with clearly different properties. The MbCO protein-ligand system has three clearly observable CS0, characterized by three CO stretch frequencies in the infrared; they are called the A substates (A_0, A_1, A_3). The A substates perform the same function (in this case, the binding of CO) but with different rates. Their conformational energies, entropies, and volumes, are different.

The structural correlation to the A states is observable by experiments that probe the position of the ligand inside the protein[31]; each A state has the CO molecule at a slightly different orientation relative to the heme plane. X-ray structures at 1.5 Å resolution have not been able to discern separate crystal structures[25] so the *overall* structure of the 3 CS0 must be roughly the same.

    iv Evidence for tier 1 comes from the rebinding kinetics of each A state below 200 K. This rebinding is non-exponential, which leads to the conclusion that there must be a distribution of conformational substates CS1, each with its own activation enthalpy, within each A state. Kinetic hole-burning measurements support this argument[32]. They reveal that different wavelengths excite different parts of the distributed population within each A state.

    v CS2 evidence comes from Mössbauer effect data[14] and by flash photolysis experiments after prolonged illumination[3]. Neutron scattering[21] and specific heat data on metmyoglobin[37] reveal the likelihood of CS3.

7. PROTEIN DYNAMICS

7.1. *Equilibrium and Non-equilibrium Motions*

In contrast to glasses, proteins perform functions. A function involves at least two different states, such as the closed and open state of a switch. Our favorite example is the binding of CO to Mb, expressed as

$$Mb + CO \rightarrow MbCO$$

There are two *equilibrium states*, Mb and MbCO. This leads to two types of motions, *equilibrium fluctuations* within an equilibrium state, (EFs) and motions between equilibrium states, or *functionally important motions* (FIMs). EFs arise from the thermodynamic fluctuations of quantities such as internal energy, entropy and volume in proteins; a particular protein in one of the equilibrium states will not remain at rest but will fluctuate from CS to CS. FIMs are the motions necessary for the protein to accomplish its job. To understand the dynamics of biomolecules, we must investigate both EFs and FIMs. These two classes of motions are not independent; they are connected by the fluctuation-dissipation theorem[13,10,26]. Spectroscopic information about the relaxation to equilibrium gives us information about the FIMs. A systematic study of the energy landscape and dynamics can be effected by using a well selected perturbation and monitoring the relaxation.

An immediate question to be asked is: *What are the temperature and time dependence of the motions in the various tiers?* In simple systems, the time dependence is exponential, and the temperature dependence of the rate is given by an Arrhenius relation

$$\Phi(t, T) = \Phi(0, T)\exp(-\kappa(T)t),$$

where

$$\kappa(T) = A'\exp\left(\frac{-H}{RT}\right) \tag{1}$$

In glasses, spin-glasses, and other disordered systems, the behavior of the system is more complex[20]. The time course becomes non-exponential[28,34], arising from the fact that there is a broad distribution of relaxation frequencies. The time dependence can be parameterized as a stretched exponential[40],

$$\Phi(t, T) = \Phi(0, T)\exp(-\{\kappa(T)t\}^{\beta}). \tag{2}$$

The temperature dependence cannot be fitted by an Arrhenius relation with a physically meaningful preexponential factor A, but $\kappa(T)$ can be described over a wide range in temperature[5,41,9,34] by

$$\kappa(T) = A'\exp\left(-H'/RT\right)^{2}. \tag{3}$$

Comparison of Eqs. (1) and (3) shows that around a given temperature T, an Arrhenius equation with a "local" activation enthalpy,

$$H(T) = \frac{2H'^{2}}{RT},$$

will fit the data.

7.2. *Motions Close to Equilibrium*

In proteins, we study the relationship between EFs and FIMs by looking at relaxation processes both close to and far away from equilibrium. In a relaxation experiment, one has a protein sample in equilibrium and subsequently introduces a small perturbation such as a pressure (P) or temperature (T) jump. The relaxation of the protein ensemble has an elastic part, and a slower, conformational part[15]. A time and temperature dependent observable $\mathcal{O}(t, T)$, can be used to define a relaxation function $\Phi(t, T)$ to characterize the conformational relaxation,

$$\Phi(t, T) = \frac{\mathcal{O}(t, T) - \mathcal{O}(\infty, T)}{\mathcal{O}(0^{+}, T) - \mathcal{O}(\infty, T)}.$$

$\Phi(t, T)$ goes from $1 \rightarrow 0$ as t progresses from $0^{+} \rightarrow \infty$.

STATE   SUBSTATE

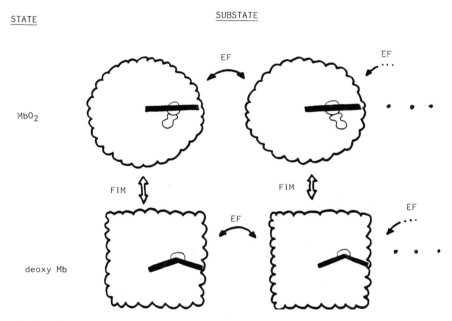

MbO$_2$

deoxy Mb

Fig. 6. EF and FIM. Proteins fluctuate between the substates of a given tier due
   to fluctuations in thermodynamic variables, such as entropy and volume.
   Transitions between states occur through non-equilibrium motions (FIM).

The relaxation phenomena in proteins show the same two important properties seen in glasses: $\Phi(t, T)$ is non-exponential in time and displays a non-Arrhenius temperature dependence. Here $\kappa(T)$ is a rate coefficient for the relaxation of the protein. The form of Eq. (2) is indicative of cooperative behavior in the protein.

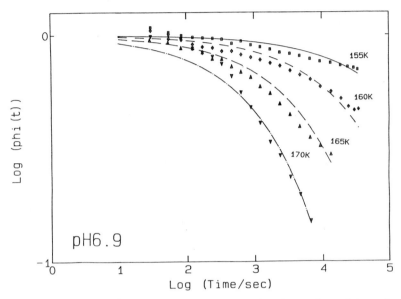

Fig. 7. Pressure Release Data. Relaxation functions for the peak position of a spectral band as a function of time after a pressure release. The relaxation is non-exponential in time.

7.3. Motions Far From Equilibrium (Proteinquakes)

Photodissociation of MbCO provides a good technique with which to examine the reaction energy landscape and the relaxation of the protein. While the details of this reaction, described by

$$MbCO + h\nu \rightarrow Mb^* + CO \rightarrow Mb + CO \rightarrow MbCO.$$

are quite complicated, there are some general features that are clear[9,4,1,38]. The absorption of a photon by the heme causes a stress that is relieved by the breaking of the Fe-CO bond. The protein is now in a state far from equilibrium and the equilibrium is restored through a *proteinquake*[2], i.e. a stress occurring at the heme

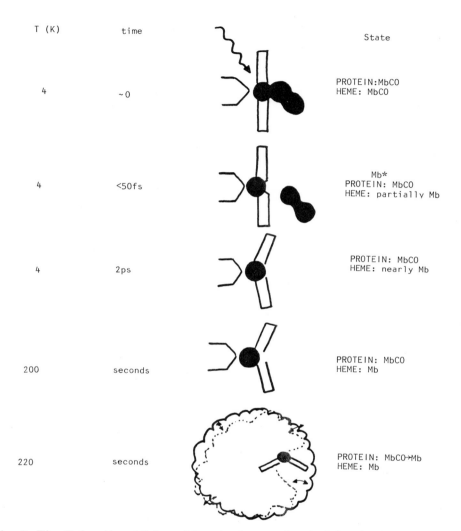

T (K)	time		State
4	~0		PROTEIN:MbCO HEME: MbCO
4	<50fs		Mb* PROTEIN: MbCO HEME: partially Mb
4	2ps		PROTEIN: MbCO HEME: nearly Mb
200	seconds		PROTEIN: MbCO HEME: Mb
220	seconds		PROTEIN: MbCO→Mb HEME: Mb

Fig. 8. The Relaxation Mb* → Mb. A photon absorbed by the heme causes the ligand to dissociate. At temperatures as low as 4 K, the iron moves toward the proximal histidine in about 50 fs and the heme begins to dome in about a few ps. Near 200 K, the iron atom moves to the deoxy position within seconds. This motion is characterized by the rate coefficient $\kappa^*(T)$. Near 200 K, the entire protein begins to move and the ligand can escape into the solvent. The relevant equilibrium fluctuations are described by the rate coefficient $\kappa_1(T)$.

iron is relieved by waves of strain that propagate from the heme and through the protein. The proteinquake is initially fast and slows as the stress is relieved. One can study the different steps of the system's equilibration as a function of temperature, pressure, solvent viscosity or pH.

Rebinding kinetics of CO to Mb show two processes. The faster process seen at all temperatures (process I for "internal") is non-exponential in time and independent of the CO concentration in the solvent. The second process (process S for "solvent") is exponential in time; the rate of rebinding for process S is proportional to the solvent CO concentration.

At temperatures above 220 K, the equilibrium fluctuations among the CS open pathways for the ligands through the protein matrix and also narrow the rate distribution. The "opening" of the protein required for the ligand to escape is goverend by the equilbrium fluctuations of tier 1 (EF1) and has a rate coefficient given by κ_1. Studies of the solvent viscosity dependence of the rate coefficients indicates that κ_1 varies over two orders of magnitude and therefore must be coupled to a large portion of the protein. These EF1s are also responsible for the narrowing of the activation rate distribution: if the characteristic time scale for rebinding at these temperatures is much longer than the time of fluctuation of the barrier, the protein-ligand system has an effective barrier height. The consequent time-average of the rebinding rate coefficients results in an effective partial collapse of the low temperature distribution. If the distribution of rate coefficients collapses to a δ function, rebinding becomes exponential.

If we consider the relaxation of the excited state of the photodissociated protein, Mb* to the relaxed state, Mb, we see that the relaxation consists of at least three steps. At low temperatures (T $\approx$ 4 K), the MbCO structure has the ligand bound and the iron is in the heme plane. The photon of energy $h\nu$ impinges upon the heme and is absorbed at time $t = 0$. By time $t \leq 50$ fs, the ligand has dissociated, but the heme iron is still partially in plane. Spectroscopically, the protein is now in the Mb* state. By $t = 1$ ns, the heme has started relaxing to its out of plane Mb structure, but has not fully reached this configuration. At higher temperatures, T $\approx$ 200 K, the heme has fully relaxed to the Mb configuration by $t = 1$ s. This relaxation is characterized by the rate coefficient $\kappa^*(T)$, which describes the relaxation at the heme.

The rebinding kinetics can be modeled by shifting the activation enthalpy distribution to higher H which decreases the rebinding rate, k, in Eq. (1). Fits using a time dependent enthalpy distribution yield a change in barrier height of approximately 9.3 kJ/mol from the Mb* state to the Mb state.

The parameters for $\kappa^*(T)$ obtained from fits of the high temperature data to Eq. (2) yield a value of $\beta = 0.24$. Fits of the relaxation function to the stretched exponential are consistent with those for glass relaxations. As expected, one sees a weak solvent viscosity dependence of κ^* suggesting that the relaxation Mb* $\rightarrow$ Mb occurs primarily at the heme.

ACKNOWLEDGMENTS

This work was supported in part by the National Science Foundation (grant DMB87-16476), the National Institutes of Health (Grants GM 18051 and 32455), and the Office of Naval Research (N00014-89-R-1300).

REFERENCES

1. Agmon, N., Hopfield, J.J. (1983) *J. Chem. Phys.* 79, 2042-2053.
2. Ansari, A., Berendzen, J., Bowne, S. F., Frauenfelder, H., Iben, I. E. T., Sauke, T. B., Shyamsunder, E., Young, R.D. (1985) *Proc. Natl. Acad. Sci. USA* 82, 5000-5004.
3. Ansari, A., Berendzen, J., Braunstein, D., Cowen, B.R., Frauenfelder, H., Hong, M.K., Iben, I.E.T., Johnson J.B., Ormos, P., Sauke, T.B., Scholl, R., Schulte, A., Steinbach, P.J., Vittitow, J., Young, R.D. (1987) *Biophys. Chem.* 26, 337-355.
4. Austin, R.H., Beeson, K.W., Eisenstein, L., Frauenfelder, H., Gunsalus, I.C. (1975) *Biochemistry* 17, 5355-5373.
5. Bässler, H. (1987) *Phys. Rev. Lett.* 58, 767-770.
6. Berendzen, J.R., Braunstein, D. (1990) *Proc. Natl. Acad. USA* 87, 1-5.
7. Brawer, S.A. (1984) *J. Chem. Phys.* f 81 (2), 954 - 975.
8. Charles L. Brooks III, Martin Karplus, B. Montgomery Pettit, *Proteins: A Theoretical Perspective of Dynamics, Structure and Thermodynamics* , Wiley, New York, 1988.
9. Bryngelson, J., Wolynes, P. G. (1989) *J. Phys. Chem.* 93, 6902-6915.
10. Callen, H. B., Welton, T. A., *Phys. Rev.* , 88, 34.
11. Chothia, C., (1984) *Ann. Rev. Biochem.* 53, 537-572.
12. R.E. Dickerson and I. Geis, *The Structure and Action of Proteins* , Benjamin-Cummings, Menlo Park, 1969.
13. Einstein, A., (1905) *A. Physik* 17, 549.
14. Frauenfelder, H. (1985) Ligand binding and protein dynamics, in *Structure and Motion: Membranes, Nucleic Acids and Proteins* , E. Clementi, G. Corongiu, M. H. Sarma, and R. H. Sarma, ed., Adenine Press, Guilderland, NY, 205-217.
15. Frauenfelder, H., Alberding, N.A., Ansari, A., Braunstein, D., Cowen, B.R., Hong, M.K., Iben, I.E.T., Johnson, J.B., Luck, S., Marden, M.C., Mourant, J.R., Ormos, P., Reinisch, L., Scholl, R., Schulte, A., Shyamsunder, E., Sorensen, L.B., Steinbach, P.J., Xie, A-H., Young, R.D., Yue, K.T. (1990) *J. Phys. Chem.* 94, 1024-1037.
16. Frauenfelder, H., Petsko, G.A., Tsernoglou, D. (1979) *Nature* 280, 558-563.
17. Frauenfelder, H., (1984) *Helv. Phys. Acta* 57, 165-187.
18. Frauenfelder, H., Parak, F., Young, R. D., (1988) *Ann. Rev. Biophys. Biophys. Chem.* 17, 451-79.
19. Frauenfelder, H., (1989) *Intern. J. of Quantum Chem.* XXXV, 711-715.
20. Derrida, B. (1980) *Phys. Rev. Lett.* 45 (2), 79 - 82.
21. Doster, W., Cusack, S. and Petry, W. (1989) *Nature (Lond.)* 337 754-756.
22. R. P. Feynman, R. B. Leighton, M. Sands, *The Feynman Lectures on Physics,* Addison-Wesley, Menlo Park, 1963.
23. Go, N. (1983) *Ann. Rev. Biophys. Bioeng.* , 12, 183-210.
24. Gratton, E., Barbieri, B. (1986) *Spectroscopy* 1, 28.
25. Kuriyan, J., Wilz, S., Karplus, M., Petsko, G. (1986), *J. Mol. Bio.* 192 133-154.
26. Kubo, R., (1966) *Rep. Prog. Phys.* 29, 225-284.
27. J. R. Lakowicz, *Principles of Fluoresence Spectroscopy* , Plenum, New York, 1983.
28. Lindsay, C.P., Patterson, G.D. (1980) *J. Chem. Phys.* 73 3348.
29. Levitt, M., (1982) *Ann. Rev. Biophys. Bioeng.*, 11, 251-71.
30. J.A. McCammon, S. C. Harver, *Dynamics of Proteins and Nucleic Acids* , Cambridge U. Press, Cambridge, 1987.
31. Ormos, P., Braunstein, D., Frauenfelder, H., Hong, M. K., Lin, S. L., Sauke, T.B., Young, R.D. (1988) *Proc. Natl. Acad. Sci. USA* 85, 8492-8496.

32. Ormos, P., Ansari, A., Braunstein, D., Cowen, B.R., Frauenfelder, H., Hong, M. K., Iben, I.E.T., Sauke, T.B., Steinbach, P.J., Young, R. D., *Biophys. J.*, 57, 191-199.

33. L. Peliti and V. Vulpiani, Eds., *Measures of Complexity*, Lecture Notes in Physics 314, Springer, New York, 1988.

34. Richert, R., Bässler, H., (1990) *J. Phys.: Condens. Matter* 2, 2273-2288.

35. Ringe, D., Petsko, G. A., (1985) *Prog. Biophy. Molec. Bio.*, 45, 197.

36. E. Schrödinger,) *What is Life?* , Cambridge Univ. Press, Cambridge, 1967.

37. Singh, G. P., Schink, H. J., Von Löhneysen, Parak, F. and Hunklinger, S. (1984) *Z. Phys.* B55 23-26.

38. Srajer, V., Reinisch, L., Champion, P. M. (1988) *J. Am. Chem. Soc.* 110, 6656-6669.

39. L. Stryer, *Biochemistry* , 3rd. ed., W.H. Freeman and Company, N.Y., 1988.

40. Williams, G., Watts, D.C. (1970) *Trans. Faraday Soc.* 66, 80.

41. Zwanzig, R. (1988) *Proc. Natl. Acad. Sci. USA* 85, 2029-2030.

SEARCH AND RECOGNITION: SPIN GLASS ENGINEERING AS AN APPROACH TO

PROTEIN STRUCTURE PREDICTION

Peter G. Wolynes

School of Chemical Sciences and Beckman Institute
University of Illinois
Urbana, Illinois 61801 USA

I. INTRODUCTION

Protein folding is perhaps the most complex process that can be usefully described in purely atomistic terms.[1,2] In order to function, a protein generally must take on a specific three-dimensional structure. It is, at present, unknown how this process occurs in the living cell. It is possible and, indeed, likely, that for many proteins, special biological machinery and catalysts are necessary for this process. However, it is also known that for some smaller proteins, this process of folding into a well defined structure can occur spontaneously in the absence of special biological machinery. In principle, this in vitro folding process can be studied using the ideas of physics and chemistry alone. A huge diversity of states of varying stabilities can occur in the folding process. It is important then to attack the problem of understanding protein folding by seeking a language that can reconcile diversity and stability.

Understanding protein folding is not an exercise motivated only by scientific curiosity about a complex process. Understanding in vivo protein folding may give clues to the nature of many diseases. Both Scrapies and Alzheimers disease are characterized diagnostically by aggregates of improperly folded proteins. An understanding of in vitro folding could lead to a scheme for predicting protein structures, knowing only the amino acid sequence or other easily obtained experimental information. Any scheme, no matter how unrealistic in mechanistic detail, for computationally predicting protein structure from sequence, would be valuable. Currently, the technology for obtaining protein sequences is considerably advanced compared with that for structure determination. Several countries are currently considering the project of sequencing the nucleic acids of the human genome completely. Among these sequences will be the information for the coding of thousands of proteins. Currently when a sequence of a protein is in hand, one could try to produce large quantities of the protein by genetic engineering and then carry out structure determinations experimentally. A structure determination currently takes, when it can be done at all, at least a month. Thus, we can see that without either a prediction scheme or improvements in structure determination technology, the human genome projects will create a huge backlog of information which cannot be

Biologically Inspired Physics, Edited by L. Peliti
Plenum Press, New York, 1991

interpreted. For all these reasons, the protein folding problem has been under study for many years by many different parts of the scientific community. Molecular biologists have been searching for the specific biological actors in _in vivo_ folding. Experimental biophysicists and physical chemists have sought new probes for investigating fleeting intermediates in the _in vitro_ folding process. Theoretical chemists have sought to describe, as precisely as possible, the molecular interactions that give rise to folding and have sought to simulate that process on the computer. For a long time too, theoretical chemists and physicists have sought to describe the folding process as a phase transition. Most recently, several of us have tried to see whether the concepts of spin glass theory are appropriate for describing aspects of the folding problem.[3-18]

In this lecture, I wish to review primarily our own efforts along these lines. I have chosen to organize the discussion around the problem of predicting protein structure from sequence. This might be called "_in machina_" protein folding. I do this for several reasons: First, I have reviewed spin glass ideas in the general context of protein folding in an article which I hope will soon see the light of day.[4] Spin glass ideas may well help in understanding some aspects of the _in vivo_ process, but considerably more biological information will be needed for definitive statements to be made. Similarly, the exact details of folding in _in vitro_ experiments depend on many detailed issues of chemical physics, such as the origin of the hydrophobic effect, and the nature of electrostatic interactions in proteins, as well as the nature of sidechain conformations. The experimental field here is also currently in a state of ferment. In truth, to solve the structure prediction problem, it may be necessary to understand the biology of _in vivo_ folding and the detailed physical chemistry of _in vitro_ folding. But it may be possible to develop an approach to structure prediction that is based on recognizing patterns in amino acid sequences since, despite the backlog in structure determinations, there are many examples of sequences whose structures are known. It is thus possible that very simple models that describe an algorithmic folding process using information about known folded proteins can be devised. Nothing completely successful in this direction has yet been achieved, but we have been able to develop very simple pattern-based Hamiltonian models for protein folding that allow us to quantify the difficulty of the problem. These models are analogous to the models introduced by Hopfield to describe neural networks.[19] Like the Hopfield neural network models, they can be analyzed fairly completely by the analytical methods of spin glass theory and by computer simulation. Even these models have a great deal of complexity, and since they can be understood in principle by the methods of theoretical physics, it seems worthwhile to study them and see what are the characteristics that would be required to carry out adequate structure prediction. This understanding of these models may allow us to engineer them so that they can work more effectively as algorithms for structure prediction. In any case, their analysis provides us with a great deal of fun as well as some insight into how molecular systems could carry out the complex pattern of information flow from one-dimensional sequence to three-dimensional structure.

The organization of the lecture is as follows: In the next section I will briefly review the phenomenology of protein structures and their folding dynamics. I will show how very simple ideas suggest that a competition between a first order transition and a spin glass transition can determine how effective folding might be carried out by simulation. Here I will use a worst-case analysis based on the random energy model.[3,5,20] In the following section, I will introduce the associative memory Hamiltonians.[6,8,9] Following this I will describe the analytical theory of the phase diagram. Then in the penultimate chapter I will show the results of various computer simulations on these models relating the mechanism of folding to these phase diagrams. Finally, I will discuss future challenges to spin glass theory inspired by the need to devise practical protein structure prediction scheme.

II. SPIN GLASS DYNAMICS AND THE SEARCH PROBLEM IN PROTEIN FOLDING

A natural analogy to bear in mind when thinking about protein folding in vitro or in machina is the analogy of the freezing of a liquid or the condensation into a crystal from a gas. One motivation for this analogy is the understanding that crystallographers have achieved of protein architecture. Although, to the untrained eye, protein structures often look random, this is very far from being the case.[2] Although there are occasionally some cavities for ligand groups, in the main, folded proteins are very nearly closely packed. For larger proteins, this close packing comes about by using some very regular structural patterns. Locally segments are arranged in helices or into extended segments called β-sheets. In addition, there are very specific ways in which infinite helices or sheets are broken up by turns. These turns often have particular local structures and have been classified. These structures which can be defined locally in sequence also form regular structures when they are packed together. Helices and sheets can only be packed together in certain ways. After understanding these patterns of packing, it has been realized that proteins can be classified into between twenty and thirty families.[21] These may be analogous to the space groups which we are familiar with for infinite systems, although little mathematical work along these lines has been done. Thus, all in all, the final folded protein is very analogous to a crystal. One knows this also because the structures can be determined by x-ray, and the Debye-Waller factors of individual atoms can be measured. The local disorder is somewhat larger than in simple amino acid crystals, but it is still far from a disorganized jumble. Since the final folded protein resembles a crystal, one would expect that Hamiltonians based on patterns seen in proteins would exhibit phase transitions like freezing. Indeed, this should be the case for in machina folding. The analogy to freezing is also relevant to in vitro protein folding. Experimentally the folded and unfolded structures are relatively balanced as far as their total free energy. This balance conceals the fact that there is a competition between entropy and energy in the folded and unfolded structures. The folded protein is very low in energy but also has little entropy. The unfolded protein is quite high in energy and very high in entropy. Thus we see that the change between the two of them would involve a latent heat, like a first order phase transition. This has been well established experimentally, using calorimetry. We should bear in mind, however, that the experimental phase diagram of proteins is quite complicated, and there are other phases potentially present that have been observed in many cases. Thus a significant question about in vitro folding is at what point is this latent heat released? This is a question to which spin glass theory, in the hands of Shakhnovich and co-workers,[14,16] has contributed some interesting insights.

Let's pursue the freezing analogy in thinking about the dynamics of folding. Freezing of simple substances occurs by nucleation. That is, a free energy barrier must be surmounted to go from the liquid state to the crystal. The gain in energy that comes from organizing the liquid into a crystal is insufficient to overcome the loss of entropy until a sufficiently large region of the material is solidified. Even in the mean field limit where critical nuclei are not localized in space, there is an effective double well which partially determines the dynamics. In the mean field limit, just at the transition, a large fraction of the system must be organized before the entropy loss is overcome by the energy gain. As the temperature is lowered, the bistable potential tilts more towards the crystalline product, and the nucleation barrier goes down. In some cases it is thought possible to approach a spinodal in which there is virtually no barrier to forming a crystal. Usually, however, another effect intervenes: glass formation. In essentially all materials, molecular motions slow down upon cooling. Supercooled liquids spend longer periods of time trapped in local minima of the potential energy. Viscosities and other measures of molecular rates of motion become exceedingly slow roughly following a Vogel-Fulcher law.

Since molecular motions are certainly required for crystallization to occur, the rate of nucleation usually shows a maximum at some intermediate temperature between the freezing point and absolute zero.[22] At the lowest temperatures, rather than having facile nucleation, there is trapping in local minima. The same situation applies to simulations of proteins which are intended to find the global minimum corresponding to the correct folded protein structure. It might be thought possible to merely write down a very accurate free energy function for a protein molecule, then by lowering the temperature in simulated annealing to make the folded structure ever more stable. By continuing to lower the temperature with a fixed free energy function, the folded state would win energetically. Trapping in local minima, however, would prevent the system from achieving this global state. Thus, the problem of local minima in simulations of protein structures is analogous to the problem of glass transitions in liquids. For simulations, aimed to simulate reality, there is another catch. If we knew the free energy function for a fixed solvent composition, something like the previous scenario could apply. In fact, most models of proteins that are folding at the atomistic level must explicitly include the solvent. A large part of the free energy association of segments in the protein arise from entropic changes in the solvent. Real proteins actually exhibit a phenomenon known as cold denaturation;[23] where proteins unfold as they are chilled in solution. The solvent entropy is certainly one contributor to this. Thus with fully atomistic models, the prediction scenario just described is impossible. Even if an exact free energy function were available, one would then be faced with the glass transition and trapping in local minimum as the rate limiting step; in fact, in most calculations that have tried to do predictions of folding, trapping is the main source of the problem. Cold denaturation has not occurred in any extant atomistic simulation of protein folding.

Spin glass theory allows a more vivid and concrete understanding of this crystallization analogy. The simplest spin glass model of folding proteins was studied by J. D. Bryngelson and me.[3,5,7] In this statistical protein model, order parameters characterizing the extent of native structure are introduced. In the simplest theory this is n_o, the fraction of residues whose conformation is consistent with that in the final folded structure. Thus, when $n_o = 1$, each amino acid residue is found in one of the ν possible states. Thus, the entropy can be written as a function of n_o:

$$\frac{S}{k_B} = -n_o \log n_o - (1-n_o) \log \frac{1-n_o}{\nu} \qquad (1)$$

In order to get a first phase transition, there must be an energy which depends on n_o. In general, this would have the form

$$E = -An_o - Bn_o^2 \qquad (2)$$

Combining this average in the entropy, one obtains the typical meanfield Hamiltonian of free energy function exhibiting two minima which we discussed in the crystallization analogy. On the other hand, not every conformational state of the protein will have this average energy. In some cases, residues will be brought together which have favorable contacts, and in others, there will be unfavorable contacts. Thus there will be a range of energies for a protein with a given fraction of native structure. To get an orientation to the problem, we can assume that the energies of different states which are only partially folded are random, as in Derrida's random energy model spin glass.[20] The variance in energy would depend on the fraction of non-native residues:

$$\Delta E^2 = B'(1-n_o)^2 \tag{3}$$

This model is entirely analogous to the ferromagnetic random energy model used by Derrida in describing spin glasses with some degree of ferromagnetism. It has a phase diagram entirely analogous with that one worked out by him in that context. This phase diagram resembles that for the associative memory models discussed in detail later. The most important point is a competition between folding and an ideal glass transition.

Using the statistical definition of the model, it is possible to get some idea about how the local minima impede the organization of the folding protein. Both Bryngelson and I and Shakhnovich and Gutin have come to similar results on the problem of trapping in local minima. One finds a very broad distribution of escape rates from individual minima reflecting the random energies of the states. The typical escape rate can be written in terms of the variance of the energies as

$$k^* = k_o N\nu \exp -\left[(\Delta E^2/2k_B T)^2\right] \tag{4}$$

The ΔE^2 in this equation depends on the fraction of native structure. If we define a local glass transition temperature by the random energy model result for systems constrained to have this value of the order parameter, k^* can be written in a more picturesque manner.

$$k^* = k_o N\nu \exp \left\{- \frac{S^*}{k_B} \left(\frac{T_g}{T}\right)^2\right\} \tag{5}$$

We see that the escape rate becomes extremely slow at the glass transition temperature. Indeed, the typical escape time at this temperature is given by

$$t^* = \frac{1}{k_o N\nu} \exp \frac{S^*}{k_B} \tag{6}$$

This is essentially the amount of time it takes to search through every possible neighboring state that preserves the fraction of native residues. Thus, if one is near the glass transition temperature with very little degree of order, the typical escape time is incredibly long. One must search essentially all configurations of the chain to make even a simple motion.

Superimposed on this trapping phenomenon is the drift down the free energy gradient. By superimposing these effects, we can obtain an approximate representation for the nucleation time for folding.

$$\bar{t} = \int^{n_o'} dn_o' \int_o^{n_o'} dn_o \; D(n_o)^{-1} \exp\left\{\frac{1}{k_B T} \{F(n_o) - F(o)\}\right\} \tag{7}$$

where the diffusion constant is given by

$$D(n_o) = \frac{1}{2} \frac{k_o}{N} \exp\left\{-S^*(n_o) + \left(\frac{1}{k_B T_g} - \frac{1}{k_B T}\right)^2 \Delta E^2(n_o)\right\} \tag{8}$$

Notice that if the folding temperature is close to the glass transition temperature, this rate is essentially the rate to search through all possible states. This is the nightmare raised by Levinthal many years ago in his thoughts about the protein structure prediction problem.[24] For a general Hamiltonian, the only way to find the global minimum is to try them all! This is what happens at the glass transition. To avoid this, one should devise energy functions for doing pattern recognition that have as low a glass transition temperature as possible in comparison to the folding temperature.

The statistical protein model raises several interesting issues about the existence of pathways for folding. Levinthal argued that the search through possible protein structures could only be carried out rapidly if there were definite intermediates along the path. The simple statistical protein model certainly argues that this need not be the case and that a large variety of intermediates can exist and still give sensible rates so long as there is sufficiency of downhill paths on the energy surface. Indeed, within the statistical protein model, the most unique pathway would apply at the glass transition where only a few possible states can be thermally populated for each value of the order parameter and this is the limit in which dynamics is extremely slow. Nevertheless, many experiments tend to suggest there may be discrete intermediates in folding in many cases in vitro. The relationship of these intermediates to possible glass transitions requires elucidation. It may well be that correlated random energy models give the possibility of unique intermediates without requiring the extreme slowing down of the folding rate. In correlated random energy models the states which exist at one value of the order parameter could be related to those at other values; it would then not be necessary to undo such deeply trapped structures. The problem of pathways has some resemblance to the problem of random directed polymers.[25] Folding paths through state space are diffusive paths, the paths for directed polymers where the paths describe tunneling entities. It is likely that in the diffusive case when the energy surface becomes rough there will be a transition to a situation in which there are fairly unique paths, just as in the random-directed polymer problem. Concrete models of this using spin glass theory would be most welcome since the pathway issue dominates the thinking of experimentalists in protein folding.

III.   RECOGNITION BY ASSOCIATIVE MEMORY HAMILTONIANS:
       A MATHEMATICAL FRAMEWORK FOR PROTEIN FOLDING CODES

Spin glass ideas in the guise of the statistical protein model illuminate many features of the search problem in protein folding. They also help illuminate the pattern recognition aspects of devising algorithms to predict protein structure. A vast effort has been expended examining and studying in artistic detail protein structures and their sequences. This study has borne "much fruit," such as the identification of secondary structure elements and the taxonomic classification of proteins into families. There are numerous observations that people have made about protein structure that can be used to decide whether a given trial structure is reasonable or not. These include constraints, such as excluded volume (the interiors of proteins are close packed), or the constraint that most hydrophobic residues are found in the interior of a protein. These insights have been obtained by qualitative examination of protein structures.

An alternative approach which has been carried through is the examination of structures in statistical ways,[26-29] for example, to obtain information about how often certain residues are found within secondary structures, or how often certain residues are found next to each other in the tertiary structure. Both the qualitative insights and the statistical results have been used to develop simple Hamiltonian models for folding proteins. A most interesting class of these simple models are lattice models, used most powerfully by Skolnick and co-workers.[29] Spin glass theory provides an organized framework in which to address the problem of what Hamiltonians have minima with properties characteristic of the known protein structures. Just as in neural network modeling where one seeks to find a set of neural connections given a known or desired set of attractors, one way of predicting patterns in proteins is to find simple Hamiltonian functions that will fold to global minima described by the known proteins. Friedrichs and I have developed a family of models which we call associative memory Hamiltonians that provide this framework for understanding the statistical mechanics of protein folding patterns.[6,8] Many of the earlier approaches to protein folding Hamiltonians can be thought of as special cases of these associative memory models in which some additional assumptions have been made. Thus, formulating questions about the associative memory models allows one to study in an organized way some of the earlier statistical approaches to protein structure prediction, and to see which assumptions in those models might be changed to give better predictions. In addition, associative memory Hamiltonians can provide a mathematical basis for examining other kinds of coding of protein folding. The statistical mechanics of these more elaborate associative memory models can tell us something about the information theoretic viability of different ideas about folding codes.

The simplest associative memory models are the direct analog in polymer physics of the associative memory spin models introduced by Hopfield. There are, of course, many ways of connecting spin models to polymer models. For proteins, there is the additional aspect that one must encode the differing identities of the amino acid residues. Our connection is made by use of the famous lattice gas analogy between magnetic and fluid systems. Density fluctuations in the fluid play the role of spin coordinates. Since the amino residues in the protein differ from each other, there are different kinds of density fluctuations. We characterize the different amino acids by charges. These "charges" might be measures of physico-chemical properties, such as hydrophobicity, or size, or, indeed, the electrical charge that is to be ascribed to each residue. These charges may also be the complete description of the identify of a residue; for example, a "cysteine-ness" charge could be introduced. In more elaborate representations that we are using now, these charges may, indeed, depend on the identity of neighboring residues. For example, we have used a preliminary secondary structure assignment based on qualitative rules such as Chou-Fasman to assign a charge indicating helical propensity, etc.

It is clear that charge density fluctuations are the appropriate analogues of the spin coordinates for the protein case. The protein should be described in a four-dimensional way, three dimensions being those of space, and one of them, the sequence number. The latter will play the role of time in various path integral treatments of the statistical mechanics of these models. If space is discretized, a protein's charge density patterns are analogous to spin patterns in a four-dimensional lattice. In the Hopfield spin model, the interaction between different sites is proportional to the correlation function of the spins in the database. Analogously, the interaction between lattice sites in the four-dimensional model representing the protein will be the correlation function of the charge densities with respect to space and sequence in a database of proteins. The lattice, however, is just an artifice. It is possible to write the energy, which is described as a function of charge density fluctuations in terms of the coordinates of the amino acid residues. We will take in simple models the amino acid residues to be represented by points located at the positions of the α-carbons. In terms of these coordinates, the interaction Hamiltonian is given by

$$H' = -\sum_{\alpha} \sum_{i,j} \sum_{m,n} q^T_{i,m} q^T_{j,n} q^\alpha_{i,m} q^\alpha_{jn} \theta(r_{ij} - r^\alpha_{ij}) \qquad (9)$$

The q^T_n in this equation are the charges of type n associated with the protein to be folded, while the q^α_n represents the charges of M proteins labeled by α that are to be memorized in the database. The θ function is taken to be either a square well or a Gaussian centered around the distance of sites i and j in the memorized protein (r^α_{ij}). As it stands, this Hamiltonian would not require the connectivity of the polymer chain. In calculations, this connectivity is ensured by imposing constraints or by adding local terms to the Hamiltonian that connect neighboring atoms. For analytical theories, this can be taken as a simple quadratic form while more elaborate forms are used in simulation. Thus, the complete interaction Hamiltonian has a form given by

$$H = H' + k_B TA \sum_i (r_i - r_{i+1})^2 \qquad (10)$$

Unlike the spin models, this Hamiltonian written in terms of r_{ij} possesses all of the appropriate physical symmetries. It is both translationally and rotationally invariant and, thus, can recognize proteins in various different orientations. There are other symmetries that the Hamiltonian does not possess. The Hamiltonian, as it stands, is not translation invariant with respect to sequence. If a protein has a few extra amino acids added at one end, in the formulation, the Hamiltonian does not necessarily possess a minimum near to a folded structure for the sequence-translated residues. Is sequence translation invariance a "biological symmetry"? We certainly expect that adding a few amino acids to a protein will not cause its folding to be very much different. In fact, there are exceptions to this statement. Thus, it is clear these biological symmetries are only approximate.

In neural network theory, we know the question of symmetry is related to the question of generalization.[30] Finding a Hamiltonian that will appropriately lead to correct folds for sequences that have never been presented to it requires making some assumptions about the possible biological symmetries. This is easy to do within the framework of the associative memory Hamiltonian models, where transformations of the sequence number indices can be introduced.

Many of the older ad hoc models for Hamiltonians make very strong assumptions as to the biological symmetries.[26-29] These are typically that interactions distant along the chain do not depend on the sequence number at all, but merely depend on the proximity of residues. Correspondingly, they often assume that only nearby residues in space contribute to the energy rather than all distant pairs. It is easy, within the context of associative memory models to evaluate the effect of making these kind of assumptions. These assumptions have been made in versions of associative memory Hamiltonians introduced by Garel and Orland,[11] as well as by Shakhnovich and Gutin. [17]

The simple associative memory Hamiltonians give a <u>delocalized global pairwise</u> code for folding. Other possibilities have been discussed in qualitative treatments of folding. For instance, it has been argued that there may be a local in sequence tetrapeptide code. Wodak has discussed statistical arguments that suggest this could be a possibility.[31] Such a tetrapeptide code could be introduced to the associative memory framework by having multiparticle interactions. Thus, one would have an interaction Hamiltonian of the form

$$H' = \sum_{\substack{local \\ quartets}} \sum_{\alpha}^{M} f(\{q_{i,}^{T}\cdots q_{i+4}^{T};q_{i,}^{\alpha}\cdots q_{i+4}^{\alpha}\})\prod_{\substack{pairs\ in \\ quartet}} e^{-\varepsilon(r_{ij}-r_{ij}^{\alpha})^2} \qquad (11)$$

The multipoint interactions in this code would be difficult to use computationally in their simple form if there were many memories, but the analytical theory of this sort of model is certainly tractable, and some progress could be made computationally as well.

Multiparticle correlations which are nonlocal in sequence have also been proposed. Ponder and Richards have introduced a template library for local multiplets in the hydrophobic core of proteins.[32] This could be implemented by another Hamiltonian with triple interactions that are local in space, rather than sequence. The statistical mechanics of these more elaborate multiplet associative memory Hamiltonians should have many parallels with the simple pairwise global one that we investigate analytically in the next section.

IV. THE PHASES OF ASSOCIATIVE MEMORY MODEL HAMILTONIANS

In order to understand the mechanisms of protein folding with associative memory Hamiltonians, it is necessary to understand their equilibrium phases. As we have seen, the ability to reach a global minimum will depend critically on how close the folding transitions are to glass transitions in the phase diagram. The resemblance of the associative memory Hamiltonians to the Hopfield neural network models suggests that a great deal can be said about these models using the analytical techniques of spin glass theory. Sasai and I have carried through such an analysis.[19] While making rather strong assumptions about the nature of the statistical memories to be encoded in the associative memory model, our analysis describes the phenomena at the molecular level without making the continuum approximations so often used in polymer physics. In this Section, I will describe the techniques and results of that analysis. There are many parallels to the techniques used by Shakhnovich and Gutin and by Garel and Orland for generic heteropolymers.

Let us imagine that the protein that we are studying has a known structure and that, in fact, one of the memory proteins exactly coincides with it. In this event, we can separate the associative memory part of the Hamiltonian into two parts. The first of these, we will call the coherent part, $\overline{V}$. It arises from pairs of distances in the database proteins which coincide to some accuracy with the target protein which we are folding. In the detailed analysis, in fact, we will assume that this term comes from precisely the target which is stored in memory. In real applications, of course, one assumes that some pairs of distances will come from one set of memories and others from other set of memories, but this greatly complicates the algebra. There are clearly contributions to the interaction which arise from pair distances which are not reflected in the target structure. This contribution will be called V', and will be treated as noise. Thus, we have for $\bar{V}$ and V' the equations

$$\bar{V} = V(r_{ij} - r_{ij}^T) \left(1 + \sum' q_i^\mu q_j^\mu q_i^T q_j^T V(r_{ij} - r_{ij}^\mu)\right)$$

$$V' = -\sum' q_i^\mu q_j^\mu q_i^T q_j^T V(r_{ij} - r_{ij}^\mu). \tag{12}$$

where the prime indicate a sum over nontarget memories.

The treatment of V' as noise is motivated by the fact that one must take into account the possibility of having different databases which include different particular proteins from nature. In addition, the sequence being studied can exist in various mutant forms and, therefore, it will only match up statistically with any particular target. Thus the statistics of carrying out averages reflect this degree of mutation and selection in a database and statements can be made precise by considering averages over all possible databases. It is the noise part of the Hamiltonian that introduces spin glass transitions.

We find the thermodynamics for a typical database by evaluating the average of the free energy over possible databases containing M memories. This average is done with the usual replica trick writing explicitly for the average of log Z

$$\langle \log Z \rangle = \mathop{Lt}_{n \to o} \frac{1}{n} \left((Z^n)'_{av} - 1 \right)$$

$$= \mathop{Lt}_{n \to o} \frac{1}{n} \left\{ \prod_\alpha \prod_i d\, \underset{\sim}{r}_i^\alpha \prod_\alpha \delta(\sum_i \underset{\sim}{r}_i^\alpha) \langle e^{-\sum_\alpha H(r_i^\alpha)} \rangle_{av} - 1 \right. \tag{13}$$

When the number of incoherent memories is small, the average over the replicated Hamiltonian is difficult to evaluate in general. However, when the number of incoherent memories is large, we expect their statistics to be approximately Gaussian. It is then possible to use the cumulant expansion Z^n to give a partition function with an effective Hamiltonian coupling different replicas.

$$Heff = k_B T \sum_\alpha \sum_i A(\underset{\sim}{r}_i - \underset{\sim}{r}_i^\alpha - r_{1+1}^\alpha)^2$$

$$- \sum_\alpha \sum_{ij} \langle \bar{V} (r_{ij}^\alpha - r_{ij}^T) \rangle_{av} - \frac{1}{2k_B T} \sum_{\alpha_1 \beta} \sum_{ij} \langle V'(r_{ij}^\alpha) V'(r_{ij}^\beta) \rangle_{av} \tag{14}$$

The interactions within a replica arise from the mean of the interaction potential over the database and arise primarily from the target sequence. The interactions between replicas reflect the variance of the noise. Since the noise terms may give rise to extraneous minima, this coupling between the replicas has the effect of causing different replicas to be attracted to the same positions given by these incorrect minima and; hence, gives an attraction between different replicas analogous to that in spin glass theory. This effective Hamiltonian has many different phases. As we have already discussed, we would expect an expanded or random coil state which is dominated by entropy and in which very few residues are interacting. There can also be a folded state in which the residues are localized in the neighborhood of the target structure. Typically, all of the memory proteins are compact structures with radius of gyration much smaller than the random coil. Thus, there can be a phase which is compact in which a variety of interactions are picked up from different memories. This phase would resemble, in many ways, a liquid state. This collapsed phase would have some similarity to the molten globule which has been seen in experiments on in vitro folding. We should be careful to point out, however, that depending on the statistics of the interactions, varying degrees of secondary struc- ture might be found in this molten globule state. The molten globule of a simple associative memory Hamiltonian may not have all of the charac- teristics of the molten globules found in in vitro folding. In both of the collapsed states, there is the possibility of glass transitions. The glass transition in the globule phase is most significant in machina.

The mean field theory for folding must reflect all the different possible orderings that may occur. It must evaluate several order parameters. We take these to be the radius of gyration of the polymer which will distinguish between the coil state and the globule and folded states, a Debye-Waller factor measuring the deviation from the correct folded structure and, finally, an overlap order parameter for the dif- ferent glassy states measuring how related the different minima are to each other. The route we have taken to the mean field theory is to construct a reference Hamiltonian in which different energy terms re- flect the different order parameters which we expect to see in the different phases. The mean field theory then uses a simple quadratic reference Hamiltonian of the form

$$H_{ref}/k_B T = A \sum_{\alpha} \sum_{i=1}^{N-1} (r_i^{\alpha} - r_{i+1}^{\alpha 2})$$

$$+ B \sum_{\alpha} \sum_{i+1}^{N} (r_i^{\alpha})^2$$

$$+ C \sum_{\alpha} \sum_{i=1}^{N} (r_i^{\alpha} - r_i^{T})^2$$

$$+ \sum_{\alpha,\beta} \sum_{i=1}^{N} D_{\alpha\beta} (r_i^{\alpha} - r_i^{\beta})^2 \qquad \qquad 15)$$

The energy term proportional to C measures the deviation from the target structure while a non zero C then reflects a finite Debye-Waller factor for the folded target structure. The term proportional to B will cause a general collapse of the structure without any particular memory structure being favored. B is related to the radius of gyration of the polymer.

The replica coupling terms reflect the degree of overlap of the different glassy states. To make progress, one must make some ansätz for the kind of replica symmetry breaking to be expected. It is argued that the typical style of replica symmetry breaking to expect in a model without any special symmetries is that of the Potts glass. This corre- sponds to one stage of replica symmetry breaking, and the matrix $D_{\alpha\beta}$

would have the form of a block diagonal matrix with zeros on the diagonal and D on the off diagonal. Other theories of random heteropolymer collapse by Shakhnovich and Gutin, as well as by Garel and Orland, also lead one to expect this sort of replica symmetry breaking.

The Potts glass replica symmetry breaking leads to a structure of the deep minima very similar to that in the statistical protein model, so with this assumption, one can directly relate the phenomena of glass transitions in the associative memory models to the dynamical search questions which we have discussed previously.

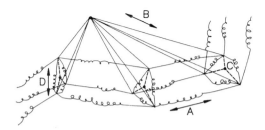

$\frac{n}{m}$ polymers of m—dim. hypertetrahedron
in random fields

Figure 1.   A diagram indicating the interactions of the reference Hamiltonian. These are conjugate to the various order parameters.

This reference Hamiltonian is rather complicated but it has a simple interpretation. It is equivalent to a set of n/m polymers in which the subunits are m dimensional hypertetrahedra. These hypertetrahedra are coupled to the minimum of the target structure harmonically. This is diagrammed in Figure 1. One can now obtain best estimates of the parameters C, B and D $\alpha\beta$ by the use of the Peierls variational principle

$$\delta\left(e^{-\langle H_{eff} - H_{ref}\rangle/k_B T} \cdot Z_{Ref}\right) = 0 \qquad 16)$$

Despite the complexity of the reference Hamiltonian, it is still harmonic. Therefore, the reference free energy and the various expectation values that are needed to compute the free energy functional can be calculated once the normal modes are known. Because of the symmetry of

Dαβ, the diagnalization, with respect to replica space, can be done. One finds one mode which involves the displacement of the entire tetrahedron and, therefore, does not contain Dαβ in the eigen frequencies, and a remaining M-1 modes which are degenerate and involve the replica symmetry breaking parameters. Once this prediagonalization is done, we see that we have two problems that involve a polymer chain with diagonal harmonic force fields. Again, the eigenmodes of this are familiar from polymer physics where they can be related to the Rouse-Zimm modes of a confined polymer.

The resulting variational free energy expression is rather complicated. In principle, it can be variationally solved given any database and encoding of the charges. The generic phase diagram can be understood more simply, making some further approximations. One takes the pair interactions as Gaussians of mean square width ϵ^{-1} and, at the same time, approximates the pair distribution of residues in the database as a Gaussian with a width of the order of the radius of gyration of typical proteins. If the width of the pair interactions are the same for all pairs, the largest part of the energy of folding comes from interactions which are distant in sequence, allowing some simplification in the formulas. In each of the different phases predicted by the model, different approximations can be made to simplify the computation of the expectation values. The resulting qualitative phase diagram is shown in Figure 2.

Several features of this phase diagram are worthy of comment. When the number of memories is rather small, there is a direct transition upon cooling from the random coil state to a folded state. This transition is first order and is analogous to the condensation of a gas directly into a crystal. When the number of memories is modest, there is a first order transition upon cooling from the coil to a molten globule state, analogous to condensation into a liquid. Now, upon further cooling, a first order transition occurs between this molten globule and the correctly folded structure. The transition temperature for this change is given by

$$kT_{MG-F}^{+} \approx \frac{v}{\log(C/A)} - MN^{-\frac{4}{3}} v \qquad (17)$$

The parameter C in this equation must be determined variationally. It is analogous to the Debye-Waller factor in forming a crystal and it has a minimum value $C \gtrsim 1.2\epsilon$. This minimum value of C reflects a Lindemann criterion for crystallization. If no glass transition intercedes, another phase transition occurs upon further cooling. The molten globule can become re-entrantly the low temperature phase. This occurs at another first order transition given by

$$kT_{MG-F}^{-} \approx MN^{-\frac{1}{3}} v \qquad (18)$$

The origin of this re-entrant phase transition is the Gaussian statistics of the misfolded memories. It can happen by chance, if their number is sufficient, that some combination state really is the true energetic ground state of the system. Typically, however, this state will have a smaller entropy than the folded one since it may require special contortions to satisfy all these unusual interactions. If the

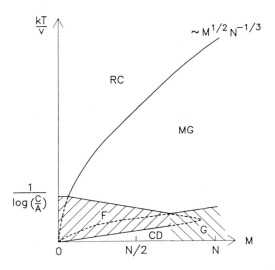

Figure 2. The phase diagram of an associative memory Hamiltonian as a function of temperature and the number of uncorrelated memories. RC designates the random coil, MG the molten globule, F the folded state, G the glassy phase and CD the cold denatured globule. The diagram is drawn for N = 100.

glass transition intervenes, these tail states will not exist. This would mean that for the associative memory Hamiltonians in this regime, the folded state is not the minimum energy state. But it still may well be the minimum free energy state under certain conditions. As remarked earlier, real proteins also have this feature in vitro, and exhibit cold denaturation; it is decidedly not the case that real proteins are always at the minimum of the energy of real systems. We should be careful, however, about taking the associative memory model too literally, for in vitro folding, since it is believed that a large fraction of the stabilization energy of the folded structure arises from hydrophobic interactions. These contain a large contribution from the entropy of the solvent. This effect is not taken into account in the associative memory Hamiltonian. In vitro cold denaturation, at least, partially, arises from the destabilization of hydrophobic interactions at low temperatures. Cold denaturation is clearly, however, also a problem in folding in machina.

Because of this re-entrant phase transition, the folded state has only a finite region of existence for uncorrelated memories. We see that if the number of memories exceeds roughly 100 for a protein of 100 residues, the folding transition will no longer exist. This is clearly one limit on the capacity of associative memory models to carry out recall. We should bear in mind, however, that since the number of protein families is finite, only a finite capacity is truly needed for practical applications.

The variational equations with respect to $D\alpha\beta$ give a glass transition. Again, there is a Lindemann criterion for the magnitude of the Debye-Waller factor of the glassy states reflecting the degree of the vibrational movement within a given minimum. This criterion indicates that $D \geq \epsilon$.

Potts glass replica symmetry breaking leads, again, to an entropy catastrophe like that in the random energy model and this gives a thermodynamic glass transition temperature

$$kT_G \sim \frac{M^{1/2} N^{-2/3} v}{\sqrt{\log(D/A)}} \tag{19}$$

The glass transition temperature increases with the number of memories and because the Lindemann criteria for glassy minima and the true minima are similar, the glass transition temperature becomes of the order of the folding temperature, roughly at the value of N, where the re-entrant phase transitions occur. Thus cold denaturation and glass transitions give very similar estimates of the capacity of these models. For proteins of length 100, this capacity is of order 70 uncorrelated memories.

Another qualitative result comes from the mean field theory. Potts glasses in the mean field limit exhibit a dynamical transition in which the individual glassy states acquire barriers which scale with the size of the system.[33] This dynamical transition occurs at a higher temperature than the obligate freezing temperature. It is related closely to the transitions obtained using mode coupling theories for glass systems.[34] For the associative memory protein models, this dynamical freezing temperature is close to the temperature at which the globule is formed, thus, the motions within the globule actually involve barriers that would be extremely high in the true thermodynamic limit. For proteins of size 100, however, these barriers may be only on the order of a few kT because of the slow scaling of barrier height with size expected in spin glasses.

As we shall see in the next Section, the qualitative features of this phase diagram give an excellent orientation to the results of computer simulations of folding. It will be worthwhile to calculate more precise phase diagrams for associative memory Hamiltonians. An important issue is the role of correlations between different memories. This is not difficult to include formally. With such a correlated theory, it should be possible to quickly examine the viability of different encodings of the protein sequence. It is clear from the dynamical theory the we have discussed that it would be ideal if the folding temperature was as large as possible compared to the glass transition temperature. This provides a criterion for discussing what makes a good encoding. Since the encoding can be very complicated and depend on context, this provides a route very analogous to back propagation for determining appropriate codes and should provide a much more efficient route to examining appropriate procedures for prediction than simply trial and error simulations with different codes.

Both the replica and variational methods that are used for the simple associative memory Hamiltonian can be used to describe the more complex multiplet codes introduced in the last Section. These same ideas may also be useful in treating more detailed models of proteins that include complete atomistic detail. Thus the development of the spin glass techniques may contribute to our understanding of in vitro folding when combined with appropriate computer simulations.

V.   COMPUTER SIMULATIONS OF ASSOCIATIVE MEMORY HAMILTONIANS

Since the primary aim of introducing associative memory Hamiltonians is to develop techniques for predicting protein tertiary structure, it is necessary to study the computational process of minimizing the energy through various schemes including simulated annealing. Thus computer simulation studies have been carried out by Friedrichs and me.[6,8] A byproduct of these studies is a qualitative confirmation of the complex phase diagram and of the associative memory Hamiltonians and the development of ideas as to the mechanism of folding in these models. Up to this point, quantitative comparisons of theory and simulation have not been carried nor have truly detailed mechanistic investigations been undertaken. Both of these are important tasks for the future.

In this Section, I will review some of the interesting features from these qualitative studies and refer the reader to the original papers for details, both of the simulation technique and of the analysis of results. Most of our studies have been on rather small proteins. Two tiny proteins, rubredoxin with 54 residues and bovine pancreatic tryposinine inhibitor with 58 residues, have been our major test cases. We have carried out a few studies with cytochromes which have roughly 100 residues, but experimentation on the larger proteins is computationally costly. At the same time, there are disadvantages to investigating small proteins. First, in the laboratory it is known that covalent interactions, such as disulfide bridges and auxiliary ligands provide a great deal of the stability for the smaller proteins. Because of this, they can tolerate an especially large degree of deviation from ideal patterns of packing of secondary structure and placement of hydrophobic residues. Thus conclusions from these computational studies may well be modified when even modestly larger proteins are studied. In addition, the dramatically larger proteins that interest many biologists often have domain structure and this may make still more dramatic modifications in the picture of folding.

Analysis of simulations of folding is difficult. Folding has often been cited as a problem in which a full movie of folding might well be too dull or too complicated to be analyzed. In fact, since we are seeking energy functions which lead to rapid folding, it may be that the problem of boredom, at least, can be solved. In any case, masses of information come out of computational studies and only limited parts of this information have been digested and analyzed.

In the early work, energy minimization was performed using a Monte Carlo based simulated annealing. Trial configurations were obtained by randomly choosing a site along the chain and then rigidly rotating that chain segment and the further segments through a random paid of dihedral bond and torsion angles. The angles were chosen from a database of angles derived from known protein structures. In the standard Metropolis way, moves which lowered the energy were always accepted and those which raised the energy were accepted with a probability that depended on a temperature. The temperature was set at an initially high value and this was then gradually lowered. The Monte Carlo procedure was rather unwieldy. It was necessary to fold structures starting with nucleated configurations; that is the first 10 or 20 residues of the protein were fixed in their crystallographic positions. The remainder of the protein was added in stages, generally two stages, and the minimization was carried out just for the new residues. These studies suggested that recall was possible when the database consisted of between 40 - 80 proteins, even for these small proteins.

In addition to giving a rather fuzzy view of the capacity, these studies had the disadvantage that the Monte Carlo procedure was very unlike the kind of motions that one might expect in real proteins. The rigid rotations lead to extremely high barriers for motions in the interior of the sequence compared to those at the end. Thus it often developed that the ends of the protein structure were rather accurate whereas the middles were rather poor.

An alternate scheme based on molecular dynamics has many nice features. First the motions are somewhat more natural. Although, one cannot argue that they explicitly imitate the detailed conformational changes, they do go on in parallel through the whole structure and complex reorganizations are possible in the MD scheme. This parallelism also makes it possible to exploit supercomputers and, in fact, the molecular dynamics approach seems to be computationally much more efficient than the simple Monte Carlo. Finally, a smooth approach to a local minimum is very easy within the MD scheme once the system has been trapped in the correct basin.

In the molecular dynamics simulated annealing approach, the system is occasionally thermalized by re-randomizing the velocities of the particles. Thus one can specify the temperature as a function of time. In most of the runs, the temperature is lowered from roughly 1 (in dimensionless units) to 5×10^{-3} where the average energy per residue is .1. We have found better annealing schedules since our first publication on this, but I will not describe those results here. The typical timesteps which are taken correspond at T=1 to a distance per timestep of roughly 10^{-1}Å. A global view of the mechanism of folding upon this annealing can be seen in Figure 3. In this figure we plot the radius of gyration of the protein as a function of time during the annealing schedule. The temperature is being continuously lowered. Also plotted

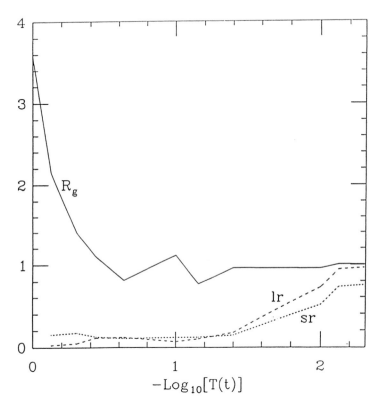

Figure 3. Time course of annealing for BPTI with a database of 45
proteins including the target. The plot shows radius of
gyration and fraction of correct short and long range
structure as a function of T(t).

in this figure are two quantities which we call the short-range structure and the long-range structure. These are the fraction of the pair distances which are either short-range in sequence, i.e., i-j<6, (this is labelled sr) or long-range in sequence (lr) that have achieved the correct values at that time. It is clear that in these runs, there is first a rapid collapse to a state which would resemble the molten globule and then, rather later in the run, at much lower temperatures, a reasonably rapid growth of both of long-range and short-range structure simultaneously. Sometimes this growth can be quite explosive, reminiscent of a highly undercooled liquid crystallizing.

The capacity, as defined by equilibrium statistical mechanics, is a fixed number. On the other hand, in molecular dynamics based simulated annealing, the effective capacity depends on the annealing schedule and, particularly, on its length. For shorter runs, less noise can be tolerated in the Hamiltonian and, so, short runs only successfully give a good final structure if a few database proteins are included in memory. For BPTI, the effective capacity is 29 dabatase proteins for runs of 1.2 x 10^6 timesteps; 40 for 2.5 x 10^6 timesteps; and 45 for 3.6 x 10^6 timesteps. These latter runs take approximately 2 hours on a CRAY 2. This is one of the reasons we have not collected elaborate statistics about these results. In runs where the effective capacity is only slightly exceeded, the errors in the final structure can usually be ascribed to defects in chirality. Hamiltonians based only on distances cannot distinguish between right-handed and left-handed structures and, therefore, domains of left-and right-handed conformations can coexist with only moderate energetic costs. Also entanglements exist in the incorrect structures.

Understanding these errors is important since they suggest ways in which the Hamiltonians can be improved. Clearly, more accurate descriptions of excluded volume and better descriptions of stereochemistry can help a lot. We are currently pursuing studies based on these.

The studies we have just discussed have used an encoding of sequence which is binary. The charge which is used is a measure of the hydrophobicity. Interestingly, when simulations fail, they still often give results that resemble one's qualitative ideas such as the idea the hydrophobic residues are found inside and hydrophilic outside. Although the simple associative memory Hamiltonian without sequence translation invariance does not do well in recognizing sequences with insertions and deletions, it can lead to robust results when only site mutations are allowed. For example, the two Rubredoxins from Desulfovibrio vulgaris and Clostridium pasteurianium have different amino acid residues at almost half their positions. With the hydrophobicity scale, only 6 positions are encoded differently. Thus it is perhaps no surprise that if the sequence of the D. vulgaris form is used with the database containing only the C. pasteurianium protein, annealing effectively goes to the correct structure.

We have also carried out studies in which random site mutations are made, changing the encoding. These studies suggest that if the hydrophobicity of 10 percent of the sites is reversed, the effective capacity is hardly reduced; whereas, if 20 percent of the sites are changed, there is a dramatic reduction of capacity of 20 - 25 percent. When the fraction of mutated sites exceeds roughly 30 percent, it is extremely difficult with large databases to obtain a correct structure. The studies of larger proteins, such as cytochrome and ribonuclease with 124 residues do show that the capacity increases with the size of the protein, at least, linearly. The computational studies suggest an effective capacity at least of the order of .5N - .7N, using the hydrophobicity code.

VI. PROTEIN STRUCTURE PREDICTION AS AN INSPIRATION FOR PHYSICS

I have tried to suggest in this lecture that spin glass theory provides a natural viewpoint for looking at some important aspects of protein folding and in particular it gives insights into the design requirements for a good protein structure prediction schemes. The protein structure prediction goal is an important one and while it is clear that our current knowledge of spin glass physics can suggest several important avenues of study, it is still clear that there is quite a ways to go.

Our understanding of the search problem in protein structure prediction is still rudimentary. The analysis, based on the random energy model, gives a picturesque result. This result may have some elements of universality because the random energy model captures many of the universal characteristics of the spectrum of deep minima of spin glasses. Nevertheless, proteins are finite systems and protein structure predictions must be made using a finite amount of computer time. Thus there are numerous questions as to how to describe the dynamics of spin glass models with finite range interactions and on finite time-scales. Perhaps such studies would lead to a more precise understanding of the concept of folding pathways, at lease for model Hamiltonians. One can also debate whether the slavish use of Monte Carlo and molecular dynamics schemes is appropriate for computational search. Perhaps other global combinatorial schemes can give better insights or genetic algorithms that search for minima by carrying out artificial evolution among structures might allow one to tunnel through barriers rather than diffusing over them.[35] Evolution on spin glassy landscapes is another topic of this workshop and the results of these studies may be of value to computational protein folding.

The problems of recognition and statistical inference also inspire further developments in the theory of spin glasses. Clearly, one aspect of protein structure prediction is to recognize structure for sequences that have mild degrees of insertion and deletion. We know from neural network approaches that this requires an understanding of the extent to which such insertions and deletions are possible and a limited set of extended symmetries these Hamiltonians might possess. We have already carried out some computational studies on this score but it is clear that we need to understand more fully, the role of correlated memories. In devising a database, what sorts of a priori constraints on sequences can be used to make better predictions? What insertions and deletion transformations take you from one structure in a family to another? Can I decide, from a sequence, which set of families a protein might lie in? If families are unequally represented in the database, how should they be weighted? These questions may be answerable if analytic estimates of capacity can be made for these cases, all of which would involve the problem correlated memories.

The associative memory Hamiltonian framework suggests that good protein folding codes would have the property that the associated Hamiltonian has a large folding temperature compared with the glass transition temperature. Both temperatures are reasonably well defined for Hamiltonians with long-range interactions. It may be necessary for generalization to limit the range of interaction. In this case an understanding of the dynamics of the short-range models will be necessary and will require going beyond the meanfield theory treatment described here. Here a scaling theory of the dynamics of short-range associative memory Hamiltonian models would be desirable. Even for the long-range models, there is the question both of the ideal encodings that we have alluded to already and the problem of ideal annealing schedules. Here the dynamics of spin glasses must be worked out in greater detail and, especially the question of how spin glass structures change with temperature or with modifications of the Hamiltonian are needed. Finally, the multiplet versions of the associative memory models raise new questions as to the nature of the phase diagram and dynamics. This is analogous to questions about the enhanced capacity of Hopfield models with multispin interactions.

I believe all of these points suggest that a deeper understanding of spin glasses will allow an organized approach to protein structure prediction. Spin glass theory may be able to change structure prediction from an art form into a discipline of engineering. Clearly, more work is in store for all of us.

ACKNOWLEDGEMENT

I have benefited from conversations with many people about the protein folding problem. I wish to single out for thanks here, J. D. Bryngelson, M. Friedrichs and M. Sasai who have worked very hard on developing the appropriate science so that spin glass engineering may come to be useful in biomolecular physics. I also wish to thank K. Bucher and H. Drickamer for linguistic suggestions. Our early work was supported by grants NSF CHE 84-18619 and NSF DMR 86-12860. Our current studies on folding are supported by NIH grant PHS 1 RO1 GM44557-01. Computational studies were carried out at the National Center for Supercomputing Applications in Urbana.

REFERENCES

1. L. M. Gierasch and J. King, "Protein Folding," AAAS, (1990).
2.   G. E. Schulz and R. H. Schimer, "Principles of Protein Structure," Springer-Verlag, New York (1979).
3. J. D. Bryngelson and P. G. Wolynes, Spin Glasses and the Statistical Mechanics of Protein Folding, Proc. Natl. Acad. Sci. 84:7524 (1987).
4. P. G. Wolynes, Spin Glass Ideas in the Protein Folding Problem, in: "Spin Glasses and Biology," D. L. Stein, ed., World Scientific Press, New York (to appear 1990).
5. J. D. Bryngelson and P. G. Wolynes, Intermediates and Barrier Crossing in a Random Energy Model (with Applications to Protein Folding), J. Phys. Chem. 93:6902 (1989).
6. M. S. Friedrichs and P. G. Wolynes, Toward Protein Tertiary Structure Recognition by Means of Associative Memory Hamiltonians, Science 246:371 (1989).
7. J. D. Bryngelson and P. G. Wolynes, A Simple Statistical Field Theory of Heteropolymer Collapse with Application to Protein Folding, Biopolymers 30:177 (1990).
8. M. S. Friedrichs and P. G. Wolynes, Molecular Dynamics of Associative Memory Hamiltonians for Protein Tertiary Structure Recognition, Tetrahedron Computer Methodology, (to appear 1990).
9. M. Sasai and P. G. Wolynes, Molecular Theory of Associative Memory Hamiltonian Models of Protein Folding, preprint.

10. T. Garel and H. Orland, Mean Field Model for Protein Folding, Europhys. Lett. 6:307 (1988).
11. T. Garel and H. Orland, Chemical Sequence and Spatial Structure in Simple Models of Biopolymers, Europhys. Lett. 6:597 (1988).
12. E. I. Shakhnovich and A. M. Gutin, The Nonergodic (Spin-glass Like) Phase of Heteropolymers with Quenched Disordered Sequence of Links, Europhys. Lett. 8:327 (1989).
13. E. I. Shakhnovich and A. M. Gutin, Frozen States of Disordered Heteropolymers, J. Phys. A22:1647 (1989); A. V. Finkelstein and E. I. Shakhnovich, Theory of Cooperative Transitions in Protein Molecules. 2. Phase Diagram for a Protein Molecule in Solution, Biopolymers 28:1681 (1989).
14. E. I. Shakhnovich and A. M. Gutin, Formation of Unique Structure in Polypeptyde Chains: Theoretical Investigation with the Aid of Replica Approach, Biophys. Chem. 34:187 (1989).
15. E. I. Shakhnovich and A. M. Gutin, Relaxation to Equilibrium in the Random Energy Model, Europhys. Lett. 9:569 (1989).
16. E. I. Shakhnovich and A. M. Gutin, Implication of Thermodynamics of Protein Folding for Evolution of Primary Sequences, Nature (to appear 1990).
17. E. I. Shakhnovich and A. M. Gutin, Protein Folding as Pattern Recognition, Studia Biophysica 132:47 (1989).
18. J. D. Honeycutt and D. Thirumalai, Metastability of the Folded States of Globular Proteins, Proc. Natl. Acad. Sci. 87:3526 (1990).
19. J. J. Hopfield, Neural Networks and Physical Systems with Emergent Collective Computational Abilities, Proc. Natl. Acad. Sci. 79:2554 (1982).
20. B. Derrida, Random Energy Model: Limit of a Family of Disordered Models, Phys. Rev. Lett. 45:79 (1980).
21. J. S. Richardson, The Anatomy and Taxonomy of Protein Structure, Adv. Protein Chem. 37:167 (1981).
22. C. A. Angell, D. R. MacFarlane and N. Oguni, The Kauzmann Paradox, Metastable Liquids and Ideal Glasses: A Summary, Ann. New York Acad. Sci., 484:241 (1987).
23. P. L. Privalov, Stability of Protein: Small Globular Proteins, Adv. Protein Chem. 33:167 (1979).
24. C. Levinthal, Are there Pathways for Protein Folding? J. Chim. Phys. 65:44 (1968).
25. B. Derrida, Mean Field Theory of Directed Polymers in a Random Medium and Beyond, in: "Complexity and Evolution," Les Honches (1990).
26. H. A. Scheraga, Calculations of Stable Conformations of Polypeptides, Proteins and Protein Complexes, Chemica Scripta 29A:3 (1989).
27. M. Levitt and A. Warshel, Computer Simulation of Protein Folding, Nature 253:694 (1975).
28. I. D. Kuntz, An Approach to the Tertiary Structure of Globular Proteins, J. Amer. Chem. Soc. 97:4364 (1975).
29. J. Skolnick, A. Kolinski and R. Yaris, Monte Carlo Simulations of the Folding of Beta Barrel Globular Proteins, Proc. Natl. Acad. Sci. 85:5057 (1988).
30. J. Denker, et al, Large Automatic Learning Rule Extraction and Generalization, Complex Systems 1:877 (1987).
31. M. J. Rooman, and S. J. Wodak, Identification of Predictive Sequence Motifs Limited by Protein Structure Data Base Size, Nature 335:45 (1988).
32. J. W. Ponder and F. Richards, Tertiary Template for Proteins - Use of Packing Criteria in the Enumeration of Allowed Sequences for Different Structural Classes, J. Mol. Biol. 193:775 (1987).

33. T. R. Kirkpatrick and P. G. Wolynes, Stable and Metastable States of Mean Field Potts and Structural Glasses, Phys. Rev. B36:8552 (1987).
34. T. R. Kirkpatrick and P. G. Wolynes, Connections Between some Kinetic and Equilibrium Theories of the Glass Transition, Phys. Rev. A35:3072 (1987).
35. M. S. Friedrichs and P. G. Wolynes, Genetic Algorithms for Model Biomolecular Optimization Problems, preprint.

RANDOM HETEROPOLYMERS FOLDING

Giorgio Parisi

Dipartimento di Fisica II Università di Roma, Tor Vergata
and INFN, sezione di Roma, Tor Vergata
Via E. Carnevale, Roma 00173, Italy

1. INTRODUCTION

In these recent years many progresses have been done in the study of the behavior of disordered systems, focusing on those features which are proper of disorder. The most interesting results have been obtained when the laws which control the evolution of the system are themselves disordered, i.e., chosen at random (see for example refs. 1,2,3,4,5).

A very rich theory has been developed, quite advanced mathematical tools have been used and the final results are very interesting. Completely unexpected mathematical structures have been discovered and these structures have many properties which recall in some sense biology. At present physicists have just started the investigations of these disordered systems. It is likely that much more precise, interesting and general results will be obtained in the future.

Many people have suggested that the time is ripe to start thinking how these new ideas can be applied in the biological sciences, and in particular to the study of the behaviour of proteins (see for example refs. 6,7).

Real proteins have characteristic time scales for different kinds of motions which go from the picosecond to the second; the same protein may exist in different foldings and sharp proteinquakes characterize the transition from one folding to an other. The time dependence of the response of proteins to external perturbations is quite similar to that of spin glasses, and a strong experimental effort has been devoted to find out the similarities and the differences among proteins and spin glasses [8,9,10].

It would be extremely nice to find out that some of the experimentally observed properties are not peculiar of real proteins, but are shared with all random heteropolymers (i.e. polymers composed of different monomers in a random order). In this case they could be understood from general principles, without having to look into the details of the chemistry of aminoacids.

It is quite clear to me that aminoacids have their peculiar chemical properties and their interaction is not random; moreover the mostly studied proteins come out from the process of natural selection. It is however worthwhile to start the investigation from the simplest theoretical hypothesis: i.e. to consider the case of completely

Biologically Inspired Physics, Edited by L. Peliti
Plenum Press, New York, 1991

random system. Eventually, when the theory of completely random system will be developed, we will see the difference between random heteropolymers and real proteins. At this stage we will incorporate in the theory the information coming from the chemistry.

Unfortunately at the present moment the theoretical analysis of protein folding from the point of view of statistical mechanics is still at its infancy; before drawing definite conclusions we must wait for the development and the refinement of a general theory of heteropolymer folding. The most interesting results have been obtained in ref.11; these results however are based on some approximations whose validity is doubtful in the real three dimensional word (it is likely that they are correct in the unphysical case where the dimensions of the space goes to infinity). A careful analysis of the theoretical predictions is still missing.

2. TWO MODELS FOR HETEROPOLYMERS

In this lecture I will introduce two models for random heteropolymers:

(a) the substrate model, in which the heteropolymers interact randomly with a substrate

(b) the self interacting model, in which the heteropolymer interacts randomly with itself.

The Hamiltonian for these two models is given by

$$H = H_0 + H_R, \tag{1}$$

where

$$H_0 = \sum_{i=1,N-1} (x_i - x_{i+1})^2, \tag{2}$$

N being the number of monomers.

H_R is given in model (a) by

$$H_R = \sum_{i=1,N} V_i(x_i), \tag{3}$$

where the V_i are independent random functions, with zero average and variance given by

$$\overline{V_i(x)V_j(y)} = \delta_{i,j} f(x - y). \tag{4}$$

In model (b) we have

$$H_R = \sum_{i,j=1,N} V_{i,j}(x_i - x_j), \tag{5}$$

where the $V_{i,j}$ are random functions. They are given by

$$V_{i,j}(x) = g_{i,j} f(x), \tag{6}$$

where the $g_{i,j}$ are quantities with zero average and variance given by

$$\overline{g_{i,j}g_{k,l}} = \delta_{i,k}\delta_{j,l} + \delta_{i,l}\delta_{j,k}. \tag{7}$$

In both cases the function $f(x)$ characterizes the range of the interaction. In many cases we can assume that the interaction is short range and that $f(x)$ may be well approximated by $\delta(x)$.

We are interested to study the most likely behaviour of these two models in the limit where the number N of monomers goes to infinity, for generic choices of the random potentials V.

The model which is really interesting for possible applications to protein physics is model (b). However it is not completely well understood for the theoretical point of view and numerical simulations are practically non existing. Model (a) is much more studied both analytically and numerically [12,13,14,15] (the usual name in the literature of this model is directed polymers in a random medium). In this lecture I will try to describe which should be the properties of model (a) and use these results to guess the properties of model (b) assuming that the difference in behaviour of these two models is small.

This educated guess may be justified if we consider the following two observation.

Let us consider a very long heteropolymer in model (b) and let us look to the interaction of a small part of it with the rest. If we do the approximation of considering the rest of the heteropolymer as fixed, we recover model (a). Model (a) can be therefore considered as a mean field approximation to model (b).

If we use the technique of broken replica symmetry and we perform a Gaussian variational approximation in both model (a) [16] and a similar approximation in model (b) [11], we find exactly the same equations. The two model differ by quantities which will appear only when we compute the corrections to the variational approximation.

In both models one finds that replica symmetry is spontaneously broken (as in spin glasses) and that different quantities behave as powers of N, with different exponents. The approximations that have been done are valid only in the limit of a space with very large dimensions and the crucial analysis of the corrections in finite dimension is under way [17].

When the heteropolymers are in the globular phase, it is crucial to study the behaviour of the following quantity:

$$d_{a,b}^2 = \frac{1}{N}\sum_{1,N}(x_i - y_i)^2, \tag{8}$$

where a and b are two equilibrium configurations of the heteropolymer with corresponding coordinates x and y. We notice that in model (b), in order avoid trivial results, the relative position and orientation of the two heteropolymers have to be adjusted in order to minimize d .

The probability distribution of d gives us crucial informations on the statistics of the folding. If the polymer has no preferential folding we expect that the value of d for two generic configurations is of the same order to the radius R of the polymer, while if it has some preferential folding it will be much smaller.

In the substrate model the most likely value of d^2 increases with N with an apparent small power law for not too large N (e.g. 0.25 in one dimension [14,15]). Configurations with too large d are suppressed. The free energy increase (relative to the ground state) for having a configuration at distance d from the ground state of order R is likely to decrease as a power of N (e.g. $N^{-\frac{1}{3}}$ in one dimension). There are however choices of the potential such that two quite different configurations with similar free energies are present; such a potential appears with a probability that decreases as a power of N at large N.

It seems that the free energy as function of d has many minima, which corresponds to different, but not too different, foldings. These minima should be separated by maxima in free energy and a detailed study of the free energy barriers should be crucial for understanding the dynamics of the system, in a similar way as it happens in spin glass like models [18,19].

3. CONCLUSIONS

We could guess that a similar behaviour (possibly with different exponents) is also present in model (b). Numerical simulations of this model have actually been done[20], but the final results are not yet available. It would be interesting to see if a more careful theoretical analysis and maybe an experimental study of the folding properties of random peptides will confirm these guesses.

At the present moment most of the theoretical effort is concentrated on model (a), also because it may be a nice testing ground of new ideas (for example replica symmetry breaking). I hope that in the future, after that the relevance of replica symmetry breaking in direct polymers will be established, the same techniques may be successfully applied also to the biologically more relevant model (b).

REFERENCES

1. Parisi G., *Field Theory and Statistical Mechanics*, ed. by Zuber J. B. and Stora R. North Holland, Amsterdam (1984).
2. Rammal R., Toulouse G. and Virasoro M.A., Ultrametricity for physicists, *Rev. Mod. Phys.* 58: 765 (1986).
3. Mezard M., Parisi G. and M. Virasoro, *Spin Glass Theory and beyond*, Word Scientific, Singapore (1987).
4. Peliti L. and Vulpiani A. Eds., *Measures of Complexity*, Springer Verlag Berlin (1988).
5. Parisi G., On the natural emergence of tree-like structures in complex systems, in Solbrig et al. *to be published* .
6. Garel T. and Orland H., Mean-Field Model for Protein Folding, *Europhys. Lett.* 6: 307 (1988).
7. Stein D. L., *Proc. Nat. Acad. Sci. USA* 82: 3670(1985).
8. Ansari A., Berendzen J., Bowne S.F., Frauenfelder H., Iben I.E.T., Sauke T.B., Shyamsunder E. and Young R., Protein states and proteinquakes, *Proc. Natl. Acad. Usa* 82: 5000 (1985).
9. Frauenfelder H., Steinbach P.J. and Young R.D. Conformational relaxation in Proteins, *Chemical Scripta* 29A: 1878 (1989).
10. Iben I.E.T., Braunstein D., Doster W., Frauenfelder H., Hong M.K., Johnson J.B., Luck S., Ormos P., Schulte A., Steinbach P.J., Xie A.H. and Young R.D., Glassy Behaviour of a Protein, *Phys. Rev. Lett.* 62:1916 (1989).

11. Shaknovich E.I. and Gutin A.M., The Nonergodic (Spin Glass Like) Phase of Heteropolymer with Quenched Disordered Sequence of Links, *Europhys. Lett.* 8: 327 (1989).
12. Kardar M. and Zhang Y.-C., *Phys. Rev. Lett.* 58: 2087 (1987).
13. Derrida B. and Spon H., *J. Stat. Phys.* 51: 817 (1988).
14. Parisi G., On the replica approach to random directed polymers in two dimensions, *J. Phys. France* 51: 1595 (1990).
15. Mezard M., *J. Physique France* 51:1423 (1990).
16. Mezard M., Parisi G., Interfaces in a random medium and replica symmetry breaking *J. Phys. A* in press.
17. Mezard M. and Parisi G., *in preparation*.
18. Vertechi A. and Virasoro M.A., Energy barriers in SK spin-glass model, *J. Phys. France* 50: 2325 (1989).
19. Macken C. and Perelson A., Protein evolution on rugged landscape, *Proc. Natl. Acad. Sci. Usa.* 86: 666 (1989).
20. Iori G., Marinari V. and Parisi G., *in preparation*.

DNA TOPOLOGY

Maxim Frank-Kamenetskii

Institute of Molecular Genetics
Academy of Sciences of USSR
Moscow, 123182 USSR

1. INTRODUCTION

The fundamental stimulus for the development of the theory to be presented in this paper was the discovery of circular DNAs. We recall that DNA molecules, which contain all the information on the structure of living organisms, consist of two polymer chains attached to one another by weak, noncovalent interactions. These chains form a double helix in which $\gamma_0 = 10$ monomer links (base pairs) occur per turn. Actual DNAs contain from several thousand to billions of monomer links. Initially the main attention was focused on studying the properties of linear DNA molecules, since this is precisely the form of DNA that could be extracted from cells and virus particles.

It was unexpectedly found in 1963 that DNA exists in certain viruses in a closed circular (CC) form. In this new state the two single chains of which the DNA consists are each closed on themselves. CC DNA is illustrated schematically in Fig.1. We see that the two complementary chains in CC DNA proved to be linked. Here they form a high-order linkage (of the order of N/γ_0, where N is the number of pairs in the DNA). Initially this discovery was not seen to be very significant, since this form of DNA was regarded as exotic. However, in this course of time, the CC form of DNA was discovered in an ever greater number of organisms. Currently it is generally acknowledged that precisely this form of DNA is typical of the simplest DNAs, and also of the cytoplasm DNAs of animals. Also most virus DNAs pass through a stage of the CC form in the course of infection of cells. Such a widespread occurrence of this form of DNA in nature has elicited the interest in its structure and properties that has been manifested in recent years.

The discovery of CC DNA has led to the formulation of fundamentally new problems, since it has turned out that many of the physical properties of the CC form differ radically from those of the linear form. The difference between the properties of these two forms of DNA is not at all due to the existence of end effects in the one case but not in the other. They involve all regions of the molecule and are caused entirely by the topological restrictions that arise in the CC form of DNA. Evidently these topological restrictions are immediately eliminated after even one of the chains has been broken. Therefore the special properties of CC DNA vanish not only when it is converted to the linear form, but also when the nicked circular form is formed, i.e., the form in which one of the strands has been broken, while the other remains closed into a ring.

Biologically Inspired Physics, Edited by L. Peliti
Plenum Press, New York, 1991

2. KNOTS

The first problem that arises in analyzing ring polymer chains consists of the following. Let a ring molecule be formed by fortuitous closure of a linear molecule consisting of n segments. What is the probability of forming a knotted chain, i.e., a nontrivial knot? This problem has been clearly formulated by Delbruck (1962) and first solved by our group (Vologodskii et al., 1974; Frank-Kamenetskii et al., 1975). To solve the problem one needs, first of all, the knot invariant. Indeed, a closed chain can be unknotted (mathematicians call such chain "trivial knot") or be a knot of different type. The very beginning of the table of knots is shown in Fig.2. However, an analytical expression for the knot invariant is unknown and most probably does not exist at all. Note that Edwards (1968) proposed such an invariant but we showed that Edwards' expression is actually not invariant towards deformations of the closed chain (Vologodskii et al., 1974). Therefore we had to use a computer and algebraic invariant elaborated in the topological theory of knots. We found that the most convenient invariant is the Alexander polynomial (for review see Frank-Kamenetskii and Vologodskii, 1981).

The next problem consisted in generating closed polymer chains. In our first calculations we simulated DNA as a freely-joint polymer chain. Several methods exist to generate exclusively closed chains for this model (see Frank-Kamenetskii and Vologodskii, 1981 and Klenin et al. 1988).
Using these methods and teaching the computer to calculate the Alexander polynomials and therefore to distinguish the knots of different types, we could calculate the knotting probability.

Analogous calculations have been performed later on by other people. All the published data on the relationship between the probability of knot formation and the number of segments in the chain, as well as the results of our recent calculations, are collected together in Fig.3. We see that the results obtained by various authors agree very well with one another. This is not surprising, since, in spite of a certain difference in the polymer models employed, to which certain differences in the results are due, the presented data in all cases fit the model of an infinitely thin polymer chain.

We see from Fig.3 that the probability of knot formation has an evident tendency to approach unity as n increases, though it has been possible to perform the calculations only up to n values such that P barely exceeds 0.5. The extensions of this relationship into the region of large n is hindered by the memory size and speed of the computer. Of course, there is a limitation in principle, which involves the fact that certain nontrivial knots have the same Alexander polynomial as an unknotted chain (trivial knot) does. However, such knots constitute such an infinitesimal fraction of all the knots, so that this restriction becomes substantial only at n values for which the probability of knot formation already has become unity.

In spite of persistent, many-years of work on the chemical synthesis of knots, this has not yet succeeded. Molecular knots were first constructed of the DNA molecules. However, before explaining how it was done, let us turn to an other DNA topological notion, the linking number.

3. DNA SUPERCOILING

From the scheme in Fig.1 it is clear that two complementary strands of DNA form a link, in topological terms. One can present a table of links similar to the table of knots in Fig. 2. However, because two complementary strands of DNA are attached to each other forming the double helix, the links which DNA can form, belong to a subclass of all possible links. Namely, the form the so called torus links because the two strands could be put into a torus. For torus links the well known Gauss integral is a strict topological invariant.

Fig.1.　Schematics of closed circular (CC) DNA.

3_1　4_1　5_1　5_2　6_1

6_2　6_3　7_1　7_2　7_3

7_4　7_5　7_6　7_7　8_1

8_2　8_3　8_4　8_5　8_6

8_7　8_8　8_9　8_{10}　8_{11}

8_{12}　8_{13}　8_{14}　8_{15}　8_{16}

8_{17}　8_{18}　8_{19}　8_{20}　8_{21}

Fig.2.　Different types of knots.

47

There is another point on the torus links. The two strands in this case could be treated as the edges of a ribbon. Therefore the theory of torus links is actually the ribbon theory.

The application of topological ideas to studying the properties of CC DNA was started by Fuller (1971), when he applied the results of ribbon theory to analyzing the properties of these molecules. According to this theory, besides the topological characteristic of a ribbon - its linking number Lk, also two differential-geometric characteristics play an important role, the twist Tw of the ribbon, and its writhing number, Wr. All three characteristics are interrelated by the condition:

$$Lk = Tw + Wr$$

CC DNA is generally not characterized by the total quantity Lk, but by the number of excess turns (the number of supercoils τ):

$$\tau = Lk - N/\gamma_0$$

We stress that the quantity N/γ_0 is rigorously fixed under given external conditions. However, upon changing the external conditions (temperature, composition of solvent, etc.), the quantity γ_0 varies. Therefore the number of supercoils, in contrast to Lk, is a topological invariant of DNA only under fixed external conditions.

Very valuable information on the energy and conformation characteristics of CC DNA has arisen from experiments in which the value of Lk could vary, and the equilibrium distribution of the CC molecules with respect to Lk was studied. The most convenient way to vary Lk is to employ special enzymes, which have been called topoisomerases.

The studies under discussion employed type I topoisomerases, which alter the topological state of CC DNA by breaking and recombining only one of the strands of the double helix. These enzymes relax the distribution of the molecules with respect to Lk to its equilibrium form. The very sensitive gel-electrophoresis method was used for analyzing the distribution of the CC DNA molecules with respect to Lk in these studies. This method can separate molecules of CC DNA that differ by unity in Lk. Naturally, the maximum of the equilibrium distribution always corresponds to $\tau=0$. We note that, although the quantity τ can only adopt discrete values that differ by no less that unity, it is not required to be an integer. Therefore, as a rule, molecules having $\tau=0$ do not appear in a preparation. It is also evident that molecules having values of τ close in absolute magnitude must have close-lying mobilities and lie close together in the gel. A distribution in which the molecules having positive and negative values of τ are separated are obtained when the electrophoresis is performed under conditions differing from those in which the reaction with the topoisomerase is conducted. The change in conditions means that we must substitute some other value γ'_0 in place of γ_0 in the equation for τ above without changing Lk. This means that the entire distribution is shifted by the amount $\delta\tau=N[(1/\gamma_0) - (1/\gamma'_0)]$. Then the molecules that possessed the value τ in the original distribution will possess the values $\tau'=\tau + \delta\tau$ in the new distribution. If $\delta\tau$ is large enough, all of the topoisomerases are well separated.

Experiments have shown that the obtained distribution is always normal. The variance of this normal distribution was measured for different DNAs. These experiments have played a very important role in studying the physical properties of CC DNA. They have made it possible for the first time to obtain a reliable estimate of the torsional rigidity of the double helix and to find what fraction of the supercoiling τ is realized in the form of a change in a axial twist Tw, and what fraction in the form of the writhing number Wr. Under conditions allowing multiple breaks and reconnections of one of the strands of DNA, i.e., in

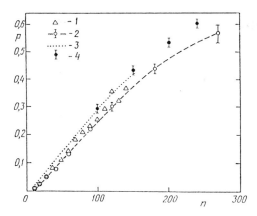

Fig.3   Probability of knot formation as a function of number of segments n for the freely-jointed model of polymer chain.

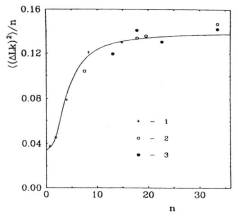

Fig.4   Dependence of the variance of the equilibrium distribution of closed DNA molecules over topoisomers, $<(\Delta Lk)^2>$, on the number of Kuhn statistical segments n. Data 1 are taken from Horowitz and Wang (1984), data 2 are from Depew and Wang (1975), and data 3 are from Pulleyblank et al. (1975) and Shure et al. (1977). The curve is calculated on the basis of Equations 3 to 7 for $C = 3^{-19} \times 10$ erg cm, $b = 100$ nm, and $d = 2$ nm (Klenin et al., 1989).

49

the presence of topoisomerases, ΔTw and Wr are independent random quantities, while the resultant quantity ΔLk equals their sum:

$$\Delta Lk = \Delta Tw + Wr .$$

Evidently the mean values are $\langle\Delta Tw\rangle = \langle Wr\rangle = \langle\Delta Lk\rangle = 0$. However, the quantities $\langle(\Delta Tw)^2\rangle$ and $\langle(Wr)^2\rangle$ differ from zero. Then, owing to the independence of the random quantities ΔTw and Wr, we have:

$$\langle(\Delta Lk)^2\rangle = \langle(\Delta Tw)^2\rangle + \langle(Wr)^2\rangle .$$

This is the master equation for our whole theoretical treatment of CC DNA (Vologodskii et al., 1979).

The quantity $\langle(\Delta Lk)^2\rangle$ is known from experiment, and the quantity $\langle(Wr)^2\rangle$ could be calculated by computer simulation of freely-jointed polymer chain because the ribbon theory offered a simple analytical formula for the Wr value provided that the shape of the chain were known. (Note that in so doing we extensively used our method of discrimination of knotted and unknotted chains.) Therefore we could find $\langle(\Delta Tw)^2\rangle$ as a function of the length of the DNA by substracting the calculated $\langle(Wr)^2\rangle$ value from the experimental $\langle(\Delta Lk)^2\rangle$ value. On the other hand the $\langle(\Delta Tw)^2\rangle$ quantity is unambiguously related to the value of the torsional rigidity of the double helix, C:

$$\langle(\Delta Tw)^2\rangle = N \langle(\Delta\varphi)^2\rangle = \frac{hkTN}{4\pi^2 C} ,$$

where h is the distance between adjacent base pairs in the double helix.

In full agreement with this equation, the value of $\langle(\Delta Tw)^2\rangle$ was found proved to be strictly proportional to N. The slope of the straight line made it possible to determine the C value.

This value of the torsional rigidity of DNA corresponds to an root-mean-square amplitude of thermal fluctuations in the value of the angle between adjacent base pairs of $4\text{-}5^o$. The obtained results indicated that in sufficiently long supercoiled DNA one third of the superhelical energy is stored in twisting and two thirds are stored in writhing. Thus the analysis of the experimental data on circular DNAs employing the topological approach made it possible to estimate one of the fundamental characteristics of the double helix.

The presented results depend in an essential manner on knowledge of another very important characteristic of the double helix: its bending rigidity. The bending rigidity of DNA is commonly characterized by the value of the persistence length a, or, which is equivalent, by the Kuhn statistical segment $b = 2a$. Studies during the past decade have yielded a reliable estimate $b = 100$ nm.

4. KNOTS AND CATENANES IN REALITY

As mathematical objects, knots and links have been studied already for more than a hundred years. The problem of the possible existence of such topological states in molecules has been raised relatively recently. It has acquired special interest since the discovery of single circular molecules of DNA, catenanes were found in certain cells, i.e., links, and even entire networks of linked circular DNAs. Catenanes are often obtained upon replication of DNA in vitro, and also upon closure into a ring of linear DNAs, having "cohesive ends", at sufficient concentration. The problem has arisen of the mechanism of replication of catenanes and networks. In fact, it is very difficult to imagine how this type of structures can duplicate itself in cell division.

The calculations of the probability of knot formation upon closing a polymer chain, the results of which are given above, have posed the problem of the possible existence of knotted DNAs. Thus, results, the equilibrium fraction of knotted DNAs must be appreciable for circular DNAs containing more than 10^4 base pairs (30 segments). In most cases DNA molecules have even a greater length, and the hypothesis has been advanced of the existence in the cell of special mechanisms that prevent the formation of knotted DNAs (Frank-Kamenetskii et al.,1975). In fact, in the replication of a knotted chain the daughter strands cannot separate. That is, the replication of knotted DNAs involves serious problems.

Knotted molecules were first detected in preparations of single-stranded circular DNAs after they had been treated under certain special conditions with a type I topoisomerase. This was the first case of a discovery of knotted polymer chains.

However, the problem of discovering knots in normal, double-stranded DNAs continued to be very intriguing. It turned out that there is a special subclass of topoisomerases, i.e., enzymes that can alter Lk in CC DNAs, called type II topoisomerases, and are capable of untying and tying knots in CC DNAs. Moreover, these enzymes catalyze the formation of catenanes from pairs or from a larger number of molecules of CC DNA. Here entire networks are formed, similarly to those observed in *vivo* in kinetoplasts. In contrast to type I topoisomerases, type II topoisomerases break, and then rejoin both strands of DNA molecules. It has been shown that the enzyme "draws" a segment of the same or of another molecule lying nearby through the "gap" that is formed in the intermediate state between the ends that arise through breakage. This operation with an individual CC DNA corresponds to a change in the writhing number by ± 2. However, it evidently does not alter Tw. Consequently, we have $\Delta Lk = \pm 2$. That is, type II topoisomerases can change the value of Lk only by an even number. In fact, experiment shows that type II topoisomerases, in contrast to type I, always alter Lk only by an even number. Thus type II topoisomerases catalyze the process of mutual penetration of segments of the double helix through one another. Consequently, these topoisomerases must lead to establishment of complete topological equilibrium, i.e., to a distribution of the molecules over the topological states that would correspond to freely penetrable strands.

As we have noted above, these molecules need not be very long for a reliable proof of the detection of knotted molecules, but then the fraction of knots, as calculated, must be small. Liu et al. (1980) have been able to overcome this contradiction by using topoisomerase II in very large concentrations in which it substantially changed the macromolecular properties of the DNA itself. Moreover, they did not add ATP to the enzyme, which is necessary for its normal operation. Precisely under these extreme conditions, they found even in short DNAs having N = 4.5 x 10^3 a considerable fraction of knotted molecules. They were able to detect them initially from the appearance of new bands in the gel electrophoretogram that corresponded to a greater mobility. Study of the properties of these fractions by various methods including electron microscopy has made it possible to show unequivocally that they correspond to knots of various types. If now one adds topoisomerase II in the normal amount and ATP to a purified preparation of knotted molecules, rapid untying of the knots takes place. That is, the system rapidly relaxes to the equilibrium state for pure DNA molecules, in which, as calculated, there should be practically no knots for the given length. As regards the reasons why the enzyme in high concentration sharply shifts the equilibrium toward knot formation, the most likely explanation consists of the data that the protein in high concentration decreases the dimensions of the polymer coil of DNA by changing the character of the interaction of segments remote along the chain. As our calculations showed (Frank-Kamenetskii and Vologodskii, 1981) even a small change in the dimensions of the polymer coil sharply increases the equilibrium fraction of knots. Knotted molecules of DNA (and also catenanes) have been obtained also by a highly refined method by employing special cell extracts that cause recombination, and preparation by gene engineering of special "chimeral" DNAs.

Thus it has been experimentally shown possible to form knots in *vitro*. Moreover, a class of enzymes has been found that can tie and untie knots in the cell. Now, after the discovery of type II topoisomerases, the existence and replication of knotted DNAs do not seem so improbable. In fact, in principle these topoisomerases eliminate all the topological problems that can arise here. Hence it becomes quite possible to assume the existence of knotted DNAs *in vivo*.

5. COMPUTER SIMULATION OF SUPERCOILED DNA

The above results were obtained more than ten years ago within the freely-joint model. This model was clearly inappropriate for short molecules comprising a few Kuhn statistical lengths. Experimental studies by Horowitz and Wang (1984) stimulated us to treat the problem within the framework of more adequate wormlike model (Frank-Kamenetskii et al., 1985). Because methods of generating closed wormlike chains are lacking we had to apply the procedure of Metropolis et al. (1953) to simulate wormlike chains. In so doing we also allowed for the excluded volume effects. Thus, our final model of DNA is as follows. We treat DNA as an isotropic homogeneous elastic rod, which is characterized by three parameters: the bending rigidity measured in terms of persistence length a, torsional rigidity C, and diameter d. Note that the d value is not necessarily equal to its geometrical value of 2 nm. Under d we assume the effective diameter which corresponds to the second virial coefficient of DNA segment under given ambient conditions. Due to the electrostatic repulsion of DNA segments, the effective d value may be much larger than the geometrical one.

Within the framework of the above sofisticated model we calculated the $<(Wr)^2>$ value as a function of the number of Kuhn statistical segments. As a result we achieved extremely good correlation between theory and experiment, as Fig.4 demonstrates. It should be emphasized that the a value is known from independent measurements and that the C value is determined from the intercept of the curve. The shape of the curve does not depend on the d value. Therefore the agreement between theory and experiment in the share of the curve is achieved without any adjustable parameters.

We consider the agreement of theory with experiment as an indication of the adequacy of our elastic model of the DNA molecule. Therefore this model may be used for calculating other DNA properties.

In its traditional form, however, the Monte Carlo approach does not permit simulating highly or even moderately supercoiled molecules because the probability of their occurrence due to thermal motion is negligible. We have extended our previous Monte Carlo calculations to make it possible to generate supercoiled DNA molecules with arbitrary supercoiling. This enables one to visualize supercoiled molecules theoretically and to calculate their various characteristics. Fig. 5 shows the results of our computer simulation of supercoiled molecules. Most significantly, we could calculate, using this approach, the dependence of superhelix energy on the DNA linking number for a wide range of superhelical energy and linking number should strongly depend on ionic strength.

Note in the conclusion that the problems discussed in the present paper, as well as related biological items, have been extensively covered in the recent volume: "DNA Topology and its Biological Effects", eds. N.R.Cozzarelli and J.C.Wang, Cold Spring Harbor Lab. Press, 1990.

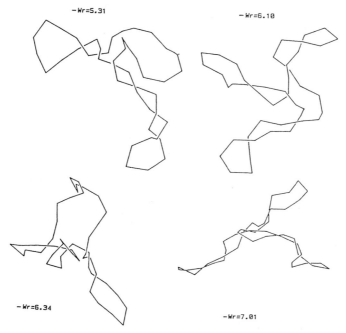

Fig.5. Results of computer simulations of supercoiled DNA molecules with different writhing number (indicated). The data are of Klenin et al. (1990).

REFERENCES

1. Delbruck, M., 1962, in: "Mathematical Problems in the Biological Sciences", Amer. Math. Soc.
2. Depew, R.E., and Wang, J.C., 1975, Conformational fluctuationg of DNA helix, *Proc. Natl. Acad. Sci. USA,* 72: 4275.
3. Frank-Kamenetskii, M.D., Likashin, A.V., and Vologodskii, A.V., 1975, Statistical mechanics and topology of polymer chains, *Nature,* 258: 398.
4. Frank-Kamenetskii, M.D., and Vologodskii, A.V., 1981, Topological aspects of the physics of polymers: the theory and its biophysical applications, *Sov. Phys. - Usp.,* 24: 679.
5. Frank-Kamenetskii, M.D., Lukashin, A.V., Anshelevich, V.V., and Vologodskii, A.V., 1985, Torsional and bending rigidity of the double helix from data on small DNA rings, *J.* Biomol . Struct. *Dyn.,* 2: 1005.
6. Fuller. F.B., 1971, The writhing number of a space curve, *Proc. Natl. Acad. Sci. USA, 68:* 815.
7. Horowitz, D.S., and Wang, J.C., 1984, The torsional rigidity of DNA and the length dependence of the free energy of DNA supercoiling *J. Mol. Biol.,* 173: 75.
8. Klenin, K.V., Vologodskii, A.V., Anshelevich, V.V., Dykhne, A.M., and Frank-Kamenetskii, M.D., 1988, Effect of excluded volume on topological properties of circular DNA, *J. Biomol. Struct. Dyn.,* 6: 1173.
9. Klenin, K.V., Vologodskii, A.V., Anshelevich, V.V., Dykhne, A.M., and Frank-Kamenetskii, M.D., 1990, Computer simulation of DNA supercoiling, *J. Mol. Biol.,* in press.
10. Klenin, R.V., Vologodskii, A.V., Anghelevich, A.V., Klighko, V.Y., Dykhne, A.M., and Frank-Kamenetskii, M.D., 1989, Variance of writhe of wormlike DNA rings with excluded volume, *J.* Biomol . *Struct. Dyn., 6:* 707 .
11. Metropolis, N., Rosenbluth, A.W., Rosenbluth, M.N., Teller, A.H., and Teller, E., 1953, *J. Chem. Phys.,* 21: 1087.
12. Pulleyblank, D.E., Shure, D.E., Tang, D., Vinograd, J., and Vosberg, H.-P., 1975, Action of nicking-closing enzyme on supercoiled and nonsupercoiled closed circular DNA: Formation of a Boltzmann distribution of topological isomers, *Proc. Natl. Acad. Sci. USA,* 72: 4280.
13. Shure, M., Pulleyblank, D.E., and Vinograd, J., 1977, The problem of eukaryotic and prokaryotic DNA packaging and *in vivo* confomation posed by supehelix density heterogeneity, *Nucleic Acids Res., 4:* 1183.
14. Vologodskii, M.D., Lukashin, A.V., Frank-Kamenetskii, M.D., and Anshelevich, V.V., 1974, The knot problem in statistical mechanics of polymer chains, *Sov. Phys. JETP* 39: 1059.

CORRELATION BETWEEN STRUCTURAL CONFORMATIONS AT DIFFERENT LEVELS IN HEMOGLOBIN DETERMINED BY XANES

A. Bianconi[1,2], A. Congiu Castellano[2], S. Della Longa[1,2]

[1] Dipartimento di Medicina Sperimentale, Università dell'Aquila,
    via S. Sisto 20, 67100 L'Aquila, Italy
[2] Biostructure Research unit of GNCB, CNR, Dipartimento di Fisica,
    Università di Roma "La Sapienza" , 00185 Roma, Italy

1. INTRODUCTION

The interaction between the structural conformations occurring at different levels of the protein organization is of key importance in understanding the dynamics of the biological molecules. Hemoglobin is studied here because it provides a good system to investigate fundamental aspects of biomolecules. In hemoglobin we can distinguish three levels of the protein organization 1) the quaternary structure of the tetramer, 2) the tertiary structure of the subunits, 3) the local structure of the Fe site in the heme .

Hemoglobin is a tetramer made up of two α–chains, each containing 141 amino acid residues, and two β-chains, each containing 146 amino acid residues. Each chain carries one heme. The α-chain (β-chain) contains 7 (8) helical segments. Hemoglobin is an allosteric protein in equilibrium between two alternative sets of structural conformations modulated by the reversible oxygen binding which are classified as: the T structure of the deoxy form with low affinity, and the R structure of the oxy form with high oxygen affinity. The co-operative binding of oxygen to hemoglobin (Hb) has been explained by the transition from the low affinity form T (tense) of the quaternary conformation of the tetramer $\alpha_2\beta_2$ in the deoxy state to the high affinity form R (relaxed) of the tetramer of the ligated state driven by the binding of the oxygen molecule in one of the subunits[1].

The oxygen affinity of the R structure is similar to that of free α and β subunits or of dimers composed of one α and β subunit, while the oxygen affinity of T structure is lower by the equivalent of 3,5 Kcal/heme, due mainly to the constraints of the additional bonds between the four subunits.

The investigation of hemoglobins from different species has pointed out that the affinity of the ligated form, measured for example by the dissociation constant, can be quite different and can be modulated by allosteric effectors. This point is of key importance because the function of hemoglobin is modulated by allosteric effectors such as the natural allosteric effector d-2,3-diphosphoglycerase (DPG) which for example facilitates the transfer of oxygen from human blood cells to the tissues by lowering the oxygen affinity of hemoglobin. The presence of allosteric effectors can change the oxygen affinity both in the R and T form and the cooperativity can be reduced[2].

Biologically Inspired Physics, Edited by L. Peliti
Plenum Press, New York, 1991

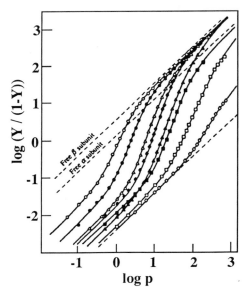

Fig.1   Oxygen equilibrium curves of hemoglobin in presence of different allosteric effectors H, C1, CO, IHP, and for Hb Milwaukee (from ref.2).

From the investigation of the crystallographic structure of the deoxygenated form (deoxy-Hb) and of the ligated forms, oxygen hemoglobin (HbO$_2$) or carbonmonoxy hemoglobin (HbCO), it has been pointed out by Perutz that the change of the tetramer from is correlated with the different displacement of Fe atom out of the heme plane.

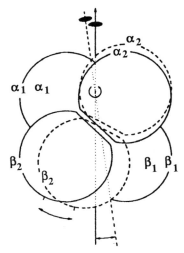

Fig.2   Rotation of the $\alpha_2\beta_2$ relative to the $\alpha_1\beta_1$ subunits with the change of the quaternary structure from the deoxy T form (solid lines) to the high affinity oxy form R (dashed lines) of hemoglobin (from ref.3)

The strain induced by the protein on the Fe active site is such that the low or high affinity forms of the tertiary structure can be distinguished by the displacement of the proximal histidine close to the Fe ion in the high affinity form, with a $Fe-N_\varepsilon$ distance of about 2.1 Å, to a low affinity conformation with the proximal histidine far away from the Fe site, with a $Fe-N_\varepsilon$ distance of about 2.7 Å[1].

The structural changes associated with the low or high affinity forms of ligated hemoglobin are not limited to the tetramer and Fe site structure but also at the intermediary level of the structure of the subunits. Two different conformations of the tertiary structure of the subunits (α or β): that can be classified in two classes[4]: the low affinity conformations t and the high affinity conformations r . Therefore the tetramer with R quaternary structure, in the ligated form can be in the R_r (where all subunits are in the high affinity form r) of R_t (where some or all subunits have the low affinity tertiary structure t).

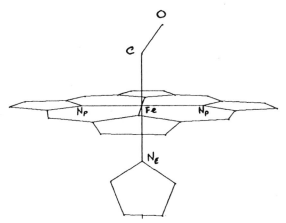

Fig.3   Fe site structure of carbonmonoxyhemoglobin, showing the heme plane with CO, above, and the proximal histidine, below.

Frauenfelder and his group have determined a correlation between the dynamics of the CO recombination and the CO bonding angle in myoglobin[5,6]. The Fe site structure in MbCO myoglobin has been found to assume different conformations A_0 , A_1 and A_3 that can be characterized by a different Fe-C-O binding angle[5]. The Fe-C-O binding angle in the most probable configuration A_1 is $155\pm4°$ in agreement with the average Fe-C-O bond angle determined by x-ray absorption near edge spectroscopy (XANES) in solution[7].

The determination of the dynamics of CO rebinding by flash photolysis mid-infrared spectroscopy has pointed out that the final state conformation for the fastest recombination process can be associated with the high affinity form of myoglobin, while the lowest process with the low affinity form. Following these results the highest affinity form A_0 is characterized by a nearly linear F-C-O bond angle $\theta = 165\pm4°$ while the lower affinity A_3 form is characterized by a bent F-C-O configuration $\theta = 148\pm4°$.

XANES [x-ray absorption near edge structure] spectroscopy of metalloproteins by using the synchrotron radiation provides a modern experimental method to investigate the structure of the metal active site and its changes in solution[8,9].

The application of this method to the investigation of the Fe site structure in hemoglobin has shown that it is possible to determine both the movement of the proximal histidine and the variation of the Fe-C-O bond angle. Therefore this technique allows to put under experimental test the two proposed correlations between the protein affinity and i) the movement of the proximal histidine found by crystallography and ii) the Fe-C-O bond angle found by mid-infrared spectroscopy.

Here we report the determination by XANES of both variation of the Fe-C-O bonding angle and the movement of proximal histidine by changing both the subunit and tetramer conformation by allosteric effectors. We report here the experimental evidence that there is a correlation between the Fe local structure and the the ligand affinity. The low affinity forms are characterized by both a lower Fe-C-O bonding angle and a larger Fe-N_ε distance. The Fe site structure has been found to be correlated with the structure of the subunit r or t but not with the tetramer conformation R or T.

2. MODULATION OF DROMEDARY HEMOGLOBIN CONFORMATIONS

We have investigated the structural changes at the Fe active site associated with the transition from the high to low affinity form in dromedary HbCO induced by binding of clofibric acid (CFA), the active hydrolysis product of the antihyperlipoprotenemia drug clofibrate which is its ethyl ester, and of inositol hexaphosphate (IHP). Since the original work of Arnone and Perutz[10] on the binding of inositol hexaphosphate (IHP) to deoxy human hemoglobin at a site between the two β chains, like DPG, the correlation between the changes of the structure of the tetramer (T or R), of the tertiary structure of the subunits (r or t) and of atomic displacements at the atomic level at the Fe site induced by the allosteric effectors have attracted the scientific interest in order to investigate the relation between structure and enzyme function. Recently the stereochemistry of drug binding is a growing field in order to establish the action of drugs at the microscopic level[11].

Camelus dromedary hemoglobin displays all the aminoacids considered relevant for the functional properties of human hemoglobin, but being characterized by the presence of two distinct binding sites for polyanions, which modulate the structural and functional properties of the protein as a function of their concentration. In particular the experimental results show that the multiplicity of interactions between hemoglobin and solvent components induces a large number of globin conformations. The IHP binding site in dromedary HbCO is assumed to be between the β-chains, and that of the clofibric acid (CFA) is assumed between the α-chains like in human deoxy hemoglobin. The CO dissociation rate, the tetramer form and the subunit form are modulated by the allosteric effectors as shown in Fig.4. The CO dissociation rate constant has ben found to $8.7 \cdot 10^3$ sec^{-1} in the native high affinity R_r form and $16 \cdot 10^3$ sec^{-1} in the R_t low affinity form obtained for the 1:1 IHP/protein ratio. The CO dissociation rate come back to the values of high affinity form by adding CFA and IHP ($8.1 \cdot 10^3$ sec^{-1}) and increases again ($23 \cdot 10^3$ sec^{-1}) by going to the low affinity form T_t adding Cl$^-$ ions.

3. RESULTS

The XANES experiments were performed at the Frascati wiggler facility using synchrotron radiation monochromatized with a Si (111) channel-cut crystal. The experimental apparatus is shown in Fig.5. The threshold energy of the Fe edge has been obtained as the first maximum of the derivative of the metal iron K edge. The energy resolution was 1 eV and energy shifts of absorption spectral features as low as 0.2 eV could be detected. Fig.6 shows the experimental Fe K-edge XANES spectrum of native dromedary HbCO and in the spectra presence of the allosteric effectors following the sequence as in Fig. 4 from bottom to top.

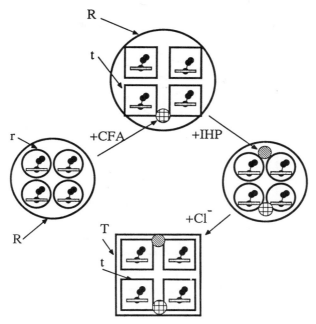

Fig.4   Changes of the tetramer R (large circles) and T (large squares) conformations and
of the subunit *r* (small circles) and *t* (small squares) conformations of dromedary
carbonmonoxyhemoglobin induced by adding CFA, CFA+IHP, and CFA+IHP+
Cl⁻ as determined by circular dichroic spectra.

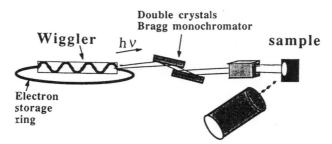

Fig.5   Experimental apparatus for XANES experiments using 1) a wiggler insertion
device in an electron storage ring as source of x- rays, 2) a double crystal
monochromator, 3) a ionization chamber for the measure of incident radiation and a
detector of x-ray flourescence emitted by the sample.

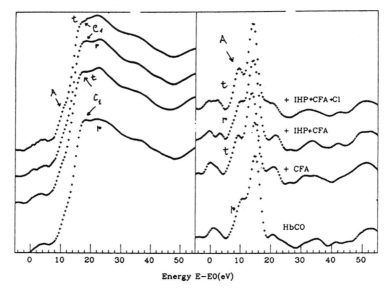

Fig.6   XANES spectra (left side) and their derivative (right side) of native dromedary
HbCO lower curves, and in the presence of allosteric effectors from bottom to top:
CFA, CFA+IHP, and CFA+IHP+Cl⁻

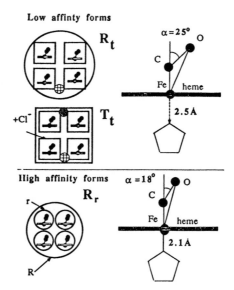

Fig.7   Fe-C-O bonding angles and the Fe-N$_\varepsilon$ distance with the nitrogen atom of the
proximal histidine as extracted from the XANES data in Fig. 5

60

The x-ray absorption spectrum probes the unoccupied electronic states in the continuum over an energy range of 50 eV. We have interpreted the XANES data in terms of multiple scattering calculations of the photoelectron wavefunction in the real space and than the absorption coefficient from the Fe(1s) level has been calculated in the dipole approximation. In the high energy continuum the modulation of the absorption coefficient from the Fe(1s) level to the states in the continuum with angular momentum $\ell=1$ are determined by the position of the atoms in a range of 5Å from the absorbing Fe atom. Therefore the XANES feature are determined by the scattering of the l=1 photoelectron by atoms around the heme iron. Attention has been focused on feature A at about 10 eV, the features C_1 at 16 eV and C_2 at about 36 eV. As reported in previous works[7] the peak C_1 is a probe of the angular bonding geometry Fe-C-O, and the feature A is sensitive to the proximal histidine distance from the heme plane[13].

We have found a correlation, shown in Fig.7, between the variation of the experimental features in Fig.6 and the variation of the of the structure of the subunits. The intensity of the peak C1 at about 22 eV is weaker for the subunits in the t forms and it is stronger for the high affinity r forms. The shoulder A in the rising edge, that is better identified in the derivative spectra, is more resolved and is shifted to lower energy in the low affinity t forms.

Multiple scattering analysis of XANES spectra has been carried out[13]. We have extracted i) the Fe-C-O bonding angle from the intensity of the peak C1 and ii) the variation of position of the proximal histidine from the shift of the feature A.

4. CONCLUSIONS

The XANES spectroscopy probes the Fe local structure in a time scale of the lifetime of the Fe(1s) core hole, about 10^{-16} sec, however because the required integration time is about 1 sec to collect spectra with a good signal-to-noise ratio, the reported spectra have to be considered as an average of the protein conformations over a large time scale. Therefore the Fe local structure determined by XANES has to be considered the structure of the conformation with higher probability. The observed structural change induced by allosteric effectors can be interpreted as the change of the relative probability of the different conformations. Therefore we think that we observe the shift of the highest probability from one conformation to another induced by allosteric effectors.

This study of the changes of the Fe local structure at the active site induced by allosteric effectors that modulate the affinity for ligand binding has allowed us to establish several aspects on the interaction between conformations at different levels of protein organization:
1) the changes of the Fe local structure are associated with changes in the tertiary structure of the subunits;
2) the change of the tetramer structure from the R to T form induced by Cl^- ions is not correlated with a change of the Fe local structure;
3) the Fe site structure of the high affinity r forms of the subunits of dromedary HbCO (CO dissociation rate ~$8\cdot10^3$ sec^{-1}) are characterized by a Fe-C-O bonding angle of 162º (the angle α, defined in Fig.7, is 18º);
4) the Fe site structure of the low affinity t forms of the subunits of dromedary HbCO (CO dissociation rate ~$20\cdot10^3$ sec^{-1}) are characterized by a Fe-C-O bonding angle of 155º (the angle α is 25º);
5) The proximal histidine moves away from Fe, with the Fe-N_ε distance going from the 2.1 Å to 2.5 Å, with the change from the high affinity r form to low affinity t form of the subunit.

The results on the correlation between the dominant configuration of the Fe-C-O bonding angle and the CO dissociation rate are in good agreement with the results of Ormos et al.[6] and Ansari et al.[5].

The movement of the proximal histidine away from the heme going from the high to low affinity from is in good agreement with the M. Perutz mechanism[1], that can explain the close correlation between the Fe site structure and the subunit conformation, found also in this work.

Finally this work shows that the tetramer structure that can be modulated by the solvent . without changing the the Fe site structure.

REFERENCES

1. M.F. Perutz, *Quat. Rev. Biophysics* **22**, 139 (1989)
2. K. Imai, *Allosteric Effects in Haemoglobin*, Cambridge Univesrsity Press, (1982)
3. J.M. Baldwin and C.Chothia, *J. Mol. Biol.* **129**, 183 (1979)
4. B.R. Gelin, A.W. Lee, and M. Karplus. *J. Mol. Biol.* **171**, 489 (1983)
5. Ansari,
6. Ormos
7. A. Bianconi, A. Congiu Castellano, P.J. Durham, S.S. Hasnain and S. Phillips, *Nature* **318**, 685 (1985)
8. A. Bianconi in *"X-ray absorption: principle, applications, techniques of EXAFS, SEXAFS, XANES"* ed. R. Prinz and D. Koningsberger, (J. Wiley and sons, New York, 1988)
9. A. Bianconi, A.Congiu-Castellano, eds, *"Biophysics and synchrotron radiation"* (Springer Verlag, Berlin, 1987)
10. A. Arnone and M.F. Perutz, *Nature* **249,** 34 (1974)
11. M. F. Perutz, G. Fermi, D. J. Abraham, C. Poyrat, and E. Bursaux, *J. Am. Chem. Soc.* **108**, 1064 (1986)
12. G. Amiconi, R. Santucci, M. Coletta, A. Congiu Castellano, A. Giovannelli, M. Dell'Ariccia, S. Della Longa, M. Barteri,E. Burattini and A. Bianconi, *Biochemistry* **28**, 8547 (1989)
13. S. Della Longa, A. Bianconi, A. Congiu Castellano, *to be published.*

SELECTIVE CONSTRAINTS OVER DNA SEQUENCE

G. Cocho[1], L. Medrano[2], P. Miramontes[2], J.L. Rius[1]

1. Instituto de Física, Universidad Nacional Autónoma de México
2. Facultad de Ciencias, Universidad Nacional Autónoma de México

Apdo. Post 20-364, México 01000 D.F. MEXICO

INTRODUCTION

After the progress derived from Watson and Crick's DNA structural model [1953], DNA has frequently been thought as a static and rigid polymer rarely disturbed by random mutations. However, the discovery of processes as transposition, hypermutability, genetic drive and gene conversion, shows that DNA is genetically more active than our first notion as being only the heredity keeping guard. From a physical point of view, studies of molecular dynamics and molecular structure have shown that DNA is far from being a rigid molecule. Indeed, there do exist several fluctuations in the molecule and the structure is not homogeneous along the polymer. At the evolutive level, the analysis of nucleotide sequence data has shown regularities that make evident selective constraints other than the protein function derived from amino acid sequence. Other selective constraints acting over the genome include protein synthesis kinetics, tRNA availability, mRNA secondary structure and DNA stability. Thus, genome evolution can be conceived as a dynamical and complex system which might be understood by the search of regularities in genomic nucleotide sequences.

NUCLEOTIDE SEQUENCE AND DNA STRUCTURE

DNA is a double-stranded nucleotide polymer made up of four kinds of bases attached to a sugar-phosphate backbone. The base sequence carries genetic information, and the sugar and phosphate groups perform a structural role. The genetic information is encoded in sequences of bases; the four bases are the purines (R), adenine (A) and guanine (G), and the pyrimidines (Y) thymine (T) and cytosine (C) (in RNA uracyl (U) replaces thymine but this fact will not be taken into account). The base elements of one strand interact with their counterparts on the other strand in a precise way: A pairs with T, and G pairs with C. This complementarity principle is the basis of replication, transcription and repair; if a segment of DNA is known, it

Biologically Inspired Physics, Edited by L. Peliti
Plenum Press, New York, 1991

serves as a template along which complementary bases are attached to form a complementary strand carrying a duplicate of the original sequence. The DNA sequence is not energetically and structuraly homogeneous; the GC pairings are linked through three hydrogen bonds while the AT pairings are held by two such bonds. Moreover, there are stacking interactions between adjacent base pairings which depend from the specific bases involved in such interactions. There is an hindrance effect when a purine and a pyirimidine are adjacent in the nucleotide sequence caused by the opposite neighboring purines (Figure 1). This hindrance causes variations in the structural angles of double stranded DNA [Dickerson 1983]. Therefore, regions with high purine-pyrimidine alternation (e.g. ...$RY RY RY R$...) have more structural variations than regions with almost only purine or pyirimidine in one strand (e.g. ...$RRRR$... or ...$YYYY$...). Thus, it is possible to define the "roughness" and "smoothness" of a particular sequence as follows: regions where purine and pyrimidine alternate will be "rough" while regions running out only one base type will be "smooth". When talking about the YR nature of one base, we will call it "0" or "1", respectively.

On the other hand, as G and C link strongly, they will be named "strong" (S) bases, while A and T will be defined as "weak" (W) bases. When talking about the WS character of the bases, they will be called "0" or "1", respectively. The energy associated to neighboring bases ("digrams") does behave as Table I [Breslauer *et al.* 1986, Freier *et al.* 1986]. So one can see that the absolute values of both, enthalpy and free energy, are smaller for 00 than for 11 digrams. 10 and 01 digrams show even smaller values. By looking at the local energy distribution, the "flexibility" of a sequence may be defined as follows: A sequence with 0 and 1 alternating will be "flexible" while another one plenty of 1's will be "rigid".

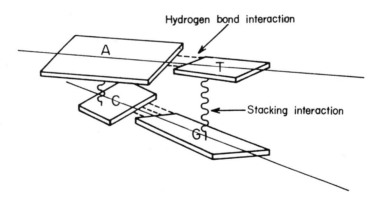

Figure 1. Diagram showing the steric hindrance effect due to neighbour purines. There are shown the different interactions that contribute to the double strand stability.

The chromosome structure and, the rather complicated, processes of replication and transcription, suggest that more flexible and smoother sequences might be favoured in eukaryotic cells. These features should be evident in non-coding regions where constraints associated to the translation into functional proteins are absent. Prokaryotic sequences should be, in consequence, more rigid and less smooth than eukaryotic ones.

Table I. Stability of double-stranded nucleic acids.*

Digrams	RNA		DNA	
	Δ H	Δ G	Δ H	Δ G
AA	6.6	0.9	9.1	1.9
AT	5.7	0.9	8.6	1.3
TA	8.1	1.1	6.0	1.9
TT	6.6	0.9	9.1	1.9
CC	12.2	2.9	11.0	3.1
CG	8.0	2.0	11.9	3.6
GC	14.2	3.4	11.1	3.1
GG	12.2	2.9	11.0	3.1
AC	10.2	2.1	6.5	1.3
AG	7.6	1.7	7.8	1.6
TC	13.3	2.3	5.6	1.6
TG	10.5	1.8	5.8	1.9
CA	10.5	1.8	5.8	1.9
CT	7.6	1.7	7.8	1.6
GA	13.3	2.3	5.6	1.6
GT	10.2	2.1	6.5	1.3

* Absolute values of ΔH and ΔG are expressed in Kcal/mol.

In order to test this hypothesis, it is necessary to measure the content of 0's and 1's for the two binary representations discussed above. We, therefore, define N_{ij}, the number of ij digrams ($ij = 00, 01, 10$ and 11) in a sequence of length N. $N_1 = N_{11} + N_{10}$ is the number of 1's in the sequence and $N_0 = N_{11} + N_{10}$ is the number of 0's.

Furthermore, we define d as:

$$d = \frac{(N_{00}N_{11} - N_{01}N_{10})}{(N_{00} + N_{01})(N_{11} + N_{10})}$$

For a random sequence, the value of d should be near to zero. If the 0's and 1's are aggregated, then $d \simeq 1$, and if they are almost completely alternated, then $d \simeq -1$. Therefore, if the binary representation is associated to the weak (W) and strong (S) base character, negative values of d (d for this WS representation) will correspond to DNA sequences more flexible than random ones and positive values to less flexible sequences. For the purine-pyrimidine YR representation, $d > 0$ will correspond to smooth sequences and $d < 0$ to rough ones.

In Figures 2 and 3, the values of d_{YR} and d_{WS} for eukaryotic, prokaryotic and viral sequences are presented. Sequence data were obtained from the GenBank genetic sequence database [Bilofsky et al. 1986]. In Figure 2, the values for eukaryotic sequences are plotted. All the sequences show $d_{YR} > 0$ (smooth) and $d_{WS} < 0$ (flexible) values. The absolute values of the d's are high for immunoglobulins and low for histones. In Figure 3, values for prokaryotes and a phage are shown. Human immunodeficency virus type 1 (HIV-1) and other eukaryotic viruses show a pattern quite similar to eukaryotes, while the prokaryote values are distributed around zero. These results are consistent with a correlation between the smoothness and flexibility characteristics and chromosome dynamics with eukaryotic and prokaryotic viruses showing a pattern similar to their corresponding hosts.

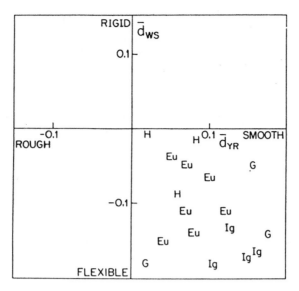

Figure 2. Distribution of eukaryotic sequences in the $d_{WS} - d_{YR}$ plane. Ig stands for immunoglobulins, H for histones, G for globins and Eu for other eukaryotic sequences.

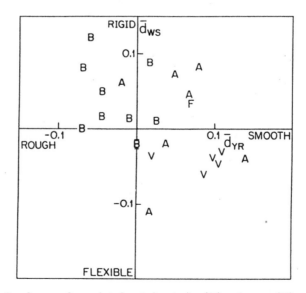

Figure 3. $d_{WS} - d_{YR}$ plot for eubacteria (B), phages (F), archaebacteria (A) and eukaryotic viruses (V).

GENE EXPRESSION AND DNA STRUCTURE

Figures 2 and 3 show a distribution of eukaryotic and prokaryotic sequences. Such a distribution is more disperse depending upon factors other than eukaryotic chromosome or prokaryotic genophore dynamics. On the other hand, Figure 4 shows the $d_{WS} - d_{YR}$ distribution of highly expressed genes (HE) and low expressed ones (LE) from *Escherichia coli*. HE genes do occur around the plane's origin while LE genes have $d_{WS} > 0$ and $d_{YR} < 0$ values. This fact is mainly due to a shortest content of GC in the third and first codon positions in HE genes. It has been shown that this difference could obey to protein synthesis kinetics [Grosjean and Fiers, 1982] but other explanation is possible too: LE genes could have higher base substitution rates, thus favouring the formation of strong ΔH digrams from weak ones since base substitutions are more frequent in sites having lower local DNA stability [Medrano, 1989].

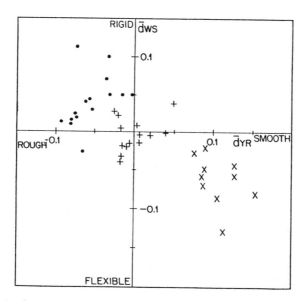

Figure 4. $d_{WS} - d_{YR}$ plot for differently expressed genes of *E. coli*. Black dots stand for low expressed genes (LE) and '+' for high expressed genes (HE). For comparison purposes, eukaryotic HE genes are shown and represented by 'X'.

SELECTIVE CONSTRAINTS OVER DNA SEQUENCE

The DNA structural constraints are neatly different from those constraints arising from the function of the resulting peptide, which are random with respect to the DNA structure. In order to relate DNA structure regularities with constraints over amino acid sequence, the euclidean distance M between the origin of the $d_{WS} - d_{YR}$ plane to any point in that plane as a measure of sequence regularity with respect to DNA structure was determined. Figure 5 shows that M decreases as the sequence is more restricted by peptide function. The constraint strength over amino acid sequence is measured using the *Unit Evolutionary Period* (*UEP*) [Wilson et al. 1977, Efstratiadis et al. 1980]. Intergenic sequences and introns have the highest M values while coding sequences show decreasing values as *UEP* increases. It can be concluded that the selection over the amino acid sequence is conflictive with the family of constraints associated to DNA structure.

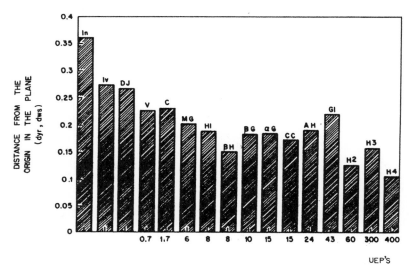

Figure 5. Average distance in the $d_{WS} - d_{YR}$ plane of several sequences. The numbers below the horizontal scale are the estimated UEP's according to Wilson *et al.* [1977] and Efstratiadis *et al.* [1980]. In stands for intergenic sequences, Iv for introns, DJ for hypervariable regions of immunoglobulins (Ig), V for Ig variable regions, C for Ig constant regions, Mg for myoglobin, H1 for histone 1, H2 for histone 2, H3 for histone 3, H4 for histone 4, βH for β-lipotropin, βG for β-globin, αG for α-globin, CC for cytochrome c, AH for adrenicorticotropin and Gl for glucagon.

FINAL COMMENTS

The regularities related to DNA structure found in nucleotide sequences, suggest the existence of selective constraints on that structure. These constraints could arise from the replication dynamics and the chromosome structure. It has been shown, in this paper, that viral sequences share the structural features of their host cell and this fact is consistent with the notion of structural constraints. The findings of Barrie et al. [1981] on the low substitution rates in intergenic sequences also support the idea of constraints in silent DNA. If exonic or coding regions are considered, the amino acid sequence constraints might be in conflict with the DNA structural ones. Thus, we suggest the term *Internal Natural Selection (INS)* to name the selective factors acting on DNA physical and geometrical structure, and the term *External Natural Selection (ENS)* for the constraints over amino acid sequence. The showed conflict between ENS and INS might explain the existence of high proportions of non-coding DNA in eukaryotes.

REFERENCES

1.- Barrie PA, Jeffreys AJ, Scott AF (1981) Evolution of the β-globin gene cluster in man and the primates. J Mol Biol 149: 319-336
2.- Bilofsky HS, Burks C, Fickett JW, Goad WR, Lewitter FI, Rindone WP, Swindell CD, Tung CS (1986) The GenBank genetic sequence databank. Nucleic Acid Res 14: 1-4
3.- Breslauer KJ, Frank R, Blöcker H, Marky L (1986) Predicting DNA duplex stability from the base sequence. Proc Natl Acad Sci USA 83: 3746-3750
4.- Dickerson RE (1983) Base sequence and helix structure variation in B and A DNA. J Mol Biol 166: 419-441
5.- Efstratiadis A, Posakony JW, Maniatis T, Lawn RM, O'Conell C, Spritz RA, DeRiel JK, Forget BG, Weissman SM, Slightom JL, Blechl AE, Smithies O, Baralle FE, Shoulders CC, Proudfoot NJ (1980) The structure and evolution of the human β-globin gene family. Cell 21: 653-668
6.- Freier SM, Kierzec R, Jaeger JA, Sugimoto N, Caruthers H (1986) Improved free energy parameters for the prediction of RNA duplex stability. Proc Natl Acad Sci USA 83: 9373-9377
7.- Grosjean H, Fiers W (1982) Preferential codon usage in prokaryotic genes: the optimal codon-anticodon interaction energy and the selective codon usage in efficiently expressed genes. Gene 18: 199-209
8.- Medrano L (1989) Consideraciones biofísicas sobre las mutaciones puntuales. M. Sc. Thesis. Facultad de Ciencias, Universidad Nacional Autónoma de México.
9.- Watson JD, Crick FHC (1953) Molecular structure of nucleic acid. A structure for deoxyribose nucleic acid. Nature 171: 737-738
10.- Wilson AC, Steven SC, White TJ (1977) Biochemical evolution. Annu Rev Biochem 46: 573-639

THE PHYSICS OF DNA ELECTROPHORESIS

Tom Duke

Cavendish Laboratory
Madingley Road,
Cambridge CB3 0HE
England

1. INTRODUCTION

An area that has seen much fruitful collaboration between biologists and physicists in recent years is the development of the gel electrophoresis technique for separating DNA fragments. To some extent, the success achieved is due to the artificial nature of the system, which makes it more immediately amenable to theoretical treatment than most of the naturally occurring systems discussed in this Workshop, where the degree of complexity is far greater. The process involved here is straightforward and essentially physical - charged DNA molecules are forced to migrate through a gel by the application of an electric field - but it is of immense importance to biologists, as the efficient separation of different fragments is an inevitable requirement for the manipulation of DNA performed in molecular genetics, cancer research and, more recently, the human genome project. The physics turns out to be rather interesting, with a rich variety of dynamical behaviour displaying some unusual and unexpected features. More generally, gel electrophoresis is a good example of a driven diffusive system, many types of which are currently under investigation.

The essence of the technique is to distinguish molecules of different size by their different mobilities as they diffuse through a gel under the influence of an eiectric field. The reason for using a gel is that, in free solution, all fragments travel at much the same speed and no separation is achieved; the gel fibres obstruct large molecules more than smaller ones, slowing them down. However, analysis by conventional gel electrophoresis turns out to be limited to fragment sizes less than one hundred thousand base-pairs - all larger molecules travel at the same velocity and cannot be segregated. This limit of resolution may sound high, but is frustratingly low compared to the size of human genomes, the biggest of which contains two hundred million base-pairs. Six years ago Schwartz and Cantor[1] made the exciting discovery that pulsing the electric field spectacularly overcomes this difficulty; by a careful choice of pulse time, fragments within a desired size range can be separated - the longer the pulse time, the larger the molecular size. This finding spawned the development of a variety of pulsed-field techniques, each using a different field configurations, proceeding

Biologically Inspired Physics, Edited by L. Peliti
Plenum Press, New York, 1991

mainly by trial and error since the molecular mechanism of separation remained a mystery. The best of these can separate molecules containing ten million base-pairs.

How do pulsed-field techniques work? The task of unravelling the mechanism has been taken up by physicists, using ideas borrowed from the field of polymer dynamics and with the help of computer simulation.

2. REPTATION AND MOLECULAR ORIENTATION

The diffusion of long chain macromolecules in a gel was first described nearly twenty years ago by de Gennes[2]. His reptation model argues that the gel fibres strongly restrict the motion of a molecule, effectively confining it to a tube-like region. Any lateral excusion out of the tube is inhibited and the main mode of motion is the longitudinal displacement of the molecule along the tube axis. The chain slithers backwards and forwards guided by the tube, and at the same time, the tube configuration is continually being modified to follow the random direction taken by the molecule as it emerges from either end.

The success of this model, which correctly predicts a diffusion coefficient inversely proportional to the square of the chain length, motivates its use as a basis for the theory of the dynamics involved in gel electrophoresis. Modifications must be made to account for the effect of the electric field, which drives the charged molecule in a particular direction. The simplest approach[3], developed into the biased reptation model (BRM)[4], treats the DNA as an inextensible rope following the tube contour. The net electric force acting longitudinally along the tube biases the diffusion towards one end, so that the molecule moves head-first through the gel, rather like a snake through grass (although it may occasionally go backwards, when thermal fluctuations overcome the field bias). The field also has another effect, which turns out to be very significant. When a small section of the chain emerges from the tube and enters a new pore, it is more likely to choose one that lowers its electrostatic energy; that is, there is an orientational bias in the field direction. Since, in reptation, the molecule always follows the lead of the end segments, this orientation is eventually transferred to the entire tube so that in the steady state, instead of having a random configuration, the molecule is aligned with the field. This is not an equilibrium effect (there is no energetic advantage for the chain as a whole), but rather a dynamic effect, resulting from the interplay of the driving field and the hinderence of the gel fibres - it would not arise if either were absent. The occurrence of this oriented non-equilibrium steady state has since been confirmed by experiment[5], most recently and unambiguously by direct optical observation[6]. It is rather striking, since it means that large fragments containing millions of base pairs can be extended over many microns.

While the BRM is a useful first step in understanding orientation effects, it is not sophisticated enough to describe the dynamics in detail. DNA is a highly flexible molecule and it is clear that the pushing and pulling of the electric force will cause local stretching or compression (depending on the way that the molecule is wound around the gel fibres), rather than inducing the entire chain to slide bodily along the tube contour. Consequently, to correctly describe the dynamics in a field, it is important to consider longitudinal fluctuations of the molecule in the tube. In fact, we find that these fluctuations, which have negligible effect in zero field, are of utmost importance in accounting for pulsed field effects.

3. A SIMPLE LATTICE MODEL

Rather than considering the DNA within the tube to be an inextensible rope, we model it as a chain of springs following the tube axis. For ease of analysis we choose a discrete, lattice-based approach which, while being a somewhat oversimplified picture of the real DNA molecule, is designed to incorporate the most important features and illuminate, at least qualitatively, the mechanism of enhanced separation in pulsed field electrophoresis. It is described in greater detail elsewhere[7,8] and here we give only a brief summary.

The complex gel structure is approximated by a regular 3-dimensional array of connected pores, each of size a. Following the reptation prescription, we imagine the DNA molecule to thread its way through the gel fibres, passing from pore to pore, in such a way that the sequence of inhabited pores defines the tube (Fig.1). The DNA itself is modelled as a connected chain of N springs, where each spring is allowed only two possible states: either extended (length a), or compressed (length 0). An extended spring represents a section of the chain that passes between adjacent pores, while a compressed spring represents a chain segment residing in a single pore.

Local longitudinal fluctuations of the molecule that lead to the translation of a segment of the chain from one pore to another may be modelled by the extension of one spring and the compression of its neighbour on the chain. The dynamic of the model, specified by a set of rules governing the springs' motion, is based on the assumption of local detailed balance - that is, we assume that there is a local thermodynamic equilibrium between Brownian impulses that make the chain move randomly and the electric force that drives it in a particular direction. This may be implemented by a Monte Carlo procedure, where the motion of a segment in the field direction is biased by an amount controlled by the parameter $\Delta = qEa/\sqrt{2}kT$: local detailed balance is preserved by choosing the ratio of probabilities of moving against the field (-) or in the field direction (+) to be $P_-/P_+ = \exp(-\Delta)$. Here, q is the effective charge carried by a spring and Δ, measuring the field strength, is a ratio of electrostatic to thermal energies.

The discretization we have used enables us to map the problem onto a 1-dimensional lattice model (Rubinstein's repton model[9]), greatly easing computation. Furthermore, it is possible to describe the dynamics in an elegant manner by a spin exchange model.

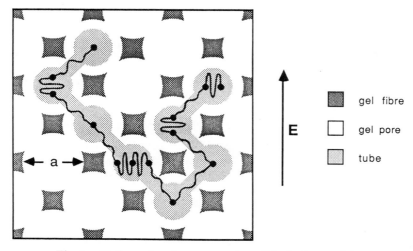

Fig.1   2-dimensional representation of the lattice model.

4. CONTINUOUS FIELD ELECTROPHORESIS

The dependence of mobility on chain length in a steady field is shown in Fig.2. This reproduces the behaviour observed experimentally, where for short molecules, the mobility decreases with increasing chain length, but at high molecular weights, the mobility becomes independent of chain length and fragments of different size cannot be distinguished. These features can be explained by molecular orientation. In the BRM, the total electric force biasing the motion along the tube contour is proportional to the distance h_x between the two ends of the molecule in the field direction, which leads to a mobility varying with the mean-square end-to-end separation[3]:

$$\mu \sim <h_x^2>/N^2 \tag{1}$$

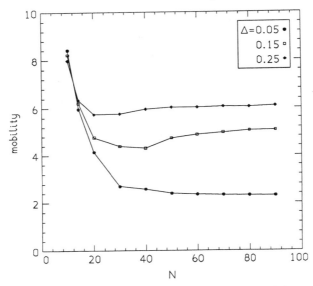

Fig.2   Dependance of mobility of chain length in continuous field electrophoresis.

Consequently, the mobility depends critically on the degree of alignment of the chain with the field. Typically, the orientation per segment in slight, so that short chains retain approximately Gaussian configurations; $<h_x^2> \sim N$ and $\sim \mu N^{-1}$. But since the orientation accumulates with chain length, long chains are aligned with the field; $<h_x^2> \sim N^2$ and $\mu \sim N^0$.

Thus it is the molecular orientation that is responsible for the loss of resolution at high molecular weights in continuous field electrophoresis - the alignment of molecules increases their speed so that the longer ones catch up with the shorter.

Fig.2 also displays a highly unexpected feature: molecules of intermediate size actually migrate more slowly than the longest molecules. This, too, can be described by the BRM. In fact, such 'band-inversion' was first predicted on the basis of this model[4] and later confirmed experimentally - previously, experimenters had assumed that the relationship between mobility and molecular weight was monotonic. The explanation of this phenomenon is rather subtle. As already stated, the degree of bias depends on the end-to-end separation h_x. For very long molecules, which in the steady state are oriented by the field, h_x is so large as to preclude the possiblity of reptation backwards, counter to the bias. But for fragments of intermediate size, the bias is slighter and reverse reptation occasionally occurs. When this happens, the chain emerging at the 'tail' end of the tube is oriented by the field so that it turns back on the rest of the tube, creating a U-shaped conformation (Fig.3). Since these have small end-to-end separation h_x, Eq.1 indicates that their occurrence leads to a small reduction in the overall mobility.

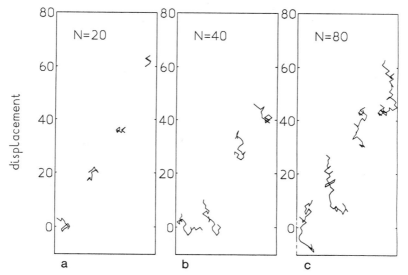

Fig.3   Time-sequence of tube configurations: (a) short chains are approximately Gaussian, (b) molecules of intermediate size are frequently U-shaped, (c) long chains are oriented along the field.

The band-inversion effect is enhanced when longitudinal fluctuations are included, allowing the chain to extend. In this case, when a U-shaped configuration is formed, the electric force acts to stretch the chain in both arms of the U and reptation is temporarily suspended. After a while, the tension in the chain exceeds the electric force and only then does reptative motion resume, with the longer arm of the U growing at the expense of the shorter. More generally, the effect of fluctuations is to destabilise the oriented steady-state of the BRM; typically, molecules go through cycles where J or U shapes form and extend before reverting to uniformly orientated configurations.

5. FIELD-INVERSION GEL ELECTROPHORESIS (FIGE)

In FIGE[10] the field is periodically inverted, but the forward and reverse pulses are of unequal duration so that there is a net drift of molecules in the forward direction. Experimentalists have observed that the mobility displays a curious non-monotonic variation with pulsing frequency; there is a sharp drop in velocity at a particular value of the pulse time that depends on the chain length. It is this feature that, by a judicious choice of pulsing frequency, enables the separation of long molecules in a given size range.

The behaviour of the model, which clearly displays dips in the mobility, is shown in Fig.4. The reduction in mobility occurs at a pulse time $t_p=t^*$ which increases roughly in proportion to chain length. On examining the centre of mass motion of the molecules at $t_p=t^*$, it is surprising to find that the decrease in mobility is not due to a slower average drift velocity, which one might have expected if the mean orientation is slightly reduced by pulsing[11]. Rather, it is a result of the chain frequently halting, typically for a number of pulse cycles, before resuming its drift in the field direction. How does this unusual behaviour arise? Fig.5 which illustrates a time-sequence of tube configurations in both the migrating and the stationary state, explains the phenomenon. In the first case, the molecule is aligned along the field, as expected, and drifts steadily forwards and backwards, following the sign of the field. But suppose that, by chance, the chain loses much of its orientation - then the bias towards the head is slighter and the field drives the chain out of *both* ends of the tube simultaneously, forming a U-shaped conformation. The molecule does not have enough time to fully orient before the field switches and under the influence of the reverse pulse, the arms of the U retract and reextend in the opposite direction. Overall, the molecule just switches between U-shaped configurations pointing alternately forwards and backwards and does not migrate. This situation is quite stable and may endure for a large number of cycles before a chance large fluctuation enables the chain to regain its oriented state.

The FME effect, then, is a consequence of abrupt conformational changes, initiated by longitudinal fluctuations, which frequently trap the chain in U-shaped tubes. The trapping happens only near a particular value t^* of the pulse time, associated with the orientation time of the chain, since for $t_p \gg t^*$ the molecule has plenty of opportunity to realign after a U occurs, while for $t_p \ll t$ there is not sufficient time for the U to form in the first place.

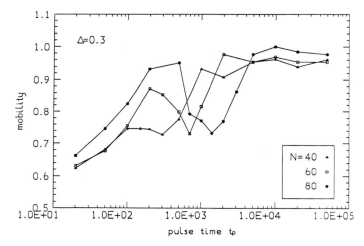

Fig.4   Variation of mobility with pulse time in FIGE, for a forward pulse twice the duration of the reverse pulse.

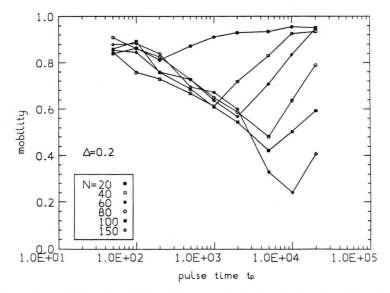

Fig.5   Snapshots of the tube configuration at $t_p=t^*$ during (a) the migrating phase and (b) the stationary phase. The field direction is indicated above.

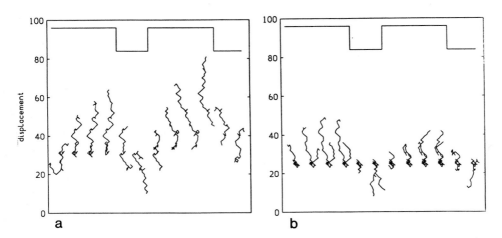

a   b

Fig.6   Variation of mobility with pulse time in OFAGE.

6. ORTHOGONAL FIELD ALTERNATION GEL ELECTROPHORESIS (OFAGE)

In OFAGE[1], the field is applied alternately, for equal durations, along two orthogonal directions. We know that continuous field electrophoresis fails to separate long molecules because of molecular orientation, so the idea behind this technique is to force the chains to change their alignment by switching the field and make use of the fact that fragments of different size take different times to reorient along the new field direction.

In general, the overall motion of the molecules is along the diagonal, but their average orientation and detailed dynamics depends on the value of the pulse time. The variation of the mobility with the pulsing frequency is shown in Fig.6. Essentially, there are three regimes, corresponding to pulse times greater than, less than, or of the same order of magnitude as the reorientation time tor, which increases roughly in proportion to molecular length.

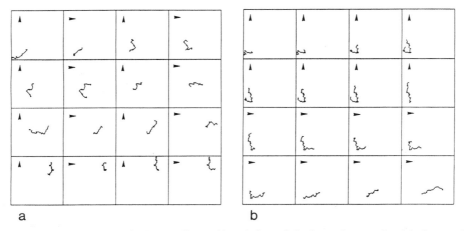

Fig.7   Time-sequence of tube configurations, left to right from the top, for (a) short pulse duration $t_p < t_{or}$ and (b) long pulse duration $t_p > t_{or}$. The field direction is indicated by the arrows.

At long pulse times $t_p > t_{or}$, a molecule has time to completely realign after each switch of field (Fig.7b) and then migrate for some distance along the field direction. Overall, it follows a broad, regular zig-zag path along the diagonal (Fig.8c). In this regime, long molecules of different size can be distinguished, provided that the pulse time is not too much greater than their reorientation time; the mobility of larger molecules is lower, since they spend longer realigning and so have less time to migrate.

When the pulse time is short $t_p << t_{or}$, only a small section of the molecule reorients along the current field direction before the field switches. As a result, the chain typically consists of a sequence of short segments aligned alternately along either field. Fluctuations somewhat confuse the picture, but very roughly the molecule has a net orientation along the diagonal (Fig.7a). The motion, too, is a fairly steady drift in this direction (Fig.8a). This regime is rather like the continuous field case (with an effective field along the diagonal) and is consequently no good for separating large molecules, which all travel at the same speed.

As in FIGE, the most interesting regime is the intermediate one, $t_p \sim t_{or}$. This value of the pulse time gives rise to the lowest mobility and is the best choice for good separation. Fig.8b indicates that the motion is rather irregular - sometimes the molecule drifts along the diagonal, sometimes it moves for many cycles along just one of the fields, remaining stationary when the other field is applied, and occasionally it halts altogether. In fact, neglecting fluctuations, the BRM predicts a broken symmetry in this regime[13]. Once the chain becomes aligned along one of the fields, it remains so; the pulse time is not long enough for it to reorient completely along the other field direction, and any small section that does realign always vanishes as the molecule rapidly drifts out of its tube during the other half of the cycle. One would expect a sample of molecules to split into two batches, one half migrating along each field. In practice, fluctuations recover the symmetry so that the observed motion is along the diagonal - but the path followed is irregular since transitions between the field directions depend on chance. This is reflected, experimentally, in the dispersion of the trace on the gel plate.

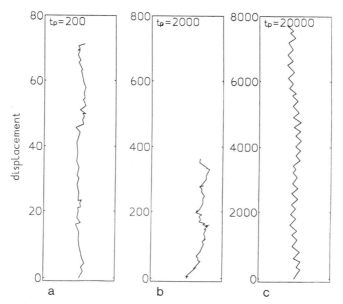

Fig.8   Path followed by a molecule over 25 field cycles: (a) short pulse duration $t_p > t_{or}$, (b) intermediate pulse duration $t_p \sim t_{or}$, (c) long pulse duration $t_p > t_{or}$. Note the scale change between figures.

7. FUTURE DIRECTIONS

The model expains many of the features observed in practice, but there are still a number of reservations about the approximations involved. At high field strengths, the entire validity of the tube picture may be called into question since hernias, or unentangled loop-like excursions of the chain through the side of the tube, although penalized entropically, may be sufficiently favoured energetically to form. This would be particularly relevant in OFAGE, where they would provide an additional mechanism for reorientation when the field switches. The role of hernias is being investigated by simulating flexible chains in a 2-dimensional array of obstacles[14,15]. A second uncertainty is the effect that randomness in the gel structure has on the dynamics. Usually, disorder profoundly modifies dynamical properties, although in this case we might expect its significance to be reduced on account of the large spatial extent of the DNA molecule. A third effect that we have not considered is the self-interaction of the chain which may, in practice, provide the fundamental limit to resolution. Extremely long molecules are statistically almost certain to form knots around gel fibres; when the electric field is applied, the force pulling on the chain tightens the knots, immobilising the molecule!

Despite this reservations, it seens that we have a picture of the basic mechanism involved in pulsed-field methods. Essentially, the idea is to confuse the molecules as much as possible, by pulsing at a frequency just faster than they can respond! To date, one of the best protocols involves alternating between fields set at an angle of 120º, and superimposing a short, sharp pulse backwards along the diagonal. Given the knowledge we have gained, it is a challenge to our ingenuity to devise yet more effective techniques in pursuit of the elusive goal of separating intact human chromosomes.

REFERENCES

1. D.C.Schwartz and C.R.Cantor, *Cell* 37, 67 (1984).
2. P.G. de Gennes, *J. Chem. Phys.* 55, 572 (1971).
3. O.J.Lumpkin, P.Dejardin and B.H.Zimm, *Biopolymers* 24, 1573 (1985).
4. J.Noolandi, J.Rousseau, G.W.Slater, C.Turmel and M.Lalande, *Phys. Rev. Lett.* 58, 2428 (1987).
5. G.Holzwarth, C.B.McKee, S.Steiger and G.Crater, *Nucl. Acids Res.* 23, 10031 (1987).
6. D.C.Schwartz and M.Koval, *Nature* 338, 520 (1989).
7. T.A.J.Duke, *Phys. Rev. Lett.* 62, 2877 (1989).
8. T.A.J.Duke, *J. Chem. Phys.*, in press.
9. M.Rubinstein, *Phys. Rev. Lett.* 59, 1946 (1987).
10. G.F.Carle, M.Frank and M.V.Olson, *Science* 232, 65 (1986).
11. J.L.Viovy, *Phys. Rev. Lett.* 60, 855 (1988).
12. J.L.Viovy, *Electrophoresis* 10, 429 (1989).
13. J.M.Deutsch, *Science* 240, 922 (1988).
14. M.Olivera de la Cruz, D.Gersappe and E.O.Schaffer, *Phys. Rev. Lett.* 64, 2324 (1990).

RECENT DEVELOPMENTS IN THE PHYSICS OF FLUCTUATING MEMBRANES

Stanislas Leibler

Service de Physique Théorique
C.E.N. - Saclay, F-91191 Gif-sur-Yvette

1. INTRODUCTION

Figure 1 shows a schematic view of an ensemble of membranes in a living cell. Each of these structures consists of a large quantity of various constituents: proteins, glycolipids, lipids and others. Each membrane has a specific composition, and plays a specific role in the functioning of the cell. For instance, the Golgi apparatus, shown in Figure 1, is mainly used by the cell to sort out proteins produced by ribosomes in the endoplasmic reticulum (another membrane structure) and send them towards different parts of the cell.

From the point of view of condensed matter physics, ensembles of biological membranes are extremely complex systems. We are very far from understanding the mechanisms of their assembly or their collective behaviour. The only thing we can try to describe at this point are the simplest properties of the membrane structures. One of them is shape and shape transformations.

Membranes are quasi-two-dimensional objects: their thickness, typically of the order of 100 Å, is much smaller than their lateral extension which can reach hundreds of microns. These surfaces, however, can have quite complicated shapes and topologies. For instance, the Golgi apparatus (Figure 1) consists of a complex network of cisternae (squashed vesicles) with holes, necks, interconnecting tubes etc. Next to this network one can observe a large quantity of small vesicles budding off the curved edges. The problem of the geometry of membranes and of the shape transformations is thus a highly non-trivial one. It is also very important even from the biological point of view: the process of protein sorting involves, for instance, a constant exchange of membrane components between various parts of the structures; this exchange takes place precisely through the budding off of the vesicles and their fusion with other membranes.

It is thus tempting for a physicist to try to clarify the basis of the collective behaviour of membranes, by starting with this simplest problem: membrane geometry. However, even then, one must restrict oneself for the moment to artificial, physical systems such as pure phospholipid bilayers, rather than more realistic complex multicomponent structures involving many lipids and proteins. Therefore, the problem

Biologically Inspired Physics, Edited by L. Peliti
Plenum Press, New York, 1991

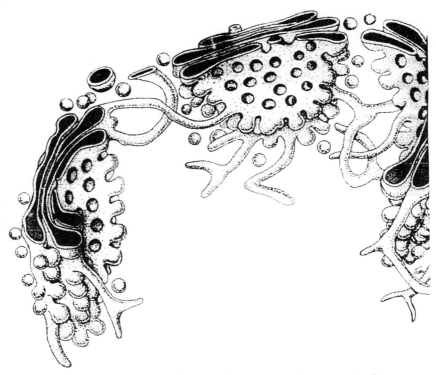

Fig. 1. The Golgi apparatus in a living cell consists of a network of interconnected membranes. A schematic view, based on electron microphotographs, shows here some characteristic structures such as flattened cisternae and small vesicles budding off from their curved borders. These vesicles transport proteins which have been produced in the endoplasmic reticulum and are sorted out in the Golgi apparatus. [Adapted from (Becker, 1986)].

enters the domain of Biologically Inspired Physics, and should be considered in this light. Such an approach consisting in trying to describe the physics of ensembles of fluctuating membranes, such as lyotropic liquid crystals, microemulsions or vesicle suspensions. One hopes that the concepts and results developed for these simple systems will help to understand some properties of biological membranes.

2. STATISTICAL MECHANICS OF A SINGLE FLUCTUATING MEMBRANE

2.1. *Membranes as semi–flexible, fluctuating sheets*

Amphiphilic molecules – which are the main constituents of many natural or artificial membranes – tend to form either *monolayers* on a fluctuating interface (e.g., on the oil/water interface of microemulsions) or freely fluctuating *bilayers*. Although in both cases, it is tempting to describe these aggregates as two-dimensional molecular systems, one has to be cautious since treating membranes as two-dimensional objects is usually an oversimplification. The typical thickness of lipid bilayers is of the order of a few nanometers and is often comparable to the separation between two sheets. Fortunately, in some systems, such as diluted solutions of big vesicles (closed objects of a size of several microns) or "hyperswollen" lamellar crystals (where the separation between neighbouring membranes can increase up to a micron), this theoretical approximation is quite adequate.

2.2. Curvature energy and the shapes of closed vesicles

An important consequence of the non-trivial molecular structure of membranes is the fact that their statistical behaviour is often governed by rigidity and not by surface tension. This can be contrasted, for instance, with interfaces between two fluids: the energy of configurations is in general dominated by surface tension, i.e., a term proportional to their area. Since the area of membranes can in most cases adjust itself freely, such a term is unimportant and rigidity or *curvature energy* dominates.

An experimental proof of the dominance of curvature energy for amphiphilic membranes is furnished, for instance, by the nontrivial shape of red blood cells and vesicles. Indeed, well-known shapes such as discocytes, stomatocytes etc., can easily be obtained through the minimization of the following elastic energy (with simple constraints of a fixed total area and of a fixed volume):

$$\mathcal{H}_{el} = \int d^2\sigma \left[\frac{\kappa}{2} (H - H_0)^2 + \bar{\kappa} K \right], \tag{1}$$

where H and K are respectively the mean and the Gaussian curvature at the point $\vec{r}(\sigma)$ of the membrane, $(\sigma = (\sigma^1, \sigma^2)$ being local coordinates of the surface). If R_1 and R_2 are the principal radii of curvature at $\vec{r}(\sigma)$, then $H = 1/2 \, (R_1^{-1} + R_2^{-1})$ and $K = (R_1 \cdot R_2)^{-1}$. H_0 is the spontaneous curvature connected to the asymmetricity of the membrane. The elastic constants κ and $\bar{\kappa}$ are the bending (rigidity) coefficient and the Gaussian bending coefficient respectively. For objects with a fixed topology, such as vesicles, the Gaussian term can be neglected since integrated over the whole surface it gives a constant contribution.

The shapes obtained on the basis of the Hamiltonian (1) are in quite good agreement with the observations in experimental systems (Figure 2). Moreover, one can calculate the spectrum of thermal fluctuations around these equilibrium shapes and compare it with the results obtained by the video microscopy or light scattering techniques. This gives an estimate of the actual value of the elastic constant κ: $\kappa \sim 10 \, k_B T$. Since κ is *not* much larger than $k_B T$, *thermal fluctuations can easily modify the behaviour of molecular membranes*. In other words, the physics of membranes is a generalization of the mechanical problem of thin shells (which are also governed by their bending energy) to the case where: (i) the internal structure can be modified (liquid, solid, etc.); (ii) the elastic constants are small enough that the thermal fluctuations can influence the global (thermodynamical) behaviour.

2.3. Internal structure of membranes. Polymerized nets

Since fluctuating membranes can undergo internal transformations and change their molecular structure, we expect the statistical behaviour of these systems to be

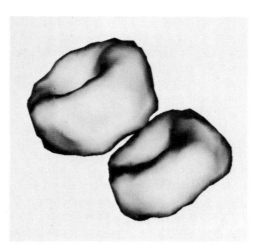

Fig. 2. A picture from a Monte-Carlo simulation of two interacting vesicles. Each
of the two vesicles consists of a closed hexagonal net of 980 triangles. The
energy used in the simulation is a sum of a discretized version of Hamiltonian
(1) and a pressure term pV, which compresses the vesicles. The pressure p
is kept constant while the volume V is allowed to fluctuate. Here $pV \approx 49\kappa$,
$\kappa \approx 5.2k_BT$, $H_0R \approx -1.16$, where $V = (4\pi/3)R^3$. The shapes obtained
for these parameters are similar to discocytes of red blood cells. Despite
the existence of an attraction between the two vesicles they can separate
due to thermal undulations. This is an example of unbinding transitions in
fluctuating membranes. For details see (Leibler and Maggs, 1990).

quite rich. Indeed, many recent theoretical studies have shown that there are several
universality classes of fluctuating membranes and films.

A striking example of how the internal structure of a fluctuating membrane
can influence its statistical behavior is the difference between fluid-like and solid-
like membranes. Fluid-like membranes, in which the molecules can diffuse and thus
change their neighbours, are much freeer to modify their shape than solid-like membranes, which have a fixed internal neighbourhood (connectivity). To understand
this difference let us imagine that we want to "wrap" a piece of membrane around an
object with a complicated shape. In general, the object has different curvatures K at
different points of its surface (Figure 3): locally flat points ($K = 0$) can coexist with
points where $K > 0$ or $K < 0$. A fluid membrane will easily adjust to this geometry;
it will introduce some local defects in its molecular structures (such defects do not
cost much energy in a fluid). A solid–like membrane, on the other hand, is locally flat
(or it has some frozen defects which cannot move), so it cannot adjust itself to the
shape of the object. In the same way, a free solid–like membrane has its fluctuations
constrained by its local molecular structure. In mathematical terms one has to add
a new elastic term to the Hamiltonian (1):

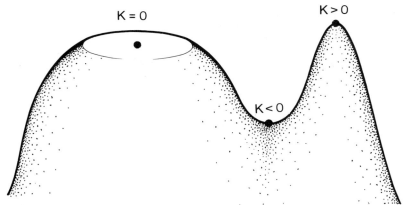

Fig. 3. The surface of a three-dimensional object does not need to be locally flat: points with $K = 0$, $K > 0$ and $K < 0$ are shown orr this simple example. If one wants to "wrap" such an object in a membrane, this can easily be done for fluid membranes, but not for polymerized membranes since the latter are locally flat.

$$\mathcal{H}_{solid} = \int d^2\sigma \left[\frac{\lambda}{2}(\mathrm{Tr}\,\mathbf{u})^2 + \mu\mathrm{Tr}(\mathbf{u}^2)\right] \tag{2}$$

where $\mathbf{u}$ is the strain tensor, connected with the solid- like order within the membrane, and λ and μ are elastic Lamé coefficients. The geometrical constraints which we have described above imply that the strain variables u_{ij} are in general *not* independent of the shape variables which enter the curvature H and K.

These geometrical constraints can have important thermodynamic consequences: for instance, non-self-avoiding solid-like membranes can undergo a finite temperature phase transition between a crumpled phase (without long- range orientational order), and a *flat, rigid phase*, characterized by a long-range orientational order, i.e., with an infinite persistence length ξ_p. Such a *crumpling transition* is quite unusual since it is an example of a transition at which a continuous symmetry (orientation) is broken in a two-dimensional system. This is not in contradiction with the Mermin-Wagner theorem, since the internal, "phonon-like" modes – when integrated out – give rise to effective, *long-range* interactions within the membrane. The crumpling transition, already observed in Monte Carlo simulations, is now being looked for in polymerized membranes. It seems that realistic *self-avoiding* solid-like, or polymerized, membranes are always in the ordered, "flat" phase. This phase has unusual elastic properties; for instance the spectrum of undulations or "phonon-like" modes is characterized by non–classical power laws. Moreover, elastic membranes, when subjected to different boundary conditions, could also undergo other phase transitions. For instance, under the action of lateral pressure these membranes might transform into a "buckled state" in which rigid regions with different orientations coexist separated by a network of defects. The *buckling transition* has been studied by several statistical and field-theoretical methods.

One has to stress again that these unusual phase transitions are closely related to the solid-like structure of fluctuating sheets: fluid membranes, the local structure of which is not coupled so strongly to configurational geometry, do not undergo a(n) (un)crumpling transition in two dimensions (at least in the absence of molecular long range forces). Thus they never become orientationally ordered, i.e. the persistence length ξ_p always remains finite. This does not necessarily mean that fluid membranes are crumpled at length scales larger than ξ_p. Indeed, based on various analytical and numerical arguments, we expect that at these length scales a single fluid membrane with a fixed topology behaves rather like a *branched polymer* made out of long tubes with a diameter of the order of ξ_p. (One should, however, keep in mind that fluid membranes can easily modify their topology, as we shall see in the next section.)

An interesting question at this stage of the development of the statistical mechanics of fluctuating membranes is to find an experimental system in which one could check various theoretical predictions for solid-like membranes. An ideal solid-like membrane should present the following properties:

its constituents should not diffuse on the time scales of thermal (shape) fluctuations;

they should be flexible enough (κ of the order of a few $k_B T$). This is not the case of lipid bilayers in solid-like phases (e.g. L_β);

if the constituents are bound together, the network should actually be a two–dimensional one. This condition seems very hard to satisfy for lipid bilayers the molecules of which are cross-linked;

if the networks include some defects one would like to be able to control their density or distribution. It seems that natural networks such as a spectrin network of red blood cells are too complex and not controlled enough (at the moment) to provide quantitative data.

In this context it is worth noting that simple "paracrystals" of proteins, such as those built out of tropomyosin (Figure 4) are very good candidates for such ideal elastic membranes. In addition, the proteins which form such "paracrystals" are abundant and relatively easy to purify. A detailed study of such systems (e.g., by light scattering methods) could make it possible to develop further the physics of fluctuating sheets, which in turn could help to understand elastic properties of biological aggregates.

As we can see, even the simplest case of an isolated membrane with fixed topology can already show a great variety of possible behaviours. We shall now show that the thermodynamic behaviour is even richer if membranes are allowed to modify their topology.

3. ENSEMBLES OF FLUID MEMBRANES WITH VARYING TOPOLOGY

3.1. *Different phases of random surfaces*

In contrast with linear polymers or polymerized membranes, fluid membranes can easily vary their topology. For instance, two membranes (e.g. two vesicles) can fuse together, a vesicle can bud off a membrane, etc. Such simple events are indeed observed in cell membranes and are of great importance for many biological processes. Here we shall give some examples of phase transitions - taking place in pure, physical

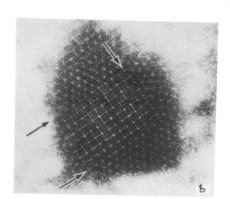

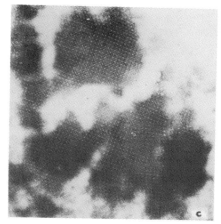

Fig. 4. Electron microscopy pictures of paracrystals of tropomyosin. (a) A two-dimensional net with a periodicity of 400Å;(b) and (c) small patches of such crystals can show both square and triangular connectivity, or even the coexistence of such structures separated by defect lines (arrows). (Yamaguchi et al., 1974).

systems - which involve nontrivial topological changes of fluctuating surfaces.

In order to study the thermodynamic behaviour of an ensemble of fluid membranes one has to add to the Hamiltonian (1) a term coupled to the total area of the membrane system, i.e. a chemical potential term for the amphiphilic molecules (we suppose that all molecules are inside the membranes and neglect their compressibility):

$$\mathcal{H} = \int d^2\sigma \left[r + \frac{\kappa}{2}H^2 + \bar{\kappa}K \right] \tag{3}$$

where r is proportional to the chemical potential. This model assumes that the bilayers are symmetric so that the spontaneous curvature H_0 vanishes. The integration is performed over many different disconnected membranes (i.e., topological

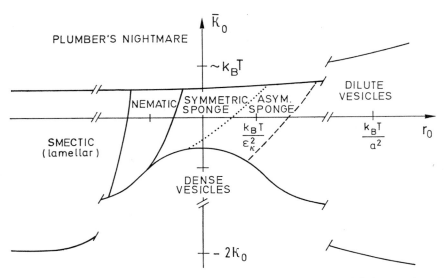

Fig. 5. A schematic phase diagram of the model defined by Eq. (3) for random fluid
    membranes. Here $\kappa > 0$, and the different phases described in the text are
    shown as function of r and $\bar{\kappa}$. (Huse and Leibler, 1987).

components); each of the latter cannot have free edges but can form many *handles*.

A schematic phase diagram of this model is depicted in Figure 5. It is based on
various theoretical arguments. Some of the phases shown here have also been studied
in the framework of other similar models. A part of this phase diagram has been
verified through Monte-Carlo simulations of random surfaces. Some details of this
diagram, however, can be altered in the future when more quantitative calculations
or simulations are carried out.

The main phases shown in this diagram are:

For small values of the Gaussian rigidity $\bar{\kappa}$, and large negative r, which favours
dense phases of membranes one expects to observe lamellar structures of parallel
membranes. These *smectic* lamellar phases will modify their topology when r
is increased. In general, one expects that at some point the defects, e.g., in the
form of necks connecting two neighbouring membranes, will proliferate and the
smectic phase will change into a *nematic* lamellar structure. For still bigger
values of r the nematic order will disappear and the ensemble of membranes
will become an isotropic fluid. The topology of such membrane structures can
be quite complex; in general, they will form a so-called *bicontinuous phase* in
which the membranes separate space into two infinite water networks (Figure
6). Such phases are now often called *sponges*, although the structure of real
sponges is quite different from a topological point of view. In Figure 5 we
show the possibility of two different types of sponge phases: symmetric and
asymmetric ones. In *symmetric sponges* the symmetry of the Hamiltonian (3)
between the two sides of the membrane is preserved. This means that one
cannot distinguish between two water networks – the interior and the exterior
of the structure look exactly the same. This geometrical symmetry can be
spontaneously broken if r is increased even further: in the so-called *asymmetric*

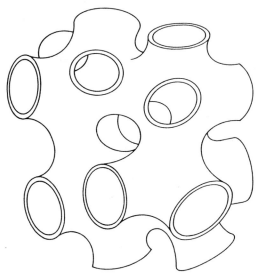

Fig. 6. A bicontinuous phase consists of a network of fluid membranes separating two water regions (an "inner" one and an "outer" one), both of which are infinite. Both regions can have the same geometrical and physical properties, in which case we call this phase symmetric; in an asymmetric bicontinuous phase the symmetry of the two sides of the membrane is broken. The bicontinuous structures are examples of the so-called sponge phase of membranes. (Drawing adapted from Strey et al., 1990).

sponge, one water network is different from the other. It has been shown that the transition between symmetric and asymmetric sponges can be continuous and then belongs to the 3d Ising universality class. Recent experiments in the so-called L_3 phases of amphiphilic solutions appear indeed to support this theoretical result. When r becomes positive and large the infinite membrane networks should break down and form an ensemble of closed vesicles. Although in such a *diluted vesicle phase* each vesicle costs some curvature energy, the membranes are diluted and their entropy is large.

For large enough values of $\bar{\kappa}$, the topological term favours a large number of disconnected objects with no handles. Therefore, we expect to observe a dense phase of spherical vesicles, such as a *crystal of densely packed vesicles*. It is possible that for small enough values of κ such crystals can be replaced by an amorphous structure. Vesicles with non- spherical topologies have been observed: one could speculate that such objects are intermediate between vesicle phases and sponge structures.

For large negative values of $\bar{\kappa}$, the topological term favours a simply-connected membrane with many handles. An example of such structures is provided by the so-called triply-periodic minimal surfaces (Figure 7). These mathematical objects are ordered crystals of surfaces for which the mean curvature H is zero at every point. We call them by the generic term *plumber's nightmare phases*. A lot of effort is now being made both by physicists and mathematicians to understand the geometry of such structures and try to use the results to describe some *cubic phases of lyotropic liquid crystals*. It seems that the main

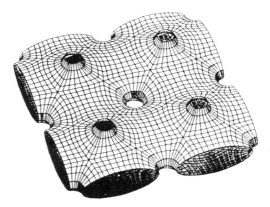

Fig. 7. An example of a triply-periodic minimal surface. This surface with a tetragonal symmetry has been obtained by a numerical minimization of the curvature energy term (Maggs and Leibler, unpublished). The solutions with $H = 0$ everywhere seem to exists for aspect ratio ("vertical" dimension to "horizontal" dimensions) smaller than 1. If this ratio is bigger than 1 these triply-periodic minimal surfaces do not exist. This conjecture by A.C. Maggs has recently been proven by D. Hoffman (D. Hoffman, 1990). The tetragonal surface shown here is one of many examples of "plumber's nightmares" studies by physicists and mathematicians.

experimental challenge consists at this point in obtaining a diluted version of these structures, so that the lattice constant should be much larger than the membrane thickness. It is only in this case that the cubic phases could be described as crystals of surfaces, and the structural details of the amphiphilic molecules could become unimportant.

3.2. *Sponge phases and lattice gauge theories*

The above examples clearly indicate that fluid membranes can often be very well described as an ensemble of random surfaces. Indeed, recent experiments on hyper-diluted amphiphilic phases (such as lamellar phases or sponge L_3 structures) have provided us with systems of membranes in which the ratio of the distance between the films to their thickness can reach a few hundred! Such systems could become experimental models to check the predictions of many theories which involve random surfaces.

Random surfaces play an important role in some theoretical problems of high energy physics. For instance, gauge theories developed in order to describe electro-weak and strong interactions between elementary particles can be formulated in a lattice approximation, in which space-time is replaced by a regular lattice of discrete points. Then, a strong coupling expansion of the physical quantities, such as the interaction potential between two quarks, can be expressed as a sum over the configurations made of many plaquettes, i.e., as a sum over two-dimensional random surfaces built on the lattice. With each configuration in this sum is associated a statistical weight factor, which depends on the details of gauge theory. In particular, it can involve a term proportional to the total area of the surface, a term analogous to the chemical potential term of the Hamiltonian (3). In this analogy between the strong coupling expansion of lattice gauge theories and the membrane phases, the lattice constant corresponds to the persistent length ξ_p.

This mathematical analogy can be used to study the phase behaviour of fluctuating membranes. For instance, the above mentioned transition between a symmetric sponge and an asymmetric sponge has been proved to belong to the Ising universality class, precisely through the analogy with $O(N)$ gauge theory on a f.c.c.lattice. It corresponds to a confining transition below which two static particles (e.g.quarks) interact through a confining (linear) potential. Inversely, the membrane systems could be used in the future to verify some non-trivial predictions of lattice gauge theories. A good example of such a possibility is provided by sponge phases *with defects*.

Let us suppose that the membranes can include some defects such as open pores (holes). Since free edges on the fluctuating membrane structure are now allowed, it is no longer possible to talk about the bicontinuous structures: there is only one infinite water network. One can therefore wonder what happens to the symmetric / asymmetric sponge phase transition the chemical potential of the edges is decreased. The answer to this question is suggested precisely by the analogy with gauge theories. It can be shown that the model of random surfaces with linear defects existing on them is equivalent to the so-called Z_2 Higgs model of lattice gauge theories. This model predicts the existence of three distinct regions of the phase diagram (Figure 8): (i) a region in which series expansion in the gauge coupling implies the phenomenon of confinement; (ii) a region where there is no confinement, separated from the confined region by a continuous Ising transition; (iii) a region where for continuous gauge groups the famous Higgs mechanism would take place. It has been shown that for the discrete group Z_2 this region is in fact continuously connected to the confinement region by a domain with no thermodynamic singularities or phase transitions. These three domains correspond to different sponge phases which we expect to exist in the presence of linear defects. These are respectively: (i) a symmetric sponge which present edges in the form of close, tense loops; (ii) an asymmetric sponge which has finite loop defects, characterized by an effective zero line tension; and (iii) a sponge with infinite edges, in which the membranes form many "sea-weed" structures. The analogy with gauge theories shows that one can go from an asymmetric sponge to a sponge with infinite edges smoothly, without passing through a thermodynamic phase transition. Numerical simulations of the Z_2 Higgs model also makes it possible to predict the nature of phase transitions between symmetric and asymmetric sponges.

It would be extremely interesting to investigate such analogies further and especially to develop quantitative experiments in amphiphilic systems which could check these predictions. The defects, such as pores (edges), could be introduced into the fluid membranes by adding a small quantity of impurities which would lower the edge energy (line tension). Good candidates for such impurities are some small polypeptides such as myletin.

4. CONCLUSIONS: HOW TO CONTROL MEMBRANE POLYMORPHISM?

We have seen that amphiphilic systems show a rich polymorphism: for instance, experimental phase diagrams of mixtures of a single surfactant and water are even more complex than the theoretical diagrams described above (since they

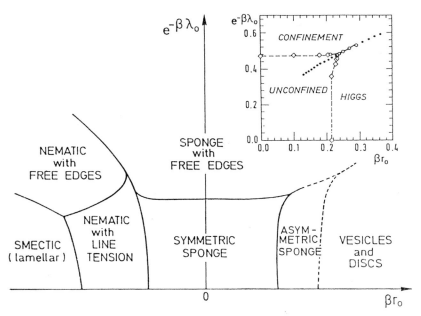

Fig. 8. A schematic phase diagram for random fluid membranes with pores. The model considered here is based on Eq. (3) with the Gaussian curvature term neglected and a new edge energy term $\lambda \int d\ell$. Here $\kappa > 0$, and the different phases described in the text are shown as function of r and of the line tension λ. The $r > 0$ part of the phase diagram with the sponge phases corresponds to the phase diagram of the Z_2 gauge-Higgs model shown in the inset (based on numerical simulations by Jongeward et al., 1980). For details see (Huse and Leibler, 1990).

include many non-bilayer phases). Therefore, much more complex mixtures of many different amphiphilic components seem to lie beyond the possibilities of systematic, quantitative studies. Moreover, some of the phases of multi-component systems can only exist in an extremely narrow range of the values of physical parameters, such as temperature, composition, pH, salinity, etc. Thus, stabilizing and studying some membrane structures seems a very hard task indeed.

From this point of view, it is astonishing to observe membrane structures in a living cell. The membranes of different organelles, such as the Golgi apparatus, the endoplasmic reticulum, or intracellular vesicles, are constantly exchanging their components (lipids, proteins,...) through frequent processes of budding, or exo- and endo- cytosis. Despite this constant exchange, each membrane (or even each mono-layer of each membrane) maintains its own characteristics. Moreover, the membrane shape and topological transformations, which make possible the exchange and transport of different components, take place in well-defined places and at well-defined moments. There can be no doubt that a living cell *controls* the geometrical and chemical structure of its membranes in a very precise way.

Of course the way a living cell regulates its structure is very different from the regulation mechanisms used in physical experiments. Not only are the cell processes out of equilibrium (which in fact is the case for many physical experiments on growth and structure formation), but – more important – they often involve some metabolic sources of energy, such as ATP or GTP hydrolysis. In addition, the regulation in the cell is usually a very local process and it can involve a very small quantity of molecules.

In my opinion, the understanding of local control mechanisms, not only of membrane systems but of other molecular assemblies, is one of the most challenging problems for future experimental and theoretical studies. Cell biology has been making rapid progress in clarifying the biochemical basis of regulation in membrane systems. Unfortunately, it seems that the regulation schemes involve many different, mutually interdependent components; in fact, many of them are still waiting to be purified or even discovered. However, the action of some proteins such as certain phospholipases or flippases, could in principle be studied *in vitro* by introducing them into artificial membranes, e.g. vesicles. Thus, parallel to biochemical or spectroscopic studies, it would then be possible to observe the consequences of the action of these enzymes on the global characteristics of membranes (shape, elasticity, ...). One could also try to control their action locally, for instance by modifying the local calcium concentration, the local ATP or GTP concentration etc. In the long run, such studies could lead to a better understanding of regulation processes. In the short term, they would probably lead to a new class of non-equilibrium phenomena in random, fluctuating surfaces.

REFERENCES

1. Alberts, B., Bray, D., Lewis, J., Raff, M., Roberts, K., Watson, J.D., 1989 *Molecular Biology of the Cell*, Garland Publishing Comp., N.Y.
2. Anderson, D.A., 1986 *Ph.D. Thesis*.
3. Aronovitz, J.A., Lubensky, T.C., 1988 *Phys. Rev. Lett.* 60, 2634.
4. Becker, W., 1986 *The World of the Cell*, Benjamin/Cummings, Reading, Mass.
5. Bensimon, D., Mutz, M., 1990 *preprint ENS Paris*.
6. Berndl, K., et al., 1990 *preprint T-U. Munich*.
7. Brochard, F., Lennon, J.-F., 1975 *J. de Physique* 36, 1035.
8. Canham, P.B., 1970 *J. Theor. Biol.* 26, 61.
9. Cates, M.C., Roux, D., Andelman, D., Milner, S.T., Safran, S.A., 1988 *Europhys. Lett.* 5, 733. Erratum: ibid 7, 94.
10. Cevc, G., Marsh, D., 1987 *Phospholipid Bilayers : Physical Principles and Models* (Wiley, N.Y.).
11. Charvolin, J., Sadoc, J.-F., 1987 *J. de Physique* 48, 1559.
12. Coulon, P., Roux, D., Bellocq A.-M., 1990 *CRPP preprint*.
13. David, F., 1989 *Europhys. Lett.* 9, 575.
14. David, F., Guitter, E., 1988 *Nucl. Phys.* B295 [FS21], 332.
15. David, F., Leibler, S., 1990 (unpublished).
16. Deuling, H.J., Helfrich, W., 1976 *Biophys. J.* 16, 861.
17. Drouffe, J-M. et al. 1981 *Nucl. Phys.* B161, 39.
18. Drouffe, J-M., Zuber, J.-B., 1983 *Phys. Rep.* 102 1.
19. Dubois-Violette, M., Pansu, B., eds., 1990 *Proceedings of the Conference on Geometry of Interfaces, Aussois*.
20. Duurhus, B., et al. 1984 *Nucl. Phys.* B240, 453.
21. Duwe, H-P., et al., 1987 *Mol. Cryst. Liq. Cryst.*, 152, 1.
22. Faucon, J-F., et al., 1989 *J. de Physique*, 50, 2389.

23. Fradkin, E., Shenker, S., 1979 *Phys. Rev.*, D19, 3682.
24. Frölich, J., 1985 in *Applications of Field Theory to Statistical Mechanics*, ed. Garrido L. *Lecture Notes in Physics* 216 (Springer, Berlin).
25. Gruner, S.M., 1985 *Proc. Natl. Acad. Sci. USA* 82, 3665.
26. Guitter, E. et al., 1988 *Phys. Rev. Lett.* 60, 2949.
27. Guitter, E. et al., 1989 *J. de Physique* 50, 1787.
28. Guitter, E. et al., 1990 *J. de Physique* 51, 1055.
29. Guitter, E., Kardar, M., 1990 *Erophys. Lett.* 13, 441.
30. Gunning, B.E.S., 1965 *Protoplasma* 60, 11.
31. Helfrich, W., 1973 *Z. Naturforsch.* 28c, 693.
32. Helfrich, W., 1978 *Z. Naturförsch.* 33a, 305.
33. Helfrich, W., 1985 *J. de Physique* 46, 1263.
34. Huse, D.A., Leibler, S., 1988 *J. de Physique* 49, 605.
35. Huse D.A. and Leibler S., *Saclay preprint Spht/90/112 (1990)* .
36. Jongeward, G.A., Stack, J.D., Jayaprakash, C., 1979 *Phys. Rev.* D21 3360.
37. Kantor, Y., Nelson, D.R., 1987 *Phys. Rev. Lett.* 58 2774.
38. Karowski, M., 1986 *J. Phys.* A19, 3375.
39. Karowski, M., Thun, H.J., 1985 *Phys. Rev. Lett.* 54, 2556.
40. Leibler S. and Maggs A.C., *Proc. Natl. Acad. Sci. USA* 87 (1990) 6433.
41. Meunier, J., Langevin, D., and Boccara, N., eds. *Physics of Amphiphilic Layers*, 1987 Springer - Verlag.
42. Milner, S., Safran, S.A., 1987 *Phys. Rev.* A36, 4371.
43. Mutz, M., Helfrich, W., 1990 *J. de Physique* 51, 991.
44. Nelson, D.R., Peliti, L., 1987 *J. de Physique* 48, 1085.
45. Nelson, D.R., Piran, T., Weinberg, S., eds. 1989 *Statistical Mechanics of Membranes and Surfaces, Proceedings of the V Jerusalem Winter School* (World Scientific, Singapore).
46. Nitsche, J.C.C., 1985 *Vorlesungen über Minimalflächen*, Springer-Verlag.
47. Peliti, L., Leibler, S., 1985 *Phys. Rev. Lett.* 54, 1690.
48. Porte, G., et al., 1988 *J. de Physique* 49, 511.
49. Roux D., et al., 1990 *Europhys. Lett.* 11, 229.
50. Scriven, L.E., 1977 in *Micellization, Solubilization, and Microemulsions*, ed. Mittal, K.L., Plenum, N.Y.
51. Silver, B.L., 1985 *The Physical Chemistry of Membranes*, Allen and Unwin, London.
52. Singer, S.J., Nicolson, G.L., 1972 *Science* 175, 720.
53. Stokke, B.T., Mikkelsen, A., Elgsaeter, A., 1986 *Eur. Biophys. J.* 13, 203 ; *ibid*, 219.
54. Strey, R., et al., 1990 *Langmuir* (in press).
55. Svetina, S., Zeks, B., 1983 *Biomed. Biochim. Acta* 42, S86.
56. Yamagushi, M., et al., 1974 *J. Ultrastructure Research* 48, 33.

VESICLE SHAPES AND SHAPE TRANSFORMATIONS: A SYSTEMATIC STUDY

Karin Berndl[1], Josef Käs[2], Reinhard Lipowsky[1], Erich Sackmann[2] and Udo Seifert[1]

1. Sektion Physik der Universität München
   Theresienstr. 37, 8000 München 2 (FRG)
2. Technische Universität München, Physik
   Department, Biophysics Group (E22)
   8046 Garching (FRG)

1. INTRODUCTION

The lipid bilayer vesicle is the simplest possible model of biological membranes. Nevertheless, it exhibits already a number of typical properties of cell membranes. The most fascinating examples are the shape transitions and shape instabilities. It has been recognized long ago that shape transitions may be induced by changing the osmotic conditions or the temperature[1]. Apart from spherical and ellipsoidal shapes more exotic shapes such as e.g. discocytes, stomatocytes[1], echinocytes[2] or a necklace of small vesicles[3] has recently been observed. Up to now, our understanding of these shape transformations has been rather limited. Indeed, all previous experiments have been performed with relatively complex systems containing, e.g. charged and unsaturated lipids, mixtures of different lipids or additional solutes such as sugar in the aqueous phase. It was generally believed that these different ingredients play an essential role in determining the vesicle shape. Therefore, no attempt has been reported so far to relate these experimentally observed shapes in a systematic way to theoretical calculations.

Biologically Inspired Physics, Edited by L. Peliti
Plenum Press, New York, 1991

In our contribution we report a systematic experimental and theoretical study on these shape transformations. In order to avoid the above mentioned complications we have investigated vesicles which consist of electrically neutral lipids (that is phosphatidylcholine) in Millipore water. We find, that even for such a simple system a change in temperature can lead to three different types of shape transformations. Theoretically, we discuss shape transformations within two well established curvature models, (i) the bilayer coupling model of Svetina and Zeks[4] and (ii) the spontaneous curvature model of Helfrich[5,6]. A comparison leads to the conclusion that the observed shape transformations can well be explained within the bilayer coupling model provided a small asymmetry in the thermal expansivities of both monolayers is assumed. In some cases, such an asymmetry is not required.

2. EXPERIMENTAL SETUP

Our experiments were performed with vesicles of dipalmitoylphosphatidylcholine (DMPC) of diameters larger than 20 μm. These were prepared in a separate test-tube. The lipid with a purity >99% was dissolved in a solvent of 2:1 chloroform/methanol to produce a 1 mM solution. Then, 60 μl of this solution were distributed as a thin film on the inner surface of the test-tube. The solvent was evaporated by placing the test-tube in a vacuum chamber for a minimum of one day. Then the vesicles were swollen by filling the test-tube with 5 ml distiled water prepared with a Milli-Q-System and heating the solution up to 40°C.

After swelling for a minimum of 12 hours, the vesicles were transferred into a special chamber which allows, for the first time, the observation of free vesicles without being driven off by thermal convection. A schematic view of this measuring chamber is shown in Fig.1. The hollow outer copper frame is used to cool and heat the chamber with a water thermostat. The inner frame is made of teflon and a temperature sensor (Pt 100) is integrated in this frame. The top and bottom of the chamber are closed with cover slides. These were fixed with vacuum grease. An inner compartment is placed onto the bottom cover slide. It is formed by a thin teflon spacer (open at one side) covered by a small cover slide. The inner

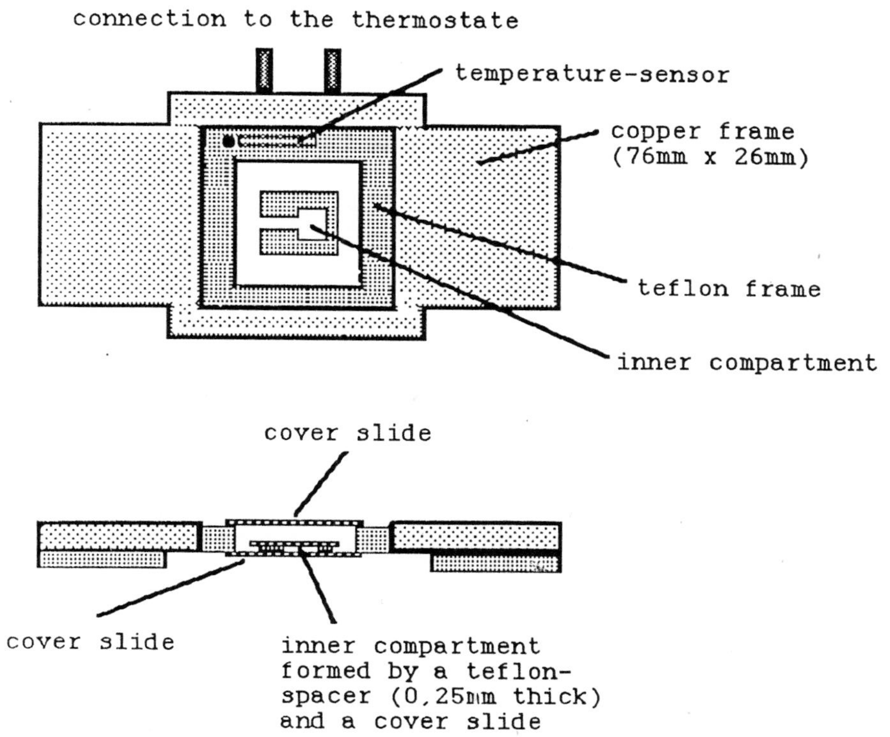

Fig.1. Measuring chamber with an inner (dead water) confinement to prevent swimming away of freely suspended vesicles by thermal convection.

compartment is essentially free of thermal convection and for this reason it is used for the observation of the vesicles. The outer compartment is required for a good thermal coupling to the thermostated frame. The vesicles were observed in phase contrast with an inverted Zeiss Axiovert 10 microscope. For the present work, a bright field air objective of magnification 40x (Zeiss) was used. During the slow increase of the temperature ≈ 0.2 K/min, the volume and the area of the observed vesicle were measured with a digital image processing system (Maxvideo, Datacube Boston, USA). The details of measuring the volume and the area will be described elsewhere[7].

3. THREE DIFFERENT TYPES OF SHAPE TRANSFORMATIONS

In our experiments we normally started with spherical or ellipsoidal vesicles of a size between 20 μm and 50 μm, which are most suitable to determine the initial values of the volume and the surface area. We found the following different types of shape transformations caused by increasing the temperature: Firstly, the budding transition exhibiting the following evolution: The sphere changes into a prolate ellipsoid and then into a pear-shaped state, which finally forms a vesicle with one bleb on the outside (see Fig.2). The second type is a reentrant shape transformation. One first obtains a prolate ellipsoid, which changes into a dumbbell shape, then into a pear-shaped state and again into a dumbbell-shaped state (see Fig.3). The third type is the discozyte-stomatozyte transition. The first step is the change from a sphere to a oblate ellipsoid. The ellipsoid changes into a discozyte and finally into a stomatozyte (see Fig.4). The last transition is completely analogous to the discocyte-stomatocyte transition of red blood cells and provides convincing evidence for our introductory remark that most simple bilayer vesicle may mimic typical behaviors of the complex biological membranes. This suggests that shape changes of biological membranes are governed by simple principles.

In Fig. 2-4, we compare the three types of experimentally observed shape transformations with shape changes, calculated with the theoretical model described below. The three theoretical sequences differ mainly in the value of a dimensionless parameter, γ, which measures the asymmetry in the thermal expansivities of the two monolayers. If the thermal expansivity of the outer monolayer is larger than the inner one and exhibits a relative difference $\gamma \geq 10^{-2}$, a small vesicle buds off the large vesicle. For $0 \leq \gamma \leq 10^{-3}$ the reentrant transitions from a dumbbell to a pear-shaped state occurs, while a larger expansivity of the inner monolayer leads to the discozyte-stomatozyte transitions and finally to the formation of a small vesicle budding towards the inside.

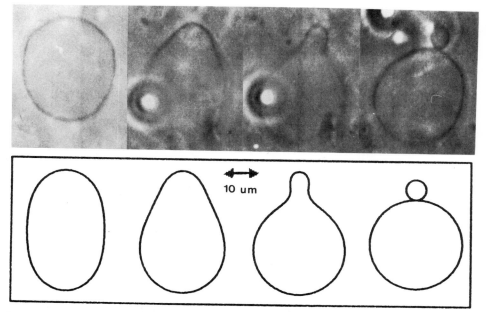

Fig.2. Demonstration of a budding transition: The shapes were measured at T = 31.4, 35.5, 35.6 and 35.8°C. The disc-like object is due to an air bubble which migrates in the outer compartment of the measuring chamber. The calculated shapes correspond to a trajectory of Eqs.(4) with initial values of the reduced volume $v_0 = 0.9446$ and reduced area difference $\Delta a_0 = 1.0305$, $\gamma = 0.057$ and b = 1500.

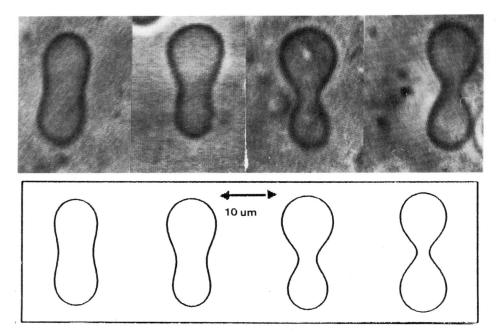

Fig.3. Symmetric-asymmetric reentrant transition: The shapes
were measured at T = 30.7, 32.6, 40.0 and 44.3⁰C. The
calculated shapes correspond to a trajectory of Eqs.(4)
with $v_0 = 0.78$, $\Delta a_0 = 1.1475$, $\gamma = 0.00166$ and b = 640.

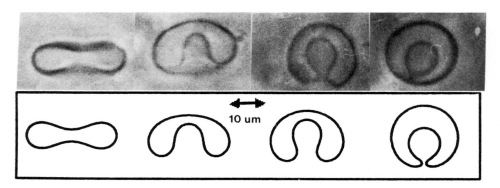

Fig.4. Discocyte-stomatocyte transition: The shapes were measured at T = 43.8, 43.9, 44.0 and 44.1⁰C. The calculated shapes correspond to a trajectory of Eqs.(4) with
$v_0 = 0.65$, $\Delta a_0 = 1.0355$, $\gamma = -0.29$ and b = 1000.

4. THEORETICAL MODELS AND THEIR AGREEMENT WITH THE EXPERIMENT

For a comparison of the experimental results with theoretical ideas, we calculated shapes and shape transitions within two variants of curvature models: (i) the bilayer coupling model of Svetina and Zeks[4], (ii) the spontaneous curvature model of Helfrich[5,6]. We first discuss the bilayer coupling model. Within this model, the two monolayers are taken to be infinite thin shells with a constant separation D, where D is about the half bilayer thickness. The shape of the vesicle is determined by the minimum of the bending energy, G_b,

$$G_b = (\kappa/2) \oint (C_1 + C_2)^2 \, dA^{in} \qquad (1)$$

which is expressed as an integral over the inner monolayer only since both monolayers are coupled. It is convenient to express all equations in terms of the surface areas A^{in} and A^{ex} rather than in terms of the neutral surface. Here, κ denotes the effective bending rigidity of the bilayer with $\kappa \approx 1.15 \times 10^{-19}$ J for DMPC[8]. The variables $C_1 = 1/R_1$ and $C_2 = 1/R_2$ are the two principal curvatures expressed by the two radii of curvature R_1 and R_2. The area difference $\Delta A \equiv A^{ex} - A^{in}$ of both monolayers is related to the total mean curvature of the inner monolayer via $\Delta A \approx D \int (C_1 + C_2) \, dA^{in}$ for small values of d. The minimization of G_b is now performed for fixed area A^{in}, fixed enclosed volume V, and fixed area difference ΔA. Thus, we assume that the exchange of lipid molecules between both monolayers can be ignored on experimentally relevant periods. This minimization leads to the shape equation

$$\delta \left[G_b + \Sigma A + PV - \frac{\kappa}{D} C_0 \Delta A \right] = 0 \qquad (2)$$

where δ denotes the variation with respect to the vesicle shape and Σ', P and C_0 denote Lagrange multipliers which can be identified with the lateral tension, the pressure difference and the spontaneous curvature, respectively.

For the model just described, we have determined the "phase diagram"[9], i.e., we have determined the axi-symmetric

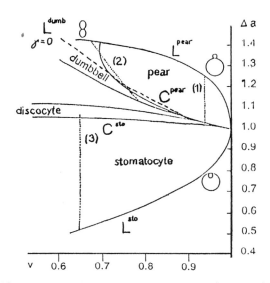

Fig.5. Phase diagram and temperature trajectories for the bilayer coupling model. This phase diagram shows the
state of lowest energy for given Δa and v. C^{pear} and
C^{sto} denote lines of continuous transitions at which
the up/down symmetry of the vesicle shape is broken.
L^{pear}, L^{sto} and L^{dumb} denote limit shapes. Note that the
dumbbell region contains for large v-values prolate
ellipsoids and that the discocyte region contains for
large v-values oblate ellipsoids. The pointed lines
(1), (2) and (3) represent the temperature trajectories for Fig.2, Fig.3 and Fig.4. The dashed line represents the temperature trajectory for the asymmetry
parameter $\gamma = 0$ with $v_0 = 0.9$ and $\Delta a_0 = 1.05$.

shapes of lowest bending energy for experimentally accessible values of the reduced volume v with

$$v \equiv V / \left[(4\pi/3) \; (A^{in}/4\pi)^{3/2} \right] \tag{3}$$

which is a measure for the excess area and the induced area difference Δa with

$$\Delta a \equiv \Delta A / \left[8\pi \; D \; (A^{in}/4\pi)^{1/2} \right] \; . \tag{4}$$

For a sphere, one has $v = \Delta a = 1$. This phase diagram is displayed in Fig.5. Some part of it has been previously described by Svetina and Zeks[4]. As two important novel features, we find (i) an instability of the dumbbell shapes with respect to the up/down symmetry which leads to the pear-shaped states, and (ii) new limiting shapes which look like two prolate ellipsoids sitting on top of each other.

In order to compare theoretical and experimental shapes, we determine the path $\Delta a = \Delta a (v)$ which corresponds to a change in temperature T. We now assume that the interior and exterior monolayer have different temperature independent relative expansivities, α^{in} and α^{ex}, as given by

$$\alpha^{in} \equiv \frac{1}{A^{in}} \frac{dA^{in}}{dT} \quad \text{and} \quad \alpha^{ex} \equiv \frac{1}{A^{ex}} \frac{dA^{ex}}{dT} \; , \quad \text{with} \; \alpha^{ex} = (1 + \gamma) \alpha^{in} \; . \tag{5}$$

For simplicity, we neglect the small thermal expansivity of the enclosed water. We assume that the thermal expansivity of D is given by $(-1/2) \alpha^{in}$ [10]. Differentiating Eqs.(3) with respect to the temperature T and inserting Eqs.(5) leads after integration to the temperature dependence of the reduced volume as given by

$$v(T) = v_0 \exp\{ (-3/2) \; \alpha^{in} (T - T_0) \} \tag{6}$$

where $v_0 = v(T_0)$ defines the initial value. A similar equation can be derived for $\Delta a (T)$. If $(T - T_0)$ is eliminated in both equations one finds the temperature trajectory

$$\Delta a(v) = \left(\frac{v_0}{v}\right)^{2/3} \left[\Delta a_0 + b \left(\left(\frac{v_0}{v}\right)^{2\gamma/3} - 1 \right) \right], \tag{7}$$

103

where $\upsilon_0 \equiv \upsilon(T_0)$ and $\Delta a_0 \equiv \Delta a(T_0)$ parametrized the initial shape at temperature $T = T_0$ and

$$b \equiv A^{ex}(T_0)/\{8\pi\ D(T_0)\ [A^{in}(T_0)/4\pi]^{1/2}\} \qquad (8)$$

For $\gamma = 0$, i.e., if the asymmetry were not present, the second term in Eqs.(7) vanishes and a temperature trajectory in the phase diagram would be given by $\Delta a(\upsilon) = (\upsilon_0/\upsilon)^{2/3}\Delta a_0$. This path is shown as a dashed line in Fig.4: Starting, e.g. with a symmetric prolate ellipsoid, it immediately crosses the phase boundary C^{pear} and enters the pear-shaped region; it then crosses again the line of continuous transitions C^{pear} and finally meets the new type of limiting shapes at L^{dumb}. Note that such a path never enters the stomatocyte region. Any $|\gamma|$ in the order of 10^{-3}, however, has already a significant influence since the parameter b as given by Eqs.(8) can be estimated to be of orders 10^3 for the typical values $A^{in}(T_0) \approx A^{ex}(T_0) \approx 1000\ \mu m^2$ and $D \approx 5$ nm.

The asymmetry γ and the initial area difference Δa_0 cannot be measured or controlled directly in our experiment. A crude estimate, however, can be obtained by comparing the experimental with theoretical shapes. The calculated shapes shown in Fig. 2-4 lie on a trajectory described by Eqs.(4) within the experimentally observed temperature intervals. For Fig.3 and Fig.4, the parameters b, υ_0, Δa_0 and γ were obtained as follows: The measured area A and volume V of the first shape at temperature T_0 determines υ_0 and b, where we used $A^{in} \approx A^{ex} \approx A$ and $D \approx 5$ nm. The area difference Δa_0 is chosen to fit the experimental shape. The measured temperature interval $T_f - T_0$ between the final and the initial shape determines $\upsilon(T_f)$ via Eqs.(6) with $\alpha^{in} \approx 6 \times 10^{-3}$ [12]. Once again, $\Delta a(T_f)$ is fitted to the experimental shape, which finally determines the value of the asymmetry parameter γ. For Fig.2, we first determined $\upsilon(T_f)$ from the last shape and then $\upsilon(T_0)$ from the measured temperature difference to the second shape.

Although the experimental and theoretical results agree very well, it is worthwhile to envisage also an explanation within the spontaneous curvature model. Within this model the shape of the vesicle with given area A and enclosed volume V is determined by the minimum of the bending energy F_b, with

$$F_b = (\kappa/2) \oint (C_1+C_2-C_0)^2\ dA , \qquad (9)$$

The microscope details of the two monolayers are described by the spontaneous curvature C_0. The minimization leads to the same shape equation as Eqs.(2) and consequently to the same extremal shapes. A shape which corresponds to a local minimum of G_b, may, however, correspond to a local maximum of F_b. Therefore the phase diagram in both models are quite different. The phase diagram for the spontaneous curvature model depends on υ given by Eqs.(3) and the reduced spontaneous curvature

$$c_0 \equiv C_0 \ (A/4\pi)^{1/2} . \tag{10}$$

We display, the phase diagram in Fig.6. Its derivation will be presented elsewhere[9]. Its main characteristics are:
1) For $C_0 \geq 2.08$, a discontinuous transition D^{pear} separates prolate/dumbbell from pear-shaped states.
2) With decreasing volume, the pear-shaped vesicles become symmetric again for $C_0 < 2\sqrt{2}$ at C^{pear}.
3) For $C_0 > 2\sqrt{2}$, however, the pear-shaped vesicles reach a limit-shape L^{pear} with decreasing volume. This limit shape consists of two spheres which are connected by an narrow neck, which contains no energy since the two curvatures have compensating signs. Such an "ideal" neck is only possible if the radii R_1 and R_2 of the two spheres fulfil the relation

$$R_1^{-1}+R_2^{-1} = C_0 \tag{11}$$

This equation, together with the conservation of area 4π $[R_1^2 +R_2^2] = A$, determines which spontaneous curvature C_0 is necessary in order to obtain budding of a smaller vesicle with radius R_1. For small R_1, we have $C_0 \approx R_1^{-1}$.
4) A discontinuous transition D^{sto} leads from oblate/discocyte shapes to the stomatocytes. For $C_0 < 0$, these shapes reach a limit shape L^{sto} given by an inverted sphere of radius $R_1 < 0$ embedded in a large sphere of radius $R_2 > 0$. Once again Eqs.(11) holds for the limit shape.

In order to compare this phase diagram and the predicted transitions with the experimental trajectory, we need the temperature dependence of C_0, which is not clear a priori. We therefore assume that C_0 remains temperature independent. For

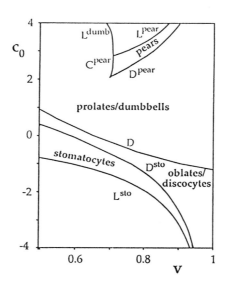

Fig.6. Phase diagram for the spontaneous curvature model. This phase diagram shows the state of lowest energy for given c_0 and v. C^{pear} denotes a line of continuous transition at which the up/down symmetry of the vesicle shape is broken. D^{pear}, D and D^{sto} denote lines of discontinuous transitions. L^{pear}, L^{sto} and L^{dumb} denote limit shapes.

the volume, Eqs.(6) remains valid. Let us now discuss which theoretical trajectories could fit the experimentally observed ones:

(1) For a reentrant trajectory as in Fig.3, the spontaneous curvature model predicts a discontinuous transition from the symmetric to the pear-shaped state. Of course, the distinction between a continuous and a discontinuous transition for a single vesicle is experimentally difficult. For a discontinuous transition, one expects a strong variation of the shape within a short time caused by an infinitely small change of the area and the occurrence of hysteresis. The observed reentrant trajectory, however, occurred over a relative large temperature interval of 14 K, which corresponds to a time interval of 70 min. A further objection against the spontaneous curvature model for this trajectory is that the neck at the C^{pear}-line is significantly narrower for the calculated shapes than for to the experimental ones.

(2) In order to obtain a budding trajectory leading to a bleb of the observed small size a spontaneous curvature $C_0 \approx 5\text{-}6$ has to be postulated. which is rather improbable for our pure system. Moreover, all budding trajectories in the spontaneous curvature model should be precursed by the first order transition from the symmetric to the pear-shaped states.

(3) Finally, the discocyte-stomatocyte transition in this model is first order and would require a negative spontaneous curvature to occur for the observed υ-value.

Summarizing, the detailed analysis of the phase diagram for both models allows to test critically whether these models apply to the observed shape transformations. We find that the bilayer coupling model augmented with an asymmetry in the monolayer expansivity fits well while the spontaneous curvature models makes qualitatively different predictions. Especially, we consider the transition from a prolate ellipsoid to a pear-shaped state in Fig.2 and the discocyte-stomatocyte transition in Fig.4 as continuous. This is in accordance with the bilayer coupling model, but in disagreement with the spontaneous curvature model.

5. POSSIBLE REASONS FOR THE ASYMMETRY IN THE THERMAL EXPANSIVITIES

What could be the origin of the different thermal expansivities of the two monolayers? One possibility could be impurities such as lyso-lipids asymmetrically distributed between the both monolayers. A second more likely explanation is that the asymmetry is induced during the swelling process or the cooling of the vesicles from 40°C to room temperature during the transfer to the measuring chamber. Experimentally we found that after a long swelling time (about 24 hours) there is a preference for γ-values close to zero indicating a relaxation process. Further more the route of shape change depends on the lipid structure. For palmitoyloleylphosphatidylcholine (POPC), for instance, the values of $|\gamma|$ are remarkable higher. If an nearly spherical vesicle is cooled, so that a lateral stress is exerted within the bilayer, a high γ value arises. Thus heating, the vesicle again leads to a budding transition owing to a high value of γ. This example is shown in Fig. 2.

6.SUMMARY

The shape changes of giant bilayer vesicles consisting of phosphatidylcholine (DMPC) in pure water were studied. They were induced by temperature variations resulting in a change of the excess surface area since the volume remains essentially constant. Three different types of shape transformations were observed: firstly, a budding of small vesicles towards the outside of large vesicles which leads to a stepwise formation of a chain of vesicles at further increasing the temperature; secondly, a reentrant transition form a dumbbell to a pear-shaped state and thirdly, an oblate ellipsoid-discocyte-stomatocyte transition.

The shape transitions and the degree of continuity (order) can be best explained in terms of the bilayer coupling approach of Zeks and Svetina[4]; by assuming that the thermal expansivities of the two monolayers are different. The type of shape change depends on the asymmetry in the thermal expansivities of both monolayers. We provide evidence that the asymmetry in the case of symmetric bilayer vesicles (equal composition of outer and inner aqueous phases) is introduced during preparation and that the route of transition depends also on the pretreatment of the vesicles. Our observations can be less well explained by the spontaneous curvature approach of Helfrich[5,6].

ACKNOWLEDGMENTS

This work was supported by the Deutsche Forschungsgemeinschaft through the Sonderforschungsbereich No.266.

REFERENCES

1. E. Sackmann, H. P. Duwe and H. Engelhardt, Membrane bending elasticity and its role for shape fluctuations and shape transformations of cells and vesicles, *Faraday Discuss.Chem. Soc.* 81:468 (1986).

2. H. Gaub, R. Rüschl, H. Ringsdorf and E. Sackmann, Phase transitions, lateral phase seperation and microstructure of model membranes composed of a polymerizable two-chain lipid and dimyristoylcholine, *Chem.Phys.Lipids* 37:19 (1985)

3. E. Evans and W. Rawicz, Entropy-driven tension and bending elasticity in condensed-fluid membranes, *Phys.Rev.Lett.* 64:2094 (1990).

4. S. Svetina and B. Zeks, Membrane bending energy and shape determination of phospholipid vesicles and red blood cells, *Eur.Biophys.J.* 17:101 (1989).

5. W. Helfrich, Elastic properties of lipid bilayers, *Z.Naturforsch.* 28c:693 (1973).

6. H. J. Deuling and W. Helfrich, The curvature elasticity of fluid membranes: A catalogue of vesicle shapes, *J.Physique* 37:1335 (1976).

7. J. Käs and E. Sackmann, to be published.

8. H. P. Duwe, J. Käs and E. Sackmann, Bending elasticity moduli of lipid bilayers: modulation by solutes, *J.Physique* 51:945 (1990).

9. U.Seifert, K.Berndl and R.Lipowsky, to be published.

10. G. Cevc and D. Marsh, Bilayer thermomechanics and thermal
    expansion, in: *Phospholipid bilayers*, G. Cevc and D.
    Marsh, ed., John Wiley & Sons, New York, Chichester,
    Brisbane, Toronto, Singapore (1987).
11. E. Evans and D. Needham, Surface-Density Transitions,
    Surface Elasticity and Rigidity, and Rupture Strength of
    Lipid Bilayer Membranes, in: *Physics of Amphiphilic
    Layers, Springer Proceedings in Physics 21,* J. Meunier,
    ed., Springer, New York, Heidelberg, Berlin (1987).

THEORETICAL ANALYSIS OF PHOSPHOLIPID VESICLES AND RED BLOOD CELL SHAPES AND

THE EFFECT OF EXTERNAL ELECTRIC FIELD

Bostjan Žekš and Saša Svetina

Institute of Biophysics, Medical Faculty, Lipičeva 2
and J. Stefan Institute, University of Ljubljana,
61105 Ljubljana, Yugoslavia

INTRODUCTION

Phospholipid vesicles and biological cells display a variety of different shapes and the question can be asked what determines the equilibrium shape of a cell and its shape changes. As the inner solutions of red blood cells (RBC) and phospholipid vesicles (PV) do not involve any structure, the shapes of these objects depend solely on the physical and chemical state of their membranes. It is commonly believed that for a given membrane the shapes that are formed correspond to the minimum value of the membrane elastic energy. This energy can, in general, be decomposed into the sum of the stretching, shear and bending energy terms[1]. It is also a general property of membranes that relatively much more energy is needed to stretch them than to cause shear deformation or bending. Consequently, the shape established by a flaccid cell or vesicle corresponds to the minimum value of the sum of the shear and bending energy terms, where its membrane area is practically constant. In particular, phospholipid membranes are two-dimensional liquids and as such do not exhibit shear elasticity. Thus their shape is determined only by the membrane bending energy. The RBC membrane is structurally more complex than the PV membrane, involving, for example, a cytoplasmic protein network and can therefore exhibit shear elasticity[2]. However, which of the above two elastic deformations is the main determinant of the RBC shape still cannot be definitely established. At least some of the shapes observed in PV and RBC systems are alike[3] which indicates a possible dominant role of the membrane bending energy. It is therefore of interest to investigate the RBC shape behavior under the assumption of a minimum value of membrane bending energy as a possible limiting case of a more general situation.

Biologically Inspired Physics, Edited by L. Peliti
Plenum Press, New York, 1991

The idea that the RBC shape is determined by the minimum value of the total membrane bending energy has been introduced by Canham[4] who, by approximating the RBC geometry by the ovals of Cassini, obtained theoretically that at constant cell volume and membrane area, the shape with the minimum value of membrane bending energy is a discocyte. Jenkins[5], by applying a general variational principle to the problem of shape determination, has shown that at a given cell volume there is a multiplicity of possible axi-symmetrical shapes, in which some involve equatorial reflection symmetry (symmetrical shapes), and others do not (asymmetrical shapes). Helfrich and Deuling[6] extended the minimum bending energy approach by also taking into consideration the spontaneous curvature, the membrane material property measuring the asymmetry of the membrane. In a continuation of their work, Deuling and Helfrich[7] gave a catalogue of vesicle shapes, the shapes having been calculated at different values of enclosed volume, membrane area and spontaneous curvature. A large variety of axisymmetrical shapes was pre-sented, allowing for indentations, cavities and contact of the membrane with itself. Deuling and Helfrich[8] have also shown the similarity of some calcu-lated shapes with discocyte and stomatocyte RBC shapes. Luke[9] contributed an efficient numerical method for calculating vesicle shapes and presented some more shapes, in particular some appearing at positive values of spontaneous curvature[10]. More recently Peterson[11] performed a stability analysis of vesicle shapes which indicated that the axisymmetrical shapes are not stable at all values of spontaneous curvature.

Another line of RBC shape research has indicated the importance of the fact that the RBC membrane is composed of layers. Evans[12], when treating the mechanical properties of layered membranes, introduced the concept of chemi-cally induced average curvature. The notion of this membrane property is very clearly represented in qualitative terms by the bilayer couple hypoth-esis[13]. In its most general form, this hypothesis is that the two leaflets of the closed membrane bilayer may respond differently to various perturba-tions while remaining in contact. The bilayer couple hypothesis was intro-duced in order to provide a qualitative explanation for RBC shape transform-ations which arise because of the interaction of the RBC membrane with a wide range of amphipathic molecules. Those of them which bind preferentially to the cytoplasmic side of the membrane were assumed to expand the inner bi-layer leaflet relative to the outer, thus causing a cell to convert from the normal biconcave disc shape (discocyte) to a cupped form (stomatocyte). Those amphipathic compounds which bind preferentially to the external half of the bilayer, were assumed to expand the outer leaflet relative to the in-ner one, thus causing a normal cell to convert into a crenated form (an echinocyte).

The idea of coupled layers can be incorporated into the quantitative treatment of RBC and PV shapes based on the membrane bending energy by considering in the minimization procedure, in addition to the constancy of the cell volume and the membrane area, the constraint of the constancy of the difference between the areas of the two bilayer leaflets[14]. The application of a strict variational approach to the problem has[15, 16] indicated a number of interesting properties of this system, such as the occurrence of symmetry instabilities and geometrical limitations of possible shapes.

In this contribution, first the bilayer couple model calculation[16] and its results will be reviewed briefly. Some newly evaluated shapes and shape sequences will be shown and some empirical rules for the general dependence of the shape on the cell geometrical parameters will be established. In the second part some preliminary results of the effect of an external electric field on PV and RBC shapes will be presented[17, 18].

BILAYER COUPLE MODEL

Minimization of the membrane bending energy

The problem is to find the extreme values of the membrane bending energy

$$W_b = \frac{1}{2}K \int (C_1 + C_2)^2 dA^*,$$ (1)

where K is the membrane bending elastic constant, C_1 and C_2 are the two principal curvatures, and integration is performed over the whole area of the neutral surface of the bilayer (A^*). The minimization procedure is to be carried out at fixed values of cell volume (V), area of neutral surface (A) and difference between the areas of the two membrane leaflets (ΔA).

The shape of a cell can therefore be obtained by minimizing the functional

$$G = W_b - \lambda(A^* - A) - \mu(V^* - V) - \nu(\Delta A^* - \Delta A).$$ (2)

Where A^*, V^* and ΔA^* are the membrane area, cell volume and the difference in leaflet areas, respectively, and the three Lagrange multipliers (λ, μ, ν) are determined from the conditions

$$A^* = A, \qquad V^* = V, \qquad \Delta A^* = \Delta A.$$ (3)

113

The difference in leaflet areas

$$\Delta A^* = \delta \int (C_1 + C_2) dA^* \tag{4}$$

is proportional to the average curvature. Here δ is the distance between the neutral surfaces of the two leaflets.

The expression for the membrane bending energy (Eq. 1) is scale invariant, i.e. it has the property that for a given shape it does not depend on the cell or vesicle size. It is therefore appropriate in the forthcoming analysis to choose the unit of length in such a way that the membrane area equals unity. If R_s is the radius of the sphere with the membrane area A

$$R_s = (A/4\pi)^{1/2} \tag{5}$$

the new variables, i.e. the two dimensionless curvatures, are

$$c_1 = R_s C_1, \qquad c_2 = R_s C_2. \tag{6}$$

It is then also convenient to define the relative volume

$$v = V/V_s, \qquad V_s = 4\pi R_s^3/3 \tag{7}$$

and relative difference between the areas of the two membrane leaflets

$$\Delta a = \Delta A/\Delta A_s, \qquad \Delta A_s = 8\pi \delta R_s \tag{8}$$

while relative area a = 1. In an analogous manner, the relative area element can be expressed as

$$da^* = dA^*/4\pi R_s^2 \tag{9}$$

and $v^* = V^*/V_s$, $\Delta a^* = \Delta A^*/\Delta A_s$.

It is clear that the result of the minimization procedure does not depend on the value of the membrane bending constant, K. It is therefore appropriate to measure also the membrane bending energy and the energy functional G (Eq. 2) relative to the bending energy of the sphere

$$w_b = W_b/8\pi K, \qquad g = G/8\pi K. \tag{10}$$

We therefore obtain for the dimensionless energy functional g the expression

114

$$g = \frac{1}{4}\int (c_1 + c_2)^2 da^* - \frac{1}{4}L(a^* - 1) - \frac{1}{6}M(v^* - v) - \frac{1}{2}N(\Delta a^* - \Delta a).$$ (11)

Here the new Lagrange multipliers L, M and N are related to λ, μ and ν as

$$L = \lambda \frac{2R_s^2}{K}, \qquad M = \mu \frac{R_s^3}{K}, \qquad N = \nu \frac{2\delta R_s}{K}.$$ (12)

It is easy to see from the structure of the energy functional G (Eq. 2) that the three Lagrange multipliers λ, μ and ν represent the thermodynamically conjugated fields to A, V and ΔA, respectively. Therefore, at equilibrium, when $W_b = W_b(A, V, \Delta A)$ the Lagrange multipliers are given by

$$\lambda = \frac{\partial W_b}{\partial A}, \qquad \mu = \frac{\partial W_b}{\partial V}, \qquad \nu = \frac{\partial W_b}{\partial \Delta A}.$$ (13)

On the other hand, one can conclude from Eq. (11) that at equilibrium, the bending energy $W_b = 8\pi K w_b(v, \Delta a)$ depends only on v and Δa. As a consequence, the Lagrange multipliers are interrelated as

$$A\lambda + \frac{3}{2}V\mu + \frac{1}{2}\Delta A\nu = 0$$ (14)

or in the dimensionless form

$$L + vM + \Delta aN = 0.$$ (15)

Following the procedure used by Deuling and Helfrich[7], for a cell with an axisymmetrical shape, the corresponding contour x = x(z), with z the dimensionless coordinate along the axis and x the dimensionless distance from the axis, can be obtained[16] by minimizing the functional g.

Possible limiting shapes

One can also pose the question[16] of what are the shapes of a cell with given A and ΔA which have maximal volume. In a dimensionless representation, this means that we are looking for a shape with extremal relative volume v^* with the condition that the relative area a^* equals one and the relative area difference Δa^* equals Δa. We therefore study the functional

$$h = v^* - \frac{1}{4}\tilde{L}(a^* - 1) - \frac{1}{2}\tilde{N}(\Delta a^* - \Delta a)$$ (16)

The Euler-Lagrange minimization procedure leads, if we restrict ourselves to shapes which are smooth at the poles, to an algebraic equation for the curvature c_p

$$\tilde{N}c_p^2 + \tilde{L}c_p - 6 = 0. \tag{17}$$

which means that the curvature c_p is constant over the surface. In such a case c_m is also constant and equal to c_p or zero. We therefore conclude that the surface which corresponds to the solution of Eq. (17) is a sphere or a section of a sphere or of a cylinder. As Eq. (17) has in general two different solutions, we conclude that the limiting shapes consist of spheres or of sections of spheres or cylinders where only two different radii are

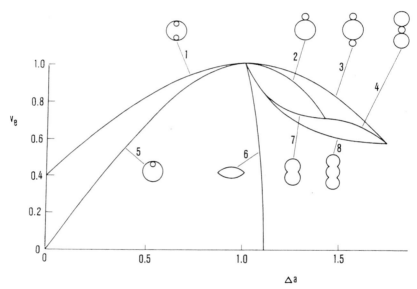

Fig. 1.   Some extreme relative volumes and the corresponding limiting shapes

allowed. Figure 1 shows some examples of limiting shapes and dependencies of their relative volumes on the relative leaflet area difference.

Some possible low energy shapes

The general result of the preceding theory is all possible shapes and the corresponding membrane bending energies at any set of values of relative volume v and relative leaflet area difference Δa. A more compact form of presenting this result is to introduce certain classes of shapes defined in

116

the following way. A class of possible PV or RBC shapes constitutes the
shapes which can be obtained in a continuous and derivable manner from a
given shape at certain values of v and Δa, if v and Δa are continuously
changed. As the previously performed analysis of limiting shapes suggests,
the classes of shapes so defined may exist within certain parts of the v/Δa
diagram only.

We first consider the behavior of two low energy classes of PV or RBC
shapes, the class of axisymmetrical shapes involving reflection symmetry
with respect to the equatorial plane of the object, which includes discocyte
shapes (designated B in the following), and the class of the axisymmetrical
shapes without such symmetry, which includes cup shapes (designated A).

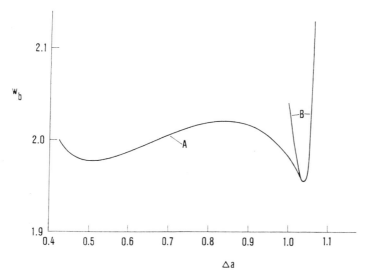

Fig. 2.   Relative membrane bending energy in dependence of Δa for discocyte
          (B) and stomatocyte (A) classes

Figure 2 shows the dependence of the relative membrane bending energy
on Δa for the value of relative cell volume v = 0.6 for classes B and A. In
Fig. 3 are given examples of some of the corresponding shapes. Symmetrical
discoidal shapes (class B) are shown within the interval $0.996 \leq \Delta a \leq 1.073$.
The lower bound is a discocyte shape at which the two poles of the object
come into contact. At values of $\Delta a < 0.996$ there is a finite amount of membrane in contact. The upper bound is the limiting shape resembling a bi-
-valved shell (see Fig. 1, line no. 6). In the interval $0.996 < \Delta a < 1.062$,
the object has a discoid shape with the distance between the poles
increasing with increasing Δa. At values $\Delta a > 1.062$ the object resembles an

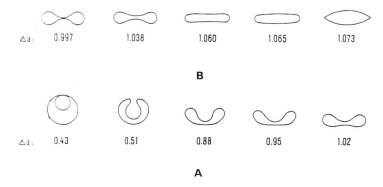

$$\Delta a: \quad 0.997 \qquad 1.038 \qquad 1.060 \qquad 1.065 \qquad 1.073$$

B

$$\Delta a: \quad 0.43 \qquad 0.51 \qquad 0.88 \qquad 0.95 \qquad 1.02$$

A

Fig. 3. Characteristic examples of discocyte (B) and stomatocyte (A) shapes

ellipsoid. The membrane bending energy has its absolute minimum at
$\Delta a = 1.038$, where the shape is discocyte. As here $N = 2\partial w_b/\partial\Delta a = 0$ this is
the shape which can be obtained by minimizing the membrane bending energy by
taking into consideration only the constraints of constant membrane area and
object volume. The present result thus confirms the results of Canham[4].

   In the interval $0.426 < \Delta a < 1.034$, there exist asymmetrical cup shapes
(class A). The upper bound represents a shape belonging to the class of symmetrical discoidal shapes described above. At this value of Δa, the symmetrical shapes become unstable and are unstable below this value. The lower
bound is the limiting spherical shape involving an inside spherical vesicle
with an inside out oriented membrane (line no. 5 in Fig. 1). The series of
shapes shown in Fig. 3 (designated by A) illustrates the dependence of these
asymmetrical shapes on the value of Δa.

   The dependence of the relative membrane bending energy on Δa as presented in Fig. 2 shows how the system would behave if the constraint
$\Delta a = $ const. were removed. If initially $\Delta a > 0.83$ an object would eventually
become a discocyte with $\Delta a = 1.038$ whereas with the initial $\Delta a < 0.83$, an
object would become a stomatocyte with $\Delta a = 0.51$.

   There exists a multitude of other solutions of the above minimization
problem, which correspond to more complicated shapes. Recently, Berndl et
al.[19] have analyzed the transition from a dumb-bell to a pear-shaped state and
have shown, that the observed temperature sequences of vesicle shapes can be
well described by the bilayer couple model. Some other examples of possible
shapes were also evaluated recently[20] and are presented for $v = 0.85$ in
Fig. 4 together with their limiting shapes. Generally, at larger Δa values
the shapes with evaginated membranes exist, while for small Δa characteristic shapes have invaginations. For intermediate Δa the shapes are evaginated
in one part and invaginated at another.

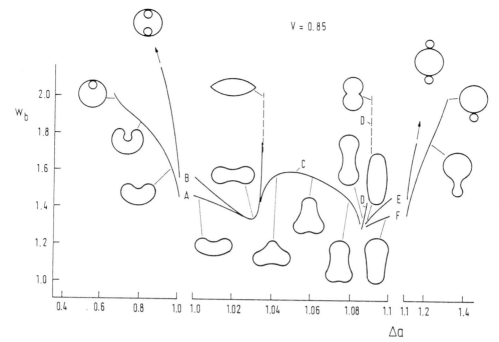

Fig. 4.   Some possible shapes for v = 0.85 for classes A to F

The other important observation is that stable shapes tend to be oblate
at smaller volumes (v) and prolate at larger v. For v = 0.6 (Figs. 2 and 3)
a discocyte, i.e. an oblate shape, represents the absolute energy minimum,
while at v = 0.85 already the prolate shape is the stable one (Fig. 4). For
v → 1 the prolate ellipsoid represents the stable shape.[16]

In Fig. 5 a case of a breaking of the mirror plane symmetry[21] in dependence of Δa is shown for v = 0.95. Fig. 5a shows the dependence of the relative membrane bending energy on Δa for symmetric and for asymmetric solutions. For 1.0219 < Δa < 1.0222 a symmetric shape (E) is stable, which
transforms with increasing Δa continuously into an asymmetric shape (F).
Some of the shapes are shown in Fig. 5b.

As the spontaneous membrane curvature is thermodynamically conjugated
parameter to Δa, the shapes obtained from the spontaneous curvature model[6-8]
are the same as those obtained from the bilayer couple model. It
nevertheless seems that the last model is physically relevant for the
description of some observed sequences of shape changes. It can be seen for
example from Figs. 2 and 3 that a monotonous decrease of Δa reproduces the
"natural" sequence of the discocyte-stomatocyte transition. To obtain the
same sequence by the spontaneous curvature model one should require a
nonmonotonous change in the spontaneous curvature. The same effect was
observed for some thermally induced sequences of vesicle shape changes[3,19].

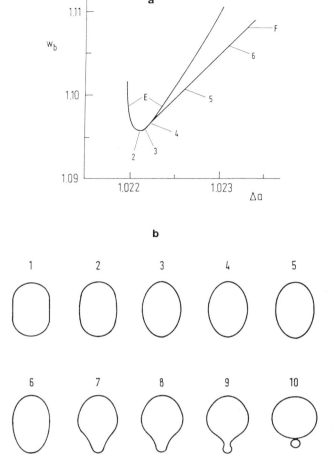

Fig. 5. (a) Relative bending energy as a function of Δa for symmetric (E) and asymmetric (F) shapes for $v = 0.95$. (b) Examples of numerically obtained shapes for Δa in the interval between $\Delta a = 1.0219$ (shape 1) and $\Delta a = 1.177$ (shape 10)

THE SHAPES OF PHOSPHOLIPID VESICLES AND RED BLOOD CELLS IN AN EXTERNAL
ELECTRIC FIELD

An external electric field deforms phospholipid vesicles because of the
electric Maxwell stresses acting on the membrane. The field tends to compress the vesicles laterally and elongate them in the direction of the
field. These field effects are stabilized by the membrane bending energy,
leading to a stable vesicle shape in the presence of the electric field.

These field induced shape changes have been studied theoretically[22,23] for almost spherical shapes, where it was assumed that the initial shape is spherical and the field induced shape change is small and can be described by the second Legendre polynomial. The area of the membrane was assumed to be constant, i.e. no lateral expansion of the membrane was allowed, while the volume of the vesicle could change because of the efflux of water. In this analysis[23] the effects of the finite electric conductivity of the phospholipid membrane were also studied and alternating electric fields were allowed. Larger electric field effects on the vesicle shape have been treated in the literature[24,17] by assuming that the equilibrium shape is an uniaxial ellipsoid and the two axes of the ellipsoid are the two parameters of the model. Bryant and Wolfe[24] allowed lateral expansion of the membrane in their analysis and showed that by increasing the electric field, not only does the excentricity of the ellipsoid increase but also its area because of the lateral stresses in the membrane induced by the field. In work by Pastushenko et al.[17] the membrane area was kept constant and so vesicle elongation only was studied as a function of the applied electric field together with the corresponding decrease of vesicle volume. A dynamic analysis has also shown[17] that because of small membrane water permeability the expected characteristic time for the volume equilibration is much larger than the measuring time, which seems to make the assumption of free volume changes questionable.

In this contribution, we shall present a method for evaluation of vesicle shapes in the presence of an electric field, which is not limited to almost spherical shapes and which is not based on a parameterization of the shapes as in previous works,[24,17] but leads to a differential equation which gives the most general stable shape at a given field by numerical integration. The method is based on the bilayer couple model of the membrane which was discussed in the previous section.

The effect of the electric field on the vesicle shape is studied here only in the case of static electric field (E_o) and with the assumption that membrane conductivity is negligible. This is a reasonable approximation, because membrane conductivity is much smaller than the conductivity of the ionic solution. In this case the vesicle disturbs the homogeneous field lines, which become tangential to the surface on the surface of the vesicle. It can be seen from the general expression for electric Maxwell stresses, that the field produces locally at the surface a force per unit area which is normal to the surface, points inward and is proportional to the square of the tangential electric field. This compression could therefore be expected to be largest on the equator, where the tangential electric field is the largest, and equal to zero at the poles, where the tangential field vanishes. This would lead to a simple elongation of the vesicle which, however, does not conserve the volume (v) and the area difference (Δa). More complicated shapes should therefore be expected to be stable in the presence of an electric field within the bilayer couple model, which requires constant v and Δa.

The problem is divided into two parts. The first is the electrostatic one. For a given shape of a vesicle, we are looking for the solution of a Laplace equation for the electric potential which gives a homogeneous field E_o far away from the vesicle, while at the surface the normal component of the field equals zero. We have developed a mathematical procedure which is based on the expansion of the potential in spherical harmonics, and which solves the potential problem with the above boundary conditions and allows the determination of the tangential field at each point on the surface of the vesicle.

The second part of the problem consists of finding the equilibrium shape, i.e. the shape for which the electric forces and the mechanical forces which originate in the membrane bending energy, are compensated. Going through the minimization procedure, which is analogous to that developed for the bilayer couple model[16] and which will be presented in detail elsewhere, one arrives at an expression which has the same structure as for a zero field case except that the Lagrange multiplier M (Eq. 11) is modified and includes the electric field term

$$M \Rightarrow M - \frac{\varepsilon\varepsilon_o E_o^2 R_s^3}{K} \frac{2}{x^2} \int_o^x e^2(x')x'dx'. \tag{18}$$

By e(x) is denoted the tangential component of the electric field relative to E_o at the point on the vesicle surface, which is at distance x from the axis. We see that the external field influences the shape only in the combination

$$\tilde{E}_o^2 = \frac{\varepsilon\varepsilon_o E_o^2 R_s^3}{K} \tag{19}$$

which represents the ratio of the electric field energy corresponding to the volume of the vesicle to the bending elastic modulus. It can be seen that the problem is no longer scale invariant and the effect of the electric field is larger for larger vesicles and smaller bending constants. For

v = 0.95 and Δa = 1.025 the shape is prolate and ellipsoid like in the absence of the electric field. Contrary to expectations, the field does not
compress the shape at the equator even though the field pressure is largest
there. At intermediate fields the shape elongates a little along the axis
but its radius on the equator also increases and an inflection develops,

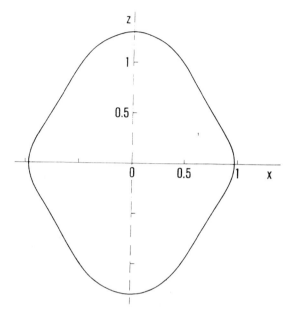

Fig. 6. The shape of a vesicle with v = 0.95 and Δa = 1.025 is shown for
$\tilde{E}_o = 5$

where the field has a plateau. A further increase of the field (Fig. 6)
makes these shape changes more pronounced, but the additional changes get
smaller and smaller as the field increases.

The calculated shapes can be understood in the following way. An electric field tends to compress the vesicle laterally and tends to make it long
and thin, but the conservation of the relative volume and relative area difference limits the elongation and produces a bulge at the equator. A careful
analysis of experimental data should decide under what experimental conditions such shapes can be observed, which means that v and Δa remain constant
during the experiment. The experiments usually do not reveal such behavior
but the simple elongation of a vesicle, which seems to mean that the field
induced lateral tension in the membrane[24] increases membrane area and/or
produces channels in the membrane which facilitate both the water flow and
the flip-flop of lipids and thus make the characteristic times for volume

equilibration and for changes in the area difference much faster than expected[17].

CONCLUSIONS

The bilayer couple model predicts a variety of different vesicle shapes, which seem to agree well with observations. In contrast to the spontaneous curvature model, it can account for observed sequences of shapes in a plausible way by assuming a monotonous change of the parameters of the model, i.e. Δa and v with varying external conditions. The bilayer couple model is strictly applicable only to phospholipid vesicles, where the membrane really consists of two leaflets. The red blood cell membrane has, in addition to the two leaflets of the bilayer part of the membrane, also a network of cytoskeletal proteins. It has been shown[25], however, that such a multilayer system can be treated in the same way as the bilayer model, though the parameters do not have purely geometrical meaning but depend on the areas and stretching constants of all the composite layers and on the distances between their neutral surfaces. It is therefore not surprising that the bilayer couple model is applicable also to red blood cells. The shapes evaluated in the presence of an external electric field do not agree with observations, which could mean that the lateral stresses which develop in the membrane are strong enough to allow the volume and area difference equilibration during measurements.

REFERENCES

1. E.A. Evans and R. Skalak, Mechanics and Thermodynamics of Biomembranes, CRC Press, Boca Raton, FL (1980).
2. R.M. Hochmuth and R.E. Waugh, Annu. Rev. Physiol. 49, 209 (1987).
3. E. Sackmann, H.-P. Duwe and H. Engelhardt, Faraday Discuss. Chem. Soc. 81, 281 (1986).
4. P.B. Canham, J. Theor. Biol. 26, 61 (1970).
5. J.T. Jenkins, J. Math. Biol. 4, 149 (1977).
6. W. Helfrich and H.J. Deuling, J. Phys. (Paris) Colloq. 36, 327 (1975).
7. H.J. Deuling and W. Helfrich, J. Phys. (Paris) 37, 1335 (1976).
8. H.J. Deuling and W. Helfrich, Biophys. J. 16, 861 (1976).
9. J.C. Luke, SIAM J. Appl. Math. 42, 333 (1982).
10. J.C. Luke and J.I. Kaplan, Biophys. J. 25, 107 (1979).
11. M.A. Peterson, J. Appl. Phys. 57, 1739 (1985).
12. E.A. Evans, Biophys. J. 14, 923 (1974).
13. M.P. Sheetz and S.J. Singer, Proc. Natl. Acad. Sci. USA 71, 4457 (1974).
14. S. Svetina, A. Ottova-Leitmannova and R. Glaser, J. Theor. Biol. 94, 13 (1982).
15. S. Svetina and B. Žekš, Biomed. Biochim. Acta 44, 979 (1985).
16. S. Svetina and B. Žekš, Eur. Biophys. J. 17, 101 (1989).
17. V. Pastushenko, A. Sokirko, S. Svetina and B. Žekš, in preparation.
18. B. Žekš, S. Svetina and V. Pastushenko, Stud. Biophys., to be published.

19. K. Berndl, J. Käs, R. Lipowsky, E. Sackmann and U. Seifert, to be published.
20. S. Svetina, V. Kralj-Iglič and B. Žekś, Proceedings of the 10th School on Biophysics of Membrane Transport, Poland, May 1990, J. Kuczera, S. Przestalski, Eds., Wroclaw (1990) Vol. II, 139.
21. S. Svetina and B. Žekś, J. Theor. Biol., to be published.
22. W. Helfrich, Z. Naturforsch. C $\underline{29}$, 182 (1974).
23. M. Winterhalter and W. Helfrich, J. Colloid Interface Sci. $\underline{122}$, 583 (1988).
24. G. Bryant and J. Wolfe, J. Membrane Biol. $\underline{96}$, 129 (1987).
25. S. Svetina, M. Brumen and B. Žekś, in Biomembranes: Basic and Medical Research, G. Benga, J.M. Tager, Eds., Springer Verlag (1988) 177.

LIPID MEMBRANE CURVATURE ELASTICITY AND PROTEIN FUNCTION

Sol M. Gruner

Department of Physics
Princeton University
Princeton, NJ 08544 USA

It is suggested that curvature elastic stress of the monolayers of biomembrane lipid bilayers can change the activity of certain imbedded membrane proteins and that this may be a rationale for the lipid compositions seen in cell membranes. Lipid monolayer curvature stress arises when lipids which are prone to exhibit nonlamellar mesomorphic phases are a large fraction of the lipids of bilayers. The stress builds as one approaches the boundry of a lamellar-nonlamellar phase transition form the lamellar side. A discussion is given of the competition governing the release of curvature stress during the formation of interfacially curved mesomorphs. The net stress energy in bilayers near to a transition is in the range of a few to perhaps ten times thermal (kT) energy per lipid molecule. Mechanisms of coupling this stress to protein conformational changes are given.

INTRODUCTION

Biological membranes consist of roughly half protein and half lipid bilayer by weight. These proteins are a very important part of the apparatus of eukaryotic cells, comprising perhaps 50% or so of all the different kinds of proteins. The literature on membrane proteins is truly massive. The literature on lipid bilayers is much smaller, but still quite large. The number of studies restricted to the interactions between proteins and lipid bilayers are fewer still. Most of this literature is concerned with either the site-specific binding of lipid molecules to proteins or the change in protein activity if the lipid environment is frozen. If attention is restricted to the physical interactions between fluid bilayer elasticity and protein conformation then the number of relevant, systematic studies can, literally, be counted on one hand.

Now this is, indeed, a telling statement of the shallowness of our understanding of protein function. Dogma has it that most proteins are conformational engines which function by distinct changes in shape or state of thermally driven motion. In the case of membrane proteins, it is also well recognized that the constraints of the amphiphilic environment imposed by the bilayer-water matrix determines the folding of the protein. Further, it has been recognized for at least a decade that lipid bilayers

Biologically Inspired Physics, Edited by L. Peliti
Plenum Press, New York, 1991

may contain a distribution of line tensions parallel to the lipid-water interface which can exert physical forces which vary over the thickness of the bilayer (Helfrich, 1980; Petrov & Bivas, 1984). Given this, a natural question is if the force distribution exerted on proteins by the bilayer can modulate conformational changes which are required for protein activity. The fact that this question is just now being addressed (Gruner, 1985; Hui, 1987; Gruner, 1989) illustrates the paucity of our understanding of the magnitudes of these forces and the conformational changes one should expect in membrane proteins. It also illustrates the degree to which the prevailing thinking about protein activity is purely chemical, as opposed to physical-chemical.

The physical interactions between bilayers and proteins have been central concerns of my laboratory for the last seven years or so. It was recognized that the structural mesomorphism of lipid-water dispersions is driven largely by a competition between a spontaneous curvature and constraints imposed by the packing of lipid hydrocarbon chains (Kirk et al, 1984). This led to some general theoretical and experimental studies of lyotropic mesomorphic behavior and quantitative procedures for determination of some of the free energies which are involved. Attention is now being focussed on applying what has been learned to the perturbation of flexible molecules imbedded in elastically stressed bilayers. Almost all of this work has been described in detail in the literature and in reviews (Gruner, 1989; Tate et al, 1991; Gruner, 1991), although little of it has appeared in the physics literature. Even so, there is no virtue to duplicating information which has already appeared, so this report will be quite short. It is intended primarily as a pointer to the existing literature.

BACKGROUND

Ultimately, we are concerned with two related sets of questions of great biological importance. The first is why are there so many chemically distinct lipid molecules in a typical biomembrane? What are the rules governing the different lipid compositions of different biomembranes? The answers to these questions are clearly intertwined with questions about how variation of lipid composition affects protein function.

To begin with, one must recognize that different combinations of roughly a dozen common lipid headgroups, chains of different lengths (say, a few to 24 carbons) and degrees of unsatuation leads to thousands of chemically distinct polar lipid entities; a given biomembrane may contain several hundred different kinds. Further, different biomembranes, even within the same cell, each tend to have distinct lipid compositional spectra. For example, the mitochondrial bilayers are quite different in composition from the nuclear membrane (Quinn & Chapman, 1980). Further, experiments with bacteria, where the membranes compositions are relatively easy to manipulate, have shown that the organism can adjust the composition of the membrane in response to changes in environmental conditions, provided that the available range of constitutents is not too limited (Wieslander, 1986; McElhaney, 1989). In other words, the organism adjusts the composition of membranes so as to conform to some set of rules, but the specific compositions appear to have a range of plasticity. The fact that some lipid types are used when available but are not absolutely required, coupled with the observation that a change in the availability of a given lipid type tends to result in a readjustment of the ratios of many other lipids suggests that the organism is adjusting collective properties of the bilayer.

Another significant observation is that biomembranes typically contain large fractions (c.a. 25-50%) of lipid types which do, by themselves, assume bilayer phases under the living conditions of the organism (Cullis et al, 1985). Rather, these lipid types form nonlamellar phases, such as the inverted hexagonal, or H_{II} phase (Figure 1). Experiments with mycoplasmas have shown that there appears to be regulation of the amount of such lipids (Wieslander, 1986; Lindblom, 1986). Our interest was in elucidating a possible role for such lipids.

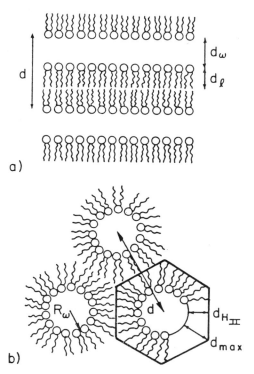

Figure 1. Cross-sections through a liquid crystalline L_α (a) and H_{II} (b) phases. In the L_α phase, the lipid monolayers, of thickness d_l, organize into bilayer lamellae which stack with intervening water layers of thickness, d_w. In the inverse hexagonal, or H_{II} phase, the water is confined to rods of radius R_w in the center of hexagonally stacked tubes of lipid. The H_{II} phase differs significantly from the L_α phase in that the thickness of the lipid monolayer has a varies, e.g., from $d_{H_{II}}$ to d_{max}. From Tate & Gruner, 1987, with permission.

Our first concern was to understand the changes in mesomorphic phase behavior which resulted when different kinds of liquid crystalline (ie, melted chain) lipids were mixed together. (Most biomembrane lipids are in a liquid crystalline state. Although phases with frozen chains are interesting from a physical point of view, they will not be discussed here. The primary biological requirement is that the biomembrane environment not be too stiff.) It had long been recognized that the curvature of the lipid-water interface may be used to characterize mesomorphic phases (Tartar, 1955;

Winsor, 1971). There also was a good understanding of the elastic behavior of lipid layers, at least in the thin layer limit (Helfrich, 1973; 1978; 1980; Evans & Skalak, 1979); in particular, it was recognized that lipid monolayers may be endowed with a spontaneous curvature. But there was only a rudimentary understanding of the way in which monolayer curvature energies compete with other free energy contributions to bring about mesomorphic transitions which resulted in discontinuous changes in the interfacial curvature. The prevailing picture (Israelachvili et al, 1980) was one in which the phase transition was the result of sharp changes in the spontaneous curvature at the phase transition temperature or composition. It was not at all obvious why the spontaneous curvature should have such discontinuities.

The point of view taken at Princeton (and, at about the same time independently in France by Charvolin and coworkers–see Charvolin, 1990) was that the phase transition involved a competition between a spontaneous curvature and constraints associated with the need to pack the hydrocarbon chains with uniform segment density (Kirk et al, 1984). For largely entropic reasons, one expects that lipid molecules diffusing along the lipid water interface should act to minimize the encounter of variations in the density or gross configurations of the chains. Yet inspection of an H_{II} phase (Figure 1) shows that the translational invariance of the chain environment is broken as lipids diffuse azimuthally around tubes because the Wigner-Seitz cell is 6-fold symmetric, ie, the distance dmax is larger than dHii. Hence, there should be a free energy cost associated with the Hii phase which is not present in the lamellar phase, where the translational invariance is not broken.

Assume that the lipid layer has a tendency to bend to a well-defined radius of curvature, as referenced to some unique point in the monolayer, and that this tendency is a smooth function of temperature and is a phase invariant. Consider, first, the thermotropic behavior when in equilibrium with a pool of bulk water. If the layer composition is such that the curvature energy is minimized when the layer bends tightly to some curvature, C_0, then a L_α (Figure 1) phase is an elastically stressed geometry for which the curvature bending energy is high. But this phase is also translationally symmetric, so costs associated with translational variations in chain packing would be relatively small. In the H_{II} phase, by contrast, the curvature energy can be made small, by bending to a curvature close to C_0, but at the cost in a rise of free energy due to a tranlationally asymmetric chain environment. Now the sum of the curvature and chain packing free energies is clearly geometrically dependent. As long as these two contributions have relatively little coupling and have different thermal dependences then it is possible for the *total* free energy of the lamellar geometry, initially at a lower free energy than the hexagonal geometry, to cross over to a higher relative free energy at some temperature. It would be this cross-over which would drive the mesomorphic transition.

Support for this transition scenario came from a series of experiments in which it was shown that the $L_\alpha - H_{II}$ phase transition temperature could be dropped dramatically by the addition of small amounts of a medium chain-length oil, such as dodecane. The oil is not anchored to the lipid-water interface and could partition differentially about the H_{II} hydophobic volume so as to remove much of the lipid chain packing stress. It was showm that the change in curvature of the lipid layer was a simple consequence of the thinning of the monolayer thickness with temperature that had long been known to be a feature of long chain amphiphiles (Luzzati, 1968).

Techniques were devised to measure the work of changing the curvature of an Hii phase. The competition between curvature and packing was shown to also have application to cubic phases. Etc. Readers interested in details are referred to the literature which is reviewed in Gruner (1989; 1991) and Tate et al, (1991).

To summarize the main conclusions, mesomorphic lipid transitions involving changes in interfacial curvature are controlled by a geometrically frustrated competition between a spontaneous tendency to bend and free energies associated with asymmetries in the hydrocarbon environment. Of course this is an oversimplified picture, but it appears to explain much of the behavior observed for electrically neutral phospholipids in excess water. Additional, geometrically dependent terms must be considered the cases of limited water or with charged lipids. Also, the curvature is not necessarily of a simple form because the radius of bend is comparable to the monolayer thickness. Clearly, there is much more work to be done in obtaining a quantitative understanding of lipid phase behavior, but these complications, while important, are somewhat incidental to the main message of this report, namely, that there is a substantial frustrated curvature stress associated with lipid bilayers just to the lamellar side of Lalpha-Hii phase transitions. The measured energy of doubling the curvature of an Hii phase may be on the order of the thermal (kT) energy (Rand et al, 1990). If this is also true for flattening the monolayers into an L_α phase, then the frustrated energy may be substantial, especially if there is a process which can interact with many lipid molecules so as to lower the net stress.

INTERACTION WITH PROTEINS

Recall that most biomembranes have large fractions of lipid constituents which are, by themselves, in nonlamellar phases at normal temperatures. In general, the effect of adding such lipids to a lipid mixture is to lower the $L_\alpha - H_{II}$ phase transition temperature by raising the spontaneous curvature. This can often be demonstrated by adding 10 to 30% dodecane to the mixture to induce the H_{II} phase and then measuring the size of the water core. What is ususally observed is that the addition of the "nonlamellar-prone" lipid decreases this size. The size of the water core in the presence of excess water and adequate dodecane may be taken as an operational measure of the spontaneous curvature. This suggests that the effect of the nonlamellar-prone lipid in the biomembrane mixture is to induce curvature stress into the native lipid bilayers. Assuming the extrapolation of the previous paragraph, the available energy might be a few tenths of kT per lipid molecule.

It is not difficult to envision ways in which proteins may tap this pool of stored elastic energy. Ultimately, the curvature stress is the result of a distribution of lateral (ie, parallel to the lipid-water interface) line tensions across the thickness of the monolayers of the bilayer. For a bilayer with an unconstrained perimeter, the integral over these tensions is zero, ie, biomembranes are not, in general, under net line tension. But the first moment of this tension distribution is, effectively, a torque which acts to increase the area of the polar part relative to the hydrocarbon part of molecules buried within the monolayer. Assume, for the moment, that the Helfrich (1980) form for the curvature energy density per unit area, E, may be taken as an approximation for the frustrated energy in the lamellae,

$$E = (K_c/2)(C - C_0)^2,$$

(1)

where K_c is the monolayer rigidity, $C = 0$ is the lamellar curvature, and C_0 is the spontaneous curvature observed upon the addition of dodecane. The measured values of K_c and C_0 (Rand et al, 1990) leads to $K_c C_o^2/2$ a few tenths kT per lipid. Equation (1) also allows an estimate of the change in energy which results when a protein imbedded within a lipid monolayer undergoes a conformational change so as to alter the ratio of its cross-sectional area at the hydrophobic core relative to the area at the lipid-water interface.

Relatively little is known about the actual conformational changes which occur when integral membrane proteins are activated. However, low resolution electron microscopy reconstructions of numerous proteins, notably pumps and channels have led to general pictures such as shown in Figure 2 (see also Unwin & Ennis, 1984). The net changes in relative protein area could plausibly be many hundreds of angstroms, corresponding to a few to perhaps 10kT of energy. This energy difference may be sufficient to change the relative free energies of the conformational states of the protein sufficiently to modulate the activity.

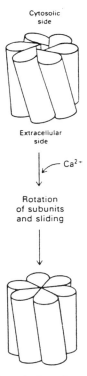

Figure 2. A model for openned and closed states of a gap junction. Reproduced with permission from *Biochemistry* by L. Stryer (W.H. Freeman, NY, 1988) after a drawing provided by Drs. N. Unwin and G. Zampighi.

The protein-lipid curvature coupling just described is purely mechanical. It may not be necessary to perform large motion mechanical changes to tap the curvature stress. A single methylation of the headgroup of the H_{II}-prone lipid dioleoylphosphatidylethanolamine results in a dramatic change in the resultant H_{II} phase radius upon the addition of dodecane (Gruner et al, 1988). The methylation appears to alter the hydrogen bonding interactions between the lipid headgroups in such a way as to significantly affect the line tension distribution. Presumably there are similar interactions between the lipid molecules immediately adjacent to an imbedded protein and the exposed protein residues. One can imagine small conformational changes, such as the rotation of single amine acid residues, which effectively alter the net chemical interactions much in the same way as lipid headgroup methylation. So little is known about the nature of these interactions even between lipid molecules, much less between lipid and protein molecules, that it is difficult to judge the importance of such a mechanism. And, of course, the lack of detailed structural information about membrane proteins would be limiting even if the strength of the interactions were known.

Is there experimental evidence that membrane protein activity correlates with the mesomorphic tendency of the imbedding lipid bilayer? The literature contains a few studies which appear tantalizing (see Gruner, 1989; 1991 for summaries). However, as is often the case in science, it is first necessary to ask the right questions before one can expect specific answers. None of the studies performed to date have explicitly attempted to correlate protein activity with a quantitative parameter such as the spontaneous curvature. And as is so often the case in the biological sciences, once the right questions are asked, the answers prove to be devilishly difficult to obtain. Ideally, one wishes to reconstitute suitable membrane proteins known to have quantitative assays of activity into lipid bilayers of different, well characterized compositions. Unfortunately, membrane protein reconstitution is a difficult art which is fraught with artefactual results and is confounded by a literature repleat with errors. Even so, the experiments are, in principle, feasible and the results would be important. One hopes that the coming years will see some unambiguous measurements which probe the coupling between protein activity and lipid monolayer curvature stress.

ACKNOWLEDGEMENTS

Research in my Princeton laboratory is supported by grants from the DOE (DE-FG02-87ER60522), NIH (GM32614) and ONR (N00014-90-J-1702). I also thank my colleagues and students for making science fun.

REFERENCES

Charvolin, J., 1990, Crystals of fluid films, *Contemp. Phys.* 31:1-17.

Cullis, P.R., Hope, M.J., de Kruijff, B., Verkleij, A.J. and Tilcock, C.P.S., 1985, Structural properties and functional roles of phospholipids in biological membranes, in: *Phospholipids and Cellular Regulations, Vol. 1*, J.F. Kuo, eds., CRC Press, Boca Raton, FL.

Evans, E.A. and Skalak, R., 1979, Mechanics and thermodynamics of biomembranes, *CRC Critical Rev. in Bioeng.* 3:180-419.

Gruner, S.M., 1985, Curvature hypothesis: Does the intrinsic curvature determine biomembrane lipid composition. A role for non-bilayer lipids, *Proc. Natl. Acad. Sci. (USA)* 82:3665-3669.

Gruner, S.M., Tate, M.W., Kirk, G.L., So, P.T.C., Turner, D.C., Keane, D.T., Tilcock, C.P.S. and Cullis, P.R., 1988, X-ray diffractions study of the polymorphic behavior of N-methylated dioleoylphosphatidylethanolamine, *Biochem.* 27:2853-2866.

Gruner, S.M., 1989, Stability of lyotropic phases with curved interfaces, *J. Phys. Chem.* 93:7562-7570.

Gruner, S.M., 1991, Nonlamellar lipid phases, in: *The Structure and Function of Cell Membranes*, P.L. Yeagle, ed., Telford Press, Caldwell, NJ (in press).

Helfrich, W., 1973, Elastic properties of lipid bilayers: Theory and possible experiments, *Z. fur Naturforschung* 28c:693-703.

Helfrich, W., 1978, Steric interaction of fluid membranes in multilayer systems, *Z. Naturforsch.* 33a:305-315.

Helfrich, W., 1980, Amphiphilic mesophases made of defects, in *Physics of Defects*, R. Balian, M. Kleman and J-P. Poirier, eds., North-Holland, Amsterdam.

Hui, S.W., 1987, Non-bilayer forming lipids: Why are they necessary in biomembranes?, *Comments Mol. Cell. Biophys.* 4:233-248.

Israelachvili, J.N., Marcelja, S. and Horn, R.G., 1980, Physical properties of membrane organization, *Quart. Rev. Biophys.* 13: 121-200.

Kirk, G.L., Gruner, S.M. and Stein, D.L., 1984, A thermodynamic model of the lamellar (L_α) to inverse hexagonal (H_{II}) phase transition of lipid membrane-water systems, *Biochem.* 23:1093-1102.

Lindblom, G., Brentel, I., Sjoland, M., Wikander, G. and Wieslander, A., 1986, Phase equilibria of membrane lipids from *Acholeplasma laidlawii*: importance of a single lipid forming nonlamellar phases, *Biochem.* 25:7502-7510.

Luzzati, V., 1968, X-ray diffraction studies of lipid-water systems, in *Biological Membranes, Vol. I*, D. Chapman, ed., Academic Press, NY.

McElhaney, R.N., 1989, The influence of membrane lipid composition and physical properties of membrane structure and function in acholeplasma laidlawii, *CRC Crit. Rev. in Microbiol.* 17:1-32.

Petrov, A.G. and Bivas, I., 1984, Elastic and flexoelectric aspects of out-of-plane fluctuations in biological and model membranes, *Prog. in Surf. Sci.* 16:389-512.

Quinn, P.J. and Chapman, D., 1980, The dynamics of membrane structure, *CRC Critical Rev. in Biochem.* 8:1-117.

Rand, R.P., Fuller, N.L., Gruner, S.M. and Parsegian, V.A., 1990, Membrane curvature, lipid segregation, and structural transitions for phospholipids under dual-solvent stress, *Biochem.* 29:76-87.

Tartar, H.V., 1955, A theory of the structure of the micelles of normal paraffin chain salts in aqueous solution, *J. Phys. Chem.* 59:1195-1199.

Tate, M.W. and Gruner, S.M., 1987, Lipid polymorphism of mixtures of dioleoylphosphatidylethanolamine and saturated and mono-unsaturated phosphatidylcholines of various chain lengths, *Biochem.* 26:231-236.

Tate, M.W., Eikenberry, E.F., Turner, D.C., Shyamsunder, E. and Gruner, S.M., 1991, Nonbilayer phases of membrane lipids, *Chem. Phys. Lipids* (in press).

Unwin, P.N.T. and Ennis, P.D., 1984, Two configurations of a channel-forming membrane protein, *Nature* 307:609-613.

Wieslander, A., Rilfors, L. and Lindblom, G., 1986, Metabolic changes of membrane lipid composition in *Acholeplasma laidlawii* by hydrocarbons, alcohols and detergents: Arguments for effects on lipid packing, *Biochem.* 25:7511-7517.

Winsor, P.A., 1971, Liquid crystallinity in relation to composition and temperature in Amphiphilic systems, *Molec. Cryst. & Liquid Crystals* 12:141-178.

OBSERVATION OF SURFACE UNDULATIONS ON THE MESOSCOPIC
LENGTH SCALE BY NMR

Myer Bloom and Evan Evans

Department of Physics, University of British Columbia
6224 Agricultural Road
Vancouver, B.C. Canada, V6T 1K9

1. INTRODUCTION

One of the striking systematic differences between materials "designed by nature" via the processes of evolution and those designed by engineers and physicists using well established principles of condensed matter physics is the predominance of ultra-soft materials in the organization of natural condensed matter structures as compared with hard materials in man-made materials. An example is that the lipid bilayer component of the membranes of virtually all cells, be they procaryotic (bacterial) or eucaryotic (animal, vegetable or fungal), are very compressible "fluids" under physiological conditions[1-3]. The word "fluid" here means, to those of the macroscopic-continuum intuitive persuasion (the "MACROS"), that the membrane does not support shear restoring forces. To those having a predominantly microscopic-molecular intuition (the "MICROS"), "fluidity" means that those spectroscopic measurements on molecules in membranes that are sensitive to translational and rotational diffusive molecular motions indicate that the membrane medium has a viscosity of moderately thick oil (≈ 1 poise).

Because of bilayer fluidity, the shapes of cells can only be understood in terms of the membrane being a composite structure [1,3-5]. A rigid, *exoskeletal* peptidoglycan cell wall determines procaryotic cell shape, while eucaryotic cell shapes are governed by soft, *cytoskeletal*, two-dimensional, rubber-like networks of proteins to which plasma membrane lipid bilayers are attached in a not yet fully understood manner.

The question arises as to whether the physical principles underlying the organization of ultra-soft natural structures will ultimately prove to be surprising in that their understanding will, in some sense, involve the discovery of "new physics". At the present time, the intuition of physical scientists attempting to investigate natural materials is limited by their experimental tools to the two regimes described in the previous paragraph. MACROS tend to "see" membranes at a length scale of slightly larger a few 1000 Å or slightly larger, as determined by the wavelengths of visible light. Most MICROS and especially biochemists who dominate this field at

Biologically Inspired Physics, Edited by L. Peliti
Plenum Press, New York, 1991

the present time, tend to focus on the structure of biologically important molecules such as proteins with a view to relating biological function to their physical structure.

The motivation for this paper is the need to identify and develop new methods for the *direct* experimental study of the *mesoscopic* length scale, i.e. intermediate between the *molecular* (≤ 10 Å) and *macroscopic* (≥ 1000 Å) length scales, in order to develop the kind of intuition that can only be done by direct confrontation with experiment. It is becoming increasingly evident that the underlying physics required to understand the mechanics of membranes is associated with the surface fluctuations going on at the mesoscopic scale[1,3,6,7]. So far, the physics of these motions at the mesoscopic length scale has had to be inferred by extrapolation from the macroscopic domain. We show here that Nuclear Magnetic Resonance (NMR) measurements *in fluid membranes* are really *mesoscopic* in character and are already providing interesting information in that regime. They are capable of providing much more.

2. FLUID MEMBRANES STUDIED WITH DEUTERIUM (^{2}H) NMR - A BRIEF INTRODUCTION

The technique of ^{2}H NMR has proven to be extremely useful for quantitative study of the L_α phase of membranes mainly because the information it provides on local orientational order and on rates of molecular reorientation is important in the quantitative characterization of membrane fluidity. In addition, the ^{2}H NMR spectrum is simple to interpret[8-10]. It is a doublet with a splitting related, for ^{2}H nuclei in hydrocarbons, to a known coupling constant of $\omega_Q/2\pi = 125$ kHz associated with the interaction of the deuteron quadrupole moment with the local electric field gradient.

Because the orientation dependence of these interactions is similar to that of anisotropic chemical shift interactions and dipolar interactions between pairs of nuclei, many of the general remarks we make here about time and length scales for quadrupolar interactions in relation to ^{2}H NMR also apply to NMR studies of membranes with other nuclei such as ^{1}H, ^{13}C and ^{31}P.

For fluid membranes, the molecular motions are axially symmetric, the axis of symmetry being the normal to the bilayer, and are rapid enough to give rise to motional averaging. As a consequence, the quadrupolar splitting $2\omega(\theta)$ is dependent only on the angle θ between the external magnetic field and the surface normal, which is the axis of symmetry for molecular reorientation in fluid bilayer membranes. One has:

$$\omega(\theta) = \omega_Q S_{CD}(3cos^2\theta - 1)/2, \tag{1}$$

where the *orientational order parameter* S_{CD} is defined by

$$S_{CD} = \langle(3cos^2\beta - 1)/2\rangle_{\text{fast motions}}, \tag{2}$$

where β is the angle between the C-^{2}H bond and the surface normal.

^{2}H NMR spectra are usually obtained using the "quadrupolar echo pulse sequence" (QEPS), $[(90)_x - \tau - (90)_y - t-]$ as the Fourier transform of the resulting free induction signal starting from the peak of the echo at $t = \tau$. The spectrum shown in Fig. 1 is a "powder spectrum", often called the "Pake doublet" in the NMR literature[9,10], corresponding to a random superposition of doublets (Eq. 1) due to an isotropic distribution of angles θ as expected for a collection of cell membranes or, as in the case in the system studied in Fig. 1, multilamellar vesicles.

The "Pake doublet" splitting in Fig. 1 is $\omega_Q|S_{CD}|/2\pi$ and hence provides a direct measure of the order parameter. More important for out present considerations, the spectrum represents a map of doublets for different local bilayer orientations. This allows one to study the orientation dependence of the longitudinal (spin-lattice) and transverse (spin-spin) relaxation times, T_1 and T_2, respectively.

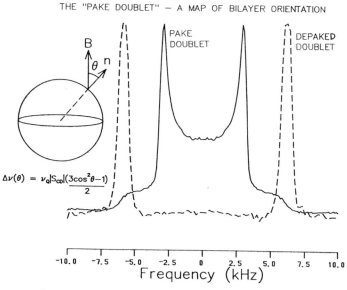

THE "PAKE DOUBLET" − A MAP OF BILAYER ORIENTATION

$$\Delta\nu(\theta) = \nu_q|S_{col}|\frac{(3\cos^2\theta-1)}{2}$$

Fig. 1. Powder ^{2}H NMR, "Pake doublet" spectrum for DPPC-d_2 deuterated in the α-position of the polar headgroup [see e.g. Akutsu and Seelig, Biochemistry 20, 7366-7373 (1981)]. Also shown in dotted lines is the doublet spectrum characteristic of the external magnetic field applied along the bilayer normal ($\theta = 0°$).

The spin-lattice relaxation time is studied by preceding the QEPS by a 180° inverting pulse and monitoring the approach of the quadrupolar echo signal to equilibrium as a function of the time between the inverting pulse and the QEPS. Since relaxation of the nuclear magnetization involves thermally induced "spin flips" requiring an exchange of energy between the spin system and the lattice, T_1 measurements are particularly sensitive to thermal motions characterized by correlation times of the order of 10^{-8}s in the vicinity of the Larmor period or shorter.

The spin-spin relaxation time may be determined directly from the decay of the quadrupolar echo as the pulse spacing τ is increased. Since decay of the echo is due to a loss of phase memory of the spins which requires no exchange of energy with the lattice, T_2 is sensitive to much slower motions than T_1.

3. MACROSCOPIC, MICROSCOPIC AND MESOSCOPIC ASPECTS OF NMR IN FLUID MEMBRANES

In analyzing spectroscopic measurements, the distinction between *molecular* and *macroscopic* features of the measurements often requires some explanation. The technique of NMR is normally considered to provide information on the *molecular* level when applied to the study of membranes. Yet, at first glance NMR appears to be a *macroscopic* technique. It uses electromagnetic radiation at frequencies in the range 10 MHz to 500 MHz, corresponding to wavelengths between 30 and 6 meters. Furthermore, signal/noise considerations dictate that large samples be used for NMR studies. At least 10^{17} deuterons are usually required to obtain interpretable ^{2}H NMR spectra in membranes. On the very first page of the "NMR BIBLE"[8], nuclear magnetism is characterized clearly as a macroscopic technique. Nevertheless, it is often correct, *to first approximation*, to identify the results for NMR measurements in membranes and many other systems with *molecular* features of the systems. The main reason for this is that the interactions between neighbouring nuclear spins are often sufficiently weak that the NMR spectral frequencies become representative of the interaction of individual spins with their local environment. In addition, coherence (interference) effects between NMR signals arising from nuclear spins far apart from each other are almost always negligible.

As may been seen from the spread of frequencies in Fig. 1, however, the establishment of a ^{2}H NMR spectrum in a fluid membrane requires a finite time of the order of 100 μs or more. During this time, an individual phospholipid molecule may diffuse randomly over a distance of a few hundred Å's, so that a proper interpretation of the information contained in such a "microscopic" measurement may require knowledge of structure at the *mesoscopic* length scale. When relaxation time measurements are carried out, the time required is even larger. Often, $T_2 \approx 1$ ms and $T_1 \approx 50$ ms, so that the length scale of the measurement may range from the *mesoscopic* to the *macroscopic* regime.

4. SUMMARY OF SOME THEORETICAL PROPERTIES OF SMALL, "WELL-BEHAVED", THERMALLY DRIVEN, SURFACE UNDULATIONS OF A SINGLE MEMBRANE [7]

In order to examine the plausibility of detecting membrane undulations in the mesoscopic length regime using NMR relaxation measurements, we summarize some of their theoretical properties.

1. The undulations are decomposed into modes of wavelength $\lambda = 2\pi/q$ and amplitude u_q. For small amplitudes corresponding to small angles of tilt $\delta\theta$ of the surface normal from its average direction, the extra surface area associated with the undulating surface is related to the area A of the flat surface for the sum of the pair of modes q and $-q$ by[7]

$$(\Delta A)_q = A\langle|\delta\theta_q|^2\rangle = Aq^2\langle|u_q|^2\rangle \tag{3}$$

2. The energy associated with these modes is related to the curvature elastic energy κ and the lateral tension σ as follows:

$$E(q) = (\kappa q^2 + \sigma)(\Delta A)_q \tag{4}$$

3. The equipartition theorem gives

$$\langle |u_q|^2 \rangle = kT/[A(\kappa q^4 + \sigma q^2)] \qquad (5)$$

4. Reasonable orders of magnitude for lipid bilayer dispersions[6] are

$$\kappa \approx (0.5 - 1.0) \times 10^{-12} \text{ergs} \qquad (6)$$

$$\sigma \ll 1 \text{dyne} - \text{cm}^{-1}, \qquad \text{probably} \leq 10^{-3} \text{dyne} - \text{cm}^{-1} \qquad (7)$$

From these orders of magnitude, it appears that one can make the approximation $\sigma \approx 0$ in the mesoscopic range ($\lambda \leq 1000\text{Å}$).

5. When the sum over all modes $\sum_q$ is replaced by an integral from a lower cut-off $q_{min} \approx \pi/\lambda_M$, where $\lambda_M \approx A^{1/2}$, to a higher cut-off $q_{max} \approx \pi/\lambda_m$, where $\lambda_m \approx$ membrane thickness ≈ 30 Å, one obtains[7]

$$(\Delta A/A)_{\sigma=0} = (kT/4\pi\kappa)\ln(\lambda_M/\lambda_m) \qquad (8)$$

The long wavelength cut-off λ_M (or equivalently $1/q_{min}$) is the effective correlation length for undulations in the membrane and is taken as $A^{1/2}$ for an isolated membrane. In many situations, however, the correlation length is reduced by geometrical constraints that create strong steric effects. For instance, in *multilamellar* dispersions or liposomes where neighbouring bilayers are separated by distances $d \ll A^{1/2}$, parallel confinement of undulations causes steric interactions between layers that reduce the effective correlation length as approximated by

$$\lambda_M \approx (2\pi^3\kappa/3kT)^{1/2}d. \qquad (9)$$

For $T = 300°K$ and $d \approx 30\text{Å}$ for lecithin multilayers, Eq. (9) predicts a value of $\lambda_M \approx 500\text{Å}$ which can be much smaller than the size of the individual lipid domains.

6. The relaxation (damping) times τ_q of undulational modes are defined by the correlation function

$$\langle u_q(0)u_r(t) \rangle = \delta_{qr}\langle |u_q|^2 \rangle exp(-t/\tau_q) \qquad (10)$$

No experimental data are available on these damping times in the mesoscopic regime.

6a. *Viscous damping:*

At optical wavelengths, measurements of shape fluctuations[11,12] of single membrane vesicles show consistency with predictions based on the well-satisfied, ultra-low Reynolds number approximation of the hydrodynamics of the coupling between a lipid bilayer and the fluid medium[12,13] of viscosity η.

$$\tau_q^{(1)} = \tau_q^{(1)}(\eta) = (4\eta)/(\kappa q^3) \approx 0.3\mu s(\lambda/1000\text{Å})^3 \qquad (11)$$

where the numerical value in Eq. (11) is based on Eq. (6) and an aqueous medium using $\eta = 10^{-2}$poise.

For multilamellar dispersions of liposomes, the hydrodynamics becomes more complicated. But, if we assume that excitations of a membrane are not correlated with adjacent membranes, then it may be shown that the viscous damping is well approximated by the relaxation time for a single membrane multiplied by the dimensionless factor $3/(qd)^3$, where d is the average water separation between membranes, i.e.:

$$\tau_q(\eta) = 3\tau_q^{(1)}(\eta)/(qd)^3 \approx 144\mu s \ (\lambda/1000\text{Å})^6 \qquad (12)$$

This expression was derived by considering the viscous damping of a single membrane undulating between two rigid boundaries separated by a distance d. Comparison of Eqs. (11) and (12) shows that the damping in a multilamellar structure can greatly increase the relaxation time for long wavelengths.

6b. *Diffusion limited damping*: For loss of correlation limited by diffusion of molecules parallel to the membrane surface, we assume

$$\tau_q = \tau_q(D) = \lambda^2/(4D) = \pi^2/Dq^2 \approx 600\mu s(\lambda/1000\text{Å})^2 \qquad (13)$$

where the numerical value is for $D = 4\text{x}10^{-8}\,\text{cm}^2\text{s}^{-1}$.

4.1. *"Adiabatic motions" and transverse relaxation*

In the quadrupolar echo (QE) experiment described earlier, the echo signal decays with a relaxation time T_2 as a result of the accumulation of random phase by the nuclear spins due to thermally-driven fluctuations in the quadrupolar splittings. For short correlation times on the "NMR time scale" (i.e. $M_2\tau_c^2 \ll 1$), this corresponds to a linear increase with time of the variance characterizing the dispersion of the nuclear spin phase distribution function, thus leading to exponential relaxation with a time constant T_2 [8]. Fast motions that contribute to T_1 also contribute to T_2 so that $T_2 \leq T_1$. In the short correlation time regime, $(\omega_0\tau_c)^2$ and hence $T_2 \approx T_1$. In systems as complex as membranes, however, there are invariably slower motions which can be classified[8] as "adiabatic motions", i.e. $(\omega_0\tau_c)^2 \gg 1$. The presence of important adiabatic motions can be recognized immediately since they contribute appreciably to T_2, but not to T_1 relaxation. Thus, $T_2 \ll T_1$ in the presence of adiabatic motions.

5. CAN SURFACE UNDULATIONS ACCOUNT FOR TRANSVERSE RELAXATION IN MODEL MEMBRANE? – AN ORDER OF MAGNITUDE CALCULATION INDICATES THAT THEY CAN

Until recently it was believed that there were no important adiabatic motions influencing ^{2}H NMR properties in the L_α phase. The presence of slow motions that influence T_2 strongly in the L phase has now been demonstrated experimentally[14], and, as described later, a technique developed for their systematic study.

If adiabatic motions, characterized by an "effective correlation time" τ_c, modulate the NMR precession frequency over a spectral distribution having a second moment M_2, the general theory of nuclear spin relaxation[8] predicts

$$1/T_2 = M_2\tau_c \qquad (14)$$

Suppose that the membrane surface normal is tilted through a small angle $\delta\theta$ from its normal orientation θ relative to the external magnetic field. This leads to a change of precession frequency which is found, using Eq. (1) and the "addition theorem for spherical harmonics" to be, to first order in $\delta\theta$

$$\delta\omega = \omega(\theta + \delta\theta) - \omega(\theta) \approx -3\omega_Q S_{CD}\sin\theta\cos\theta\cos\phi\,\delta\theta \tag{15}$$

where ϕ is the azimuthal angle of the tilt. Using $\langle\cos^2\phi\rangle = 1/2$, the second moment associated with an equilibrium ensemble of surface undulations is given by

$$M_2 = (9/2)\omega_Q^2 S_{CD}^2\sin^2\theta\cos^2\theta\langle|\delta\theta|^2\rangle \tag{16}$$

Using Eqs. (3) and (8), this may be written

$$M_2 = [(9kT)/8\pi\kappa)]\ln(\lambda_M/\lambda_m)\omega_Q^2 S_{CD}^2\sin^2\theta\cos^2\theta \tag{17}$$

Bloom and Sternin[14] measured T_2 for the fluid phase of a powder sample of DPPC-d$_{31}$ (i.e. dipalmitoylphosphatidylcholine with one chain perdeuterated) and obtained a value of $T_2(eff) = \langle 1/T_2\rangle^{-1} \approx 5\times10^{-4}s$. This is the usual order of magnitude of T_2 obtained in our laboratory for ^{2}H-labeled acyl chains in fluid membranes. We ask whether an appropriately averaged M_2 i.e. $M_2(eff) = \langle M_2\rangle$ and Eq. (14) gives the experimental value of $T_2(eff)$ for a "reasonable" value of τ_c, keeping in mind that M_2 in Eq. (17) is certainly an overestimate, since very long wavelength modes contribute appreciably to M_2, but have their spin-spin contribution suppressed because of correlation times longer than T_2.

TABLE1

Numerical values used to estimate $M_2(eff)$ and τ_c from Eqs. (14) and (17).

$T = 300K$ and $\kappa = 5\times10^{-13}$ ergs yields $kT/\kappa = 0.082$ and $\lambda_M = 500$Å.
$\lambda_M/\lambda_m \approx 17$ (a guess, but the prediction not sensitive to the number)
$\langle\sin^2\theta\cos^2\theta\rangle = 2/15$,
$\langle S_{CD}^2\rangle^{1/2} = 0.156$ (Bloom & Sternin[14] for DPPC-d$_{31}$),
$T_2(eff) = 5\times10^{-4}s$.

The result of this calculation using numerical values shown in Table 1 is that $\tau_c \approx 10^{-5}s$. A smaller value of $M_2(eff)$ would require a longer correlation time. This short value of τ_c tells us that the molecular motions associated with surface undulations represent a potent mechanism for spin-spin relaxation.

6. A BETTER THEORY OF RELAXATION BY SURFACE UNDULATIONS

The general theory of spin relaxation[8] predicts that if the Larmor frequency is modulated by a set of independent, adiabatic motions labeled by the index q, each motion being characterized, respectively, by a correlation time τ_q and a range of frequencies having a second moment $M_2^{(q)}$ then for $M_2^{(q)}\tau_q^2 \ll 1$, the spin-spin relaxation time is given by

$$1/T_2 = \sum_q M_2^{(q)}\tau_q \tag{18}$$

where

$$M_2^{(q)} = (9/4)(kT/\kappa)[1/(q^2 A)]\omega_Q^2 S_{CD}^2 \sin^2\theta\cos^2\theta \tag{19}$$

Replacing the sum in Eq. (18) by an integral[7], i.e. $\sum_q = (A/2\pi)\int q\,dq$, gives

$$1/T_2 = \Omega^2 S_{CD}^2 \sin^2\theta\cos^2\theta \int dq q^{-1}\tau_q \tag{20}$$

where

$$\Omega^2 = (9/8\pi)(kT/\kappa)\omega_Q^2 \tag{21}$$

For the values in Table 1, $\Omega = 0.17\omega_Q = 1.35\times 10^5\,s^{-1}$.

For the forms of τ_q given by Eqs. (11) and (13) and with the approximation $q_{min} \ll q_{max}$, T_2 is given by the following expressions:
(a) *Viscous damping*:

$$1/T_2 = \{\Omega^2 \tau_m(\eta)/6\} S_{CD}^2 \sin^2\theta\cos^2\theta \tag{22}$$

where the viscous damping correlation time $\tau_m(\eta)$ at the long wavelength cut-off, $q = q_{min}$, is given by

$$\tau_m(\eta) = 3[4\eta/(\kappa q_{min}^3)]/(q_{min}d)^3 = 12\eta/[\kappa d^3 q_{min}^6] \tag{23}$$

(b) *Diffusion — limited damping*:

$$1/T_2 = \{\Omega^2 \tau_m(D)/2\} S_{CD}^2 \sin^2\theta\cos^2\theta \tag{24}$$

where

$$\tau_m(D) = (Dq_{min}^2)^{-1} \tag{25}$$

Numerical values of $T_2(eff)$ and τ_m are predicted for each of these mechanism, using Eqs. (22) - (25), assuming values of parameters in Table 1, $\lambda_M = 500\text{Å}$,

$$\eta = 10^{-2}\text{poise}, \quad D = 4\times 10^{-8}\text{cm}^2\text{s}^{-1}, \quad q_{min} = \pi/\lambda_M \approx 6\times 10^5\text{cm}^{-1}$$

and $d = 30\text{Å}$.

The predictions are as follows:
Viscous damping:
$\tau_m(\eta) = 190\mu s$ and $T_2(eff) = 83\mu s$
Diffusion — limited damping:
$\tau_m(D) = 70\mu s$ and $T_2(eff) = 75\mu s$

Clearly, either relaxation process can yield reasonable values for $T_2(eff)$. However, the dependence on q_{min} is much stronger for the viscous damping mechanism. Furthermore, the longest correlation times, $\tau_m(\eta)$ and $\tau_m(D)$, predicted for the viscous-damping and diffusion-damping mechanisms, respectively, violate the short correlation time approximation, $M_2^{(q)}\tau_q^2 \ll 1$, used in deriving Eq. (18). In order to distinguish between these mechanisms and to examine the relationship between theory and experiment more critically, it is necessary to have a way of measuring the spectrum of correlation times associated with transverse spin relaxation. Fortunately, such a method has been found[14].

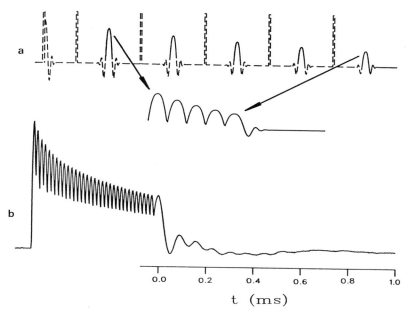

Fig. 2. (a) Schematic representation of the selective data acquisition system used in the Q-CPMG experiments. Only the data points near the peaks of the echo were recorded as indicated by the solid line. (b) An example of an actual experimental data set acquired in (a) is shown. Here, $\tau = 100\mu s$ and 32 pulses were used. The time scale is only for the time region following the last echo, which was the only one recorded completely in order to measure the shape of the spectrum following the 32 pulses.

7. METHOD OF DISTINGUISHING BETWEEN SHORT AND LONG CORRELATION TIMES

It is conventional to measure $T_2 = T_2^e$ in ^{2}H NMR from the decay of the QE signal. The refocussing effect of the second pulse via reversal of the phase accumulated by the spins between the two pulses ensures that the QE signal decays only as a result of randomly fluctuating interactions and not due to any static distribution of quadrupolar splittings. A useful technique for making a further separation of the hierarchy of motions into those whose correlation times are much greater or much less than some pre-selected time τ, is to replace the QEPS by a train of pulses[14] $[90_x - \tau - (90_y - 2\tau-)_N]$. The second pulse is replaced by a sequence of N such pulses and the rate of decrease of the amplitude of the train of echoes at times $2n\tau$, $n = 1, 2, 3......$etc., resulting from the refocussing of the spins by successive pulses is measured to determine a transverse relaxation time denoted by T_2^{CPMG}. This Q-CPMG, "Carr - Purcell - Meiboom - Gill" technique is a well known NMR method that has been adapted by Bloom and Sternin[14] to ^{2}H NMR. It is illustrated in Figures 2 and 3. The virtue of this technique is that the periodic refocussing of the nuclear spins effectively multiplies τ_q in Eq (20) by $[1 - (\tau_q/\tau)\tanh(\tau/\tau_q)]$ so that[14]

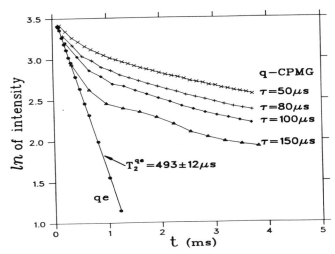

Fig. 3. Comparison of transverse relaxation rates in DPPC-d$_{31}$ membranes at 44°C. The intensity of the peak of the signal is plotted vs the actual time it occurs for the QE measurements and for the Q-CPMG measurements for various values of τ.

$$1/T_2^{CPMG} = \Omega S_{CD}^2 \sin^2\theta\cos^2\theta \int dq\; q^{-1}\tau_q[1 - (\tau_q/\tau)\tanh(\tau/\tau_q)] \qquad (26)$$

This allows the NMR spectroscopist to classify the hierarchy of correlation times τ_q into those $(q = q')$ for which $\tau_{q'} \ll \tau$ and those $(q = q'')$ for which $\tau_{q''} \gg \tau_q$. Effectively, the latter can be progressively suppressed by using shorter and shorter values of τ. If the spectrum of correlation times can be approximated as a superposition of two classes corresponding to very short and very long correlation times, then Eq. (26) may be replaced by

$$1/T_2^{CPMG} = \sum_{q'} M_2^{(q')}\tau_{q'} + \sum_{q''} M_2^{(q'')}\tau^2/(3\tau_{q''}) \qquad (27)$$

Note that the long correlation time contributions are reduced by a factor of $\tau^2/(3\tau_{q''}^2)$.

The results in Figure 3 were interpreted by Bloom and Sternin[14] in terms of the slow motions being associated with molecular diffusion on a smooth spherical surface. In a recent application of the Q-CPMG method, Stohrer et al [15] demonstrated that a similar decomposition into short and long correlation times obtains for oriented bilayers. They ascribe this spectrum to "order director fluctuations" (see also Watnick et al[16]). Order director fluctuations and surface undulations both are predicted to give rise to the observed $\sin^2\theta\cos^2\theta$ angular dependence[14-17]. They should be distinguishable from each other from a study of the spectrum of τ_q values using the CPMG method described above and from the variation of the spectrum with position of the deuteron in the lipid molecule. In one study, still at the preliminary stage (Clare Morrison, personal communication), the characteristic angular dependence of surface undulations (and order director fluctuations) was observed for DPPC deuterated in the head group. It will be interesting to compare results of that system with those of chain deuterated DPPC.

146

ACKNOWLEDGMENTS

We wish to thank Clare Morrison and Gerd Kothe for telling us about their unpublished results. This work was supported by the Natural Sciences and Engineering Research Council of Canada (M.B.), the Medical Research Council of Canada (E.E.) and the Canadian Institute for Advanced Research (M.B. and E.E.).

REFERENCES

1. M. Bloom, E. Evans and O.G. Mouritsen, *Q. Rev. Biophys.* (to be published).
2. M. Bloom and O.G. Mouritsen, *Can. J. Chem.* 66, 705–712 (1988).
3. S. Leibler, in Statistical Mechanics of Membranes and Surfaces, *Proc. of the Jerusalem Winter School for Theoretical Physics*, eds. D.R. Nelson, T. Piran and S. Weinberg (World Scientific, Singapore, 1989).
4. E. Evans and R.M. Hochmuth, *J. Membr. Biol.* 30, 351–362 (1977).
    A. Alberts, D. Bray, J. Lewis, M. Raff, K. Roberts and J.D. Watson, Molecular Biology of the Cell, Second Edition. (Garland, New York, 1989).
5. E. Evans and W. Rawicz, *Phys. Rev. Lett.* 64, 2094–2097 (1990).
6. W. Helfrich and R.-M. Servuss, *Nuovo Cimento* D3, 137–151 (1984).
7. A. Abragam, *The Principles of Nuclear Magnetism.* Oxford University Press, London, 1961.
8. J.H. Davis, *Biochim. Biophys. Acta* 737, 117–171 (1983).
9. M. Bloom, in *Physics of NMR Spectroscopy in Biology and Medicine,* Ed. B. Maraviglia. pp 121 –157. (North Holland, Amsterdam, 1988).
10. J.F. Faucon, M.D. Mitov, P. Meleard, I Bivas and P. Botherel, *J. Phys. France* 50, 2389–2413 (1989).
11. F. Brochard and J.F. Lennon, *J. Phys. France* 36, 1035 (1975).
12. S.T. Milner and S.A. Safran, *Phys. Rev.* A36. 4371–4379 (1987).
13. M. Bloom and E. Sternin, *Biochemistry* 26, 2101–2105 (1987).
14. J. Stohrer, G. Grobner, D. Reimer, K. Weisz, C. Mayer and G. Kothe, *J. Chem. Phys. (submitted).*
15. P.I. Watnick, P. Dea and S.I. Chan, *Proc. Natl. Acad. Sci. U.S.A.* 87, 2082–2086 (1990).
16. J.S. Blicharski, *Can. J. Phys.* 64, 733–45, (1986).

A POSSIBLE CORRELATION BETWEEN FLUCTUATIONS AND FUNCTION OF A MEMBRANE PROTEIN

Fritz Jähnig and Klaus Dornmair*

Max-Planck-Institut für Biologie
Corrensstr. 38, D7400 Tübingen, FRG

INTRODUCTION

It is well known that proteins are not static structures but dynamic entities which undergo conformational transitions in fulfillment of their catalytic function. The characteristic times of such transitions are given by the turnover numbers of the proteins and range from below milliseconds to seconds. In addition to these large and slow conformational transitions, proteins undergo small and fast conformational fluctuations, their characteristic times extending from picoseconds to microseconds. This implies that each of the few conformational states consists of a large number of substates and transitions between substates occur frequently. The existence of substates and fluctuations has been demonstrated by a number of different techniques[1].

What is still under debate is the relevance of such fluctuations to protein function. Are they just thermal motions unavoidable to the proteins, or are they correlated in a specific way to the slow conformational transitions involved in the catalytic processes? A model has been proposed in which the fluctuations and conformational transitions obey a hierarchical order[2]: The fastest fluctuations enable the protein to overcome low energy barriers and thus give rise to slower fluctuations, which in turn permit to overcome higher energy barriers leading to still slower fluctuations and so on until, finally, the slowest fluctuations lead to a conformational change. In an attempt to give an experimental answer to the above question, we investigated the internal fluctuations of the transport protein lactose permease (LacP) under conditions where the protein is functionally either active or inactive[3].
LacP is a protein of the cytoplasmic membrane of Escherichia coli which catalyzes the transport of galactosides

*Present address: Max-Planck-Institut für Psychiatrie, Abt. Neuroimmunologie, 8032 Martinsried, FRG.

Biologically Inspired Physics, Edited by L. Peliti
Plenum Press, New York, 1991

in symport with a proton across the membrane[4,5]. Such a process is called active transport, because due to the proton gradient across the cytoplasmic membrane the sugar is accumulated against its concentration gradient in the cell. The protein consists of 417 amino acid residues and acts as a monomer[6]. Spectroscopic data, the construction of fusion proteins, and structure predictions indicate that the polypeptide chain is folded into 12 membrane-spanning helices with the sugar binding site in the center. The cylinder should not act as a pore and let sugars and protons pass freely, because this would render active transport impossible. In a simple model, therefore, the helices are assumed to be slightly tilted, as illustrated in Fig.1. Transport would then proceed via a change in the orientation of the helices. Obviously, a transport cycle must involve two such conformational changes, one with substrates (of rate k_c) and one without (of rate k_o). The intervening steps of substrate binding and dissociation are faster, so that the two conformational changes determine the transport rate. There is experimental evidence that they occur with similar rates[4]. Hence, since the transport rate is of the order of 10 s^{-1} in the absence of a proton-motive force, the rates k_c and k_o are also of this order of magnitude. This means that the characteristic time for a conformational change is of the order of 100 ms.

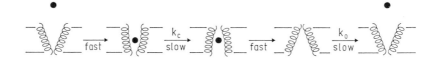

Fig.1. Schematic representation of a transport cycle of LacP. The protein is drawn as two membrane-spanning helices, the substrate as a dot.

The above question can then be stated more specifically: Are there fast orientational fluctuations of the helices which might give rise to the slow conformational changes involved in transport?

The technique we used to detect fast orientational fluctuations was fluorescence anisotropy decay (FAD)[10]. The fluorophores in the sample are excited by short laser pulses of vertically polarized light. The intensities $I_\parallel(t)$ and $I_\perp(t)$ of the fluorescence light emitted at right angle and polarized parallel or perpendicular to the exciting light are recorded as a function of time. The total intensity results as $I_{tot}(t) = I_\parallel(t) + 2 I_\perp(t)$ and the anisotropy as $r(t) = [I_\parallel(t) - I_\perp(t)]/I_{tot}(t)$. This technique permits to observe orientational fluctuations in the range from picoseconds to nanoseconds. As fluorophores one may employ either intrinsic Trp and Tyr residues or extrinsic labels. We started our studies detecting the fluorescence of the 6 Trp residues of LacP, but recognized very soon that the most interesting fluctuations have characteristic times in the range of 20-100 ns which is not easily accessible with the Trp fluorescence because of its short lifetime of a few ns. Therefore, LacP was labeled with pyrene-maleimide (MalNPyr) which has a lifetime component of 100 ns. The labeling site was Cys148 located on one of the putative membrane-spanning helices, hence, the MalNPyr label may reflect helix fluctuations.

LacP was isolated, purified, labeled, and reconstituted into vesicle membranes of dimyristoylphosphatidylcholine (DMPC) at a lipid/protein molar ratio of 2500. The vesicles are devoid of a proton-motive force across the membranes. All measurements were performed in the absence of galactoside, since labeling of LacP at Cys148 with MalNPyr inhibits galactoside binding.

RESULTS AND DISCUSSION

A typcial result of a FAD measurement is shown in Fig.2. The temperature was 28°C, i.e. the DMPC vesicle membranes were in the fluid phase. The intensity $I_{tot}(t)$ of the pyrene label decreases slowly with two lifetimes of 22 ns and 109 ns. The

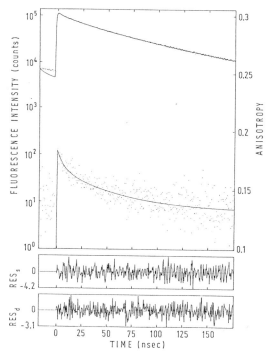

Fig.2. Intensity $I_{tot}(t)$ (upper curve) and anisotropy r(t) (lower curve) of the pyrene fluorescence of MalNPyr-labeled LacP reconstituted in vesicle membranes of DMPC and measured at 28°C. The experimental data are represented by dots, and the curves are least-square fits, with the goodness of the fits given by the residuals for the sum and for the difference of the parallel and perpendicular polarized fluorescence light.

anisotropy r(t) exhibits a fast and a slow relaxation process and levels off at a finite value r_∞. A fit of r(t) by a sum of two exponentials and a constant term yields the relaxation times $\emptyset_1$ = 4.2 ns and $\emptyset_2$ = 39 ns with the amplitudes b_1 = 0.023 and b_2 = 0.037, respectively, and the limiting value r_∞ = 0.128. If the anisotropy is fitted by three exponentials and a constant term, the long relaxation time is simply obtained twice. Furthermore, measurements extended over a larger time window of 470 ns demonstrate that up to 1 μs no further relaxation process exists. These results indicate that the pyrene label attached to LacP undergoes two relaxation processes. By contrast, when MalNPyr alone is dissolved in vesicle membranes of DMPC, it relaxes with $\emptyset_1$ = 5.3 ns and does not exhibit a slow relaxation process.

Therefore it is most obvious to assign the fast relaxation process of the pyrene label to orientational fluctuations of the label relative to the backbone of the polypeptide chain. In turn, the slow relaxation process may be assigned to fluctuations of the backbone. The relaxation time for rotational diffusion of the entire LacP molecule (about an axis parallel to the membrane normal) has been determined experimentally[6] as 25 μs, and from this result the relaxation time for wobbling of the entire molecule about the membrane normal may be estimated as a few μs. Thus, the orientational fluctuations of the entire LacP molecule seem to be much slower than the observed relaxation process with about 40 ns. Therefore, in view of the fact that the pyrene label attached to LacP is located on one of the putative membrane-spanning helices, the slow relaxation process is tentatively assigned to orientational fluctuations of the helix about the membrane normal. Orientational fluctuations of single membrane-spanning helices have indeed been observed[11,12], their relaxation times ranging from 10 to 40 ns.

A theoretical estimate for the relaxation time $\emptyset_h$ of a membrane-spanning helix may be obtained using the relation for a random walk $\langle\vartheta^2\rangle$ = $2D_\perp\emptyset_h$, with ϑ denoting the angle between the helix axis and the membrane normal and $D_\perp$ the coefficient for rotational diffusion about a short axis of the helix. Modeling a membrane-spanning helix as an ellipsoid, $D_\perp$ can be calculated[3] as $D_\perp$ = 5 . 10^5 s^{-1}. Analysis of the measured amplitude of the slow relaxation process leads to rms fluctuations of $\langle\vartheta^2\rangle^{1/2}$ = $\pm15°$. Thus, the random walk equation yields $\emptyset_h$ = 60 ns, in astonishingly good agreement with the experimental result.

The observed helix fluctuations are postulated to give rise to the conformational changes involved in transport (Fig.1). To test whether such a mechanism is compatible with the available data, one may employ the Arrhenius relation k_o = $k_o'\exp(-\Delta G_o/RT)$ for the rate k_o of the conformational change in the absence of substrate. The activation energy for sugar transport by LacP in vesicle membranes of DMPC has been measured[13] as 30-40 kJ/mol and the same value may be used for ΔG_o. The rate k_o = 10 s^{-1} then requires a prefactor of k_o' in the range of 10^6 - 10^8 s^{-1}. On the other hand, the prefactor k_o' may be expressed in terms of the helix fluctuations by applying a relation derived by Kramers[14]. He modeled a chemical reaction as a Brownian motion of a particle along the energy profile and obtained for the case of high viscosity which pertains to proteins, k' = $2\pi m\nu_o^2/\gamma$.

Here, m denotes the mass of the particle, ν_o the oscillation frequency in the energy well, and γ the friction coefficient. Introducing the Einstein relation $D_t = kT/\gamma$ for the coefficient of translational diffusion D_t and the equipartition theorem $m(2\pi\nu_o)^2<x^2>/2 = kT/2$ with $<x^2>$ denoting the mean-square fluctuations of the position of the particle, the Kramers relation may be transformed into $k' = D_t/(2\pi<x^2>)$. In this form, it can easily be adapted to reactions controlled by rotational diffusion as in our case leading to $k' = D_\perp/(2\pi<\vartheta^2>)$. If the above expression for the relaxation time of a membrane-spanning helix is employed, one finally obtains $k_o' = 1/(4\pi\phi_h)$. This is the desired relation between the prefactor k_o and the helix fluctuations. Insertion of our experimental value $\phi_h = 40$ ns yields the value $k_o' = 2 \cdot 10^6$ s^{-1} which lies well within the limits set by the transport rate and the activation energy. This analysis shows that the available experimental data are compatible with the postulated correlation between helix fluctuations and transport, it does not represent a proof.

The above picture is fully consistent with the model of a hierarchical order in proteins[2]. The two conformational states of LacP (Fig.1) would consist of a large number of substates corresponding to the many different helix orientations which occur when the helices fluctuate. One might have expected the helix fluctuations to differ in the two conformational states leading to two different relaxation processes. Actually, however, only one slow relaxation process was observed, so that the two conformational states must be similar with respect to their fluctuations.

Further support for the postulated correlation between helix fluctuations and transport may be obtained by studying the helix fluctuations under conditions where transport is known to be blocked or at least reduced. This is the case if the temperature is reduced below the lipid phase transition which for DMPC membranes occurs at $T_t = 24°C$. In the ordered phase, the transport rate is reduced by several orders of magnitude[13,15]. When FAD measurements of the kind described above were performed at temperatures below T_t, the fast relaxation process was not much altered but the slow relaxation time increased considerably and became as large as 300 ns at 13°C. The amplitude of the slow relaxation process decreased by a factor of two upon passing below T_t. Thus, the helix fluctuations became slower which is in accordance with a slower transport process. The slowing down of the helix fluctuations by one order of magnitude, however, is not sufficient to explain the decrease of the transport rate by several orders of magnitude. In addition, the energy barrier between the two conformational states must become higher. Experimental results on transport indicate[13] that ΔG_o is indeed higher at $T < T_t$. Our FAD data provide indirect evidence for a higher ΔG_o at $T < T_t$. The decrease of the amplitude of the slow relaxation process upon passing below T_t implies a decrease of the mean-square amplitude of the helix fluctuations. As illustrated in Fig.3, a smaller amplitude of the fluctuations in the two conformational states requires a higher barrier ΔG_o. Thus, the FAD data at $T < T_t$ are also consistent with the postulated correlation between fluctuations and transport.

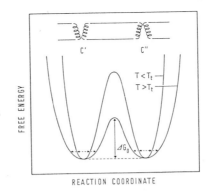

FREE ENERGY

REACTION COORDINATE

Fig.3. Schematic plot of the free energy of LacP as a function
of the reaction coordinate. The states C' and C"
represent the two conformations involved in transport.
The arrows indicate the amplitude of the helix
fluctuations. If the amplitude decreases at $T < T_t$,
while the conformations remain unaltered, the energy
barrier ΔG_o must increase.

Another possibility to reduce the transport rate is offered
by point mutations of LacP[5],[16]. Mutants have already been
constructed with unaltered galactoside binding but much slower
transport rates. FAD measurements on such mutants are under way
and hopefully will help to clarify the question of whether the
internal dynamics of a membrane protein such as LacP is just
thermal noise or the basis of transport.

REFERENCES

1. H. Frauenfelder, F. Parak and R.D. Young, Conformational
   substates in proteins, Ann. Rev. Biophys. Chem. 17:451
   (1988).
2. A. Ansari, J. Berendzen, S.F. Bowne, H. Frauenfelder, I.E.T.
   Iben, T.B. Sauke, E. Shyamsunder and R.D. Young, Protein
   states and proteinquakes, Proc. Natl. Acad. Sci. USA
   82:5000 (1985).
3. K. Dornmair and F. Jähnig, Internal dynamics of lactose
   permease, Proc. Natl. Acad. Sci. USA 86:9827 (1989).
4. J.K. Wright, R. Seckler and P. Overath, Molecular aspects of
   sugar:ion cotransport, Ann. Rev. Biochem. 55:225 (1986).
5. H.R. Kaback, E. Bibi and P.D. Roepe, ß-Galactoside transport
   in E. coli: a functional dissection of lac permease, Trends
   Biochem. Sci. 15:309 (1990).
6. K. Dornmair, A.F. Corin, J.K. Wright and F. Jähnig, The size
   of the lactose permease derived from rotational diffusion
   measurements, EMBO J. 4:3633 (1985).
7. D.L. Foster, M. Boublik and H.R. Kaback, Structure of the
   lac carrier protein of E. coli, J. Biol. Chem. 258:31
   (1983).
8. H. Vogel, J.K. Wright and F. Jähnig, The structure of the
   lactose permease derived from Raman spectroscopy and
   prediction methods, EMBO J. 4:3625 (1985).

9. J. Calamia and C. Manoil, lac permease of E. coli: Topology and sequence elements promoting membrane insertion, Proc. Natl. Acad. Sci. USA 87:4937 (1990).

10. J.M. Beecham and L. Brand, Time-resolved fluorescence of proteins, Ann. Rev. Biochem. 54:43 (1985).

11. E. John and F. Jähnig, Dynamics of melittin in water and membranes as determined by fluorescence anisotropy decay, Biophys. J. 54:817 (1988).

12. H. Vogel, L. Nilsson, R. Rigler, K.P. Voges and G. Jung, Structural fluctuations of a helical polypeptide traversing a lipid bilayer, Proc. Natl. Acad. Sci. USA 85:5067 (1988).

13. K. Dornmair, P. Overath and F. Jähnig, Fast measurement of galactoside transport by lactose permease, J. Biol. Chem. 264:342 (1989).

14. H.A. Kramers, Brownian motion in a field of force and the diffusion model of chemical reactions, Physica 7:284 (1940).

15. J.K. Wright, I. Riede and P. Overath, Lactose carrier protein of E. coli: Interaction with galactosides and protons, Biochemistry 20:6404 (1981).

16. P. Overath, U. Weigel, J.M. Neuhaus, J. Soppa, R. Seckler, I. Riede, H. Bocklage, B. Müller-Hill, G. Aichele and J.K. Wright, Lactose permease of E. coli: Properties of mutants defective in substrate translocation, Proc. Natl. Acad. Sci. USA 84:5535 (1987).

MICROPHASE SEPARATION IN BILAYER MEMBRANES

Kell Mortensen

Physics Department
Risø National Laboratory
DK-4000 Roskilde
Denmark

1. INTRODUCTION

A biological membrane is a complicated multicomponent, two-dimensional alloy of amphipathic molecules. The varied purposes that natural membranes serve require a high degree of order at the molecular level simultaneously with high flexibility and molecular mobility. Nature achieves these goals by using combinations of suitable lipid components and proteins.

The basic building blocks consists of various phospholipids, to form a liquid crystalline phase. A very important additive to the membrane is cholesterol, which appears to have dramatic effects on the membrane properties. Cholesterol serves as a precursor for various acids and hormones, however, it also has major influence on the membrane function itself. Numerous studies have been performed during the past decades in order to learn about the role of cholesterol. These include both experiments and theoretical model calculations. It has been speculated that the biological advantage associated with cholesterol may be due to a reduced fluidicity or increased microviscosity which the addition of cholesterol is claimed to impart to the liquid crystalline state of phospholipid bilayer membranes[1]. Other additives includes for example various proteins. Spectrin and clathrin, for example, certainly have major influence on the membrane structure, as they by polymerization form a network covering the bilayer membrane. Clathrin coated vesicles serve the purpose of transporting specific hormones into the cell.

In many model systems of biological membranes, one makes artificially phospholipid bilayer membranes of well characterized molecules, as for example DMPC (dimyristoylphosphatiddylcholine), DPPC (dipalmitoylphosphatidylcholine), DMPE (dimyristoylethanolamine) etc. These molecules are commercially accessible in pure form, even when perdeuterated. When mixed with water, the lipids form instantaneously bilayer systems of different phases, depending on water and temperature. In Fig. 1 is shown the schematic phase diagram of lecithin-water[2]. The absolute temperature-scale depends on the specific lecithin. At low water concentration, be-

Biologically Inspired Physics, Edited by L. Peliti
Plenum Press, New York, 1991

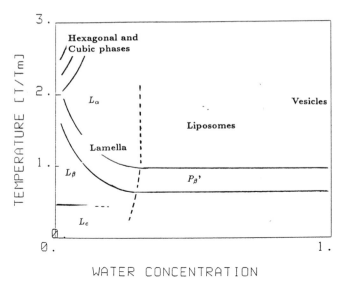

Fig. 1. Schematic phase diagram of lecithin-water system. The temperature scale
depends on the lecithin system.

low approximately 20%, the system forms a variety of structural phases, including
hexagonal, cubic and lamellar states. In high degree of excess water, vesicles are
formed. In the crystalline states formed below approximately 20% H_2O, there is
significant coupling between different bilayers. This is most probably also true for
the lamellar states. The physical properties observed in lamellar bilayer membranes
may therefore deviate quite significantly from the behavior of vesicles which are more
close to the biological situation. Still, however, many features observed in the lamellar
states seems also to be relevant for the vesicles.

In the present paper, we will describe some structural investigations performed
on model membranes incorporated with additives. The studies are all done on powder
samples of 17 mol.% water, which gives well defined lamellar structure, and still the
lowest possible inter bilayer interaction. The studies have been performed using the
Risø small angle neutron scattering facility.

2. EXPERIMENTAL

The model membranes used for the study were made at the Technical University of Munich on the basis of lecithin from Avanti (Birmingham, AL). Cholesterol
was obtained from Serva (Heidelberg, FRG), and the E34-lipids were provided by
Prof. Ringsdorf (Mainz). In order to get good contrast for the neutron studies, the
fatty acid chains of phospholipids were fully deuterated[3]: DMPC-d_{54} and DPPC-d_{62}.
(Since the conclusions of this paper are not specific to the perdeuterated lecithins,
we will in the following not differentiate between DMPC-d_{54} and DMPC and between DPPC-d_{62} and DPPC). The bilayer membranes were prepared to form well

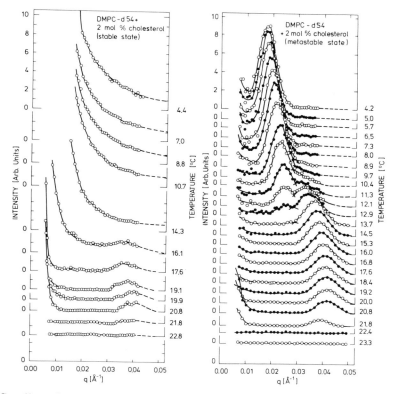

Fig. 2. Small angle neutron scattering data on DMPC-d$_{54}$ with 2 mol.% cholesterol.

characterized lamellar phases including 17 mol.% D$_2$O. The structural investigation was made using the Risø-Small Angle Neutron Scattering Facility. The spectrometer setting was 2.5 meter collimation, 3 meter sample to detector distance, 16 mm and 8 mm diameter pinholes of the collimator, and neutrons monochromatized using a velocity selector of 18% wavelength resolution. Various neutron wavelengths were used, ranging from 3 to 20Å. With these settings, the Bragg peaks observed are all resolution limited[4].

3. DMPC INCORPORATED WITH CHOLESTEROL

A variety of information on the lecithin-cholesterol phasediagram is obtained from the small angle scattering data[3].

First of all, the studies clearly demonstrate very slow kinetics involved in the formation of some of the phases. The exact history of the sample is therefore crucial for the observations obtained. This is demonstrated in Fig. 2, showing (DMPC-d$_{54}$) with 2 mol.% cholesterol. After long time storage at low temperature, below approximately 15°C, the (a) sub-phase is stabilized. This phase may be related to the L$_c$ sub-phase known from pure lecithin membranes. However, cholesterol clearly also has active influence in characterizing this phase, as seen from a quite simple relationship between the amount of cholesterol and the small angle scattering which

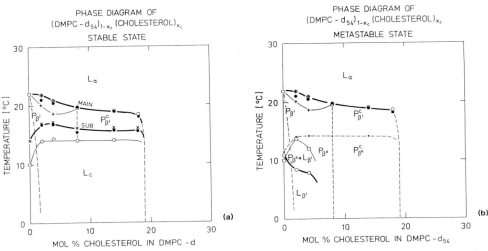

PHASE DIAGRAM OF
$(DMPC - d_{54})_{1-x_c} (CHOLESTEROL)_{x_c}$
STABLE STATE

PHASE DIAGRAM OF
$(DMPC - d_{54})_{1-x_c} (CHOLESTEROL)_{x_c}$
METASTABLE STATE

Fig. 3. Phase diagram of the DMPC-d_{54}-cholesterol system, as observed by small angle neutron scattering experiments.

dominates this regime. The small angle scattering follows a *power law* with exponent close to 3, indicating a surface fractal, probably related to *crumpled* structure within this phase. The absolute intensity of this small angle scattering increases linearly with the amount of cholesterol. For pure DMPC, the time involved in the formation is of the order of minutes, whereas for the low temperature phase of cholesterol enriched samples it is of the order of months.

After annealing above the sub-transition of approximately 15°C, an other phasediagram appear, including *ripple* phases, the L_β and the L_α phase (Fig. 3). The inclusion of cholesterol clearly stabilizes the rippled structure. The Bragg peak in the P'_β phase is believed to correspond to well-defined ripple structures, whereas the Bragg peak in the L_β phase probably relates to the step-like defect-lines between planar L_β-domains, as also observed in electron microscopy studies[5]. The ripple structure parameters, e.g. ripple periodicity and Bragg scattering intensity, correlate simply with the amount of cholesterol. It has therefore been argued that the cholesterol phase-separates on a microscopic level into *defect-lines*[3]. Above cholesterol concentrations of approximately 8 mol.%, the defect lines are fully covered with cholesterol or cholesterol complexes, and at the order of 20 mol.%, the cholesterol rich areas merge with the consequence of absence of rippled structure[3]. Ipsen et al. have demonstrated phase separation in the same area of the phase diagram[6]on the basis of Monte Carlo calculations, and also seen indication of cholesterol aggregation near possible phase boundaries[7]. Rippled structure was, however, not included in their model.

4. POLYMERIZATION OF LIPIDS WITHIN THE BILAYER MEMBRANE

As in the case of cholesterol incorporated in phospholipid bilayer membranes, it seems that also other types of lipid additives to the the membrane will phase separate on a microscopic level when the system is within a gel-state. This seems for example to be the case for glycosphingolipids, which clearly stabilizes the ripple structure as observed by electron microscopy[8]. Experiments performed on the molecule E34, give also evidence of the E34 molecules being concentrated within the defect lines of the ripple structure. This is seen from the very pronounced *ripple* structure observed in the scattering experiment (Fig. 4). E34 is a phospholipid of 15 CH_2 units in the acyl chain, and with a polymerizable head-group. The polymerization can be initiated by UV-light or by chemical initiators heated to a given temperature. Polymer formation within bilayer membranes has also great biological interests. For example with the relationship to the clathrin network, coating biological vesicles when transporting e.g., hormones in to the cell[9,10].

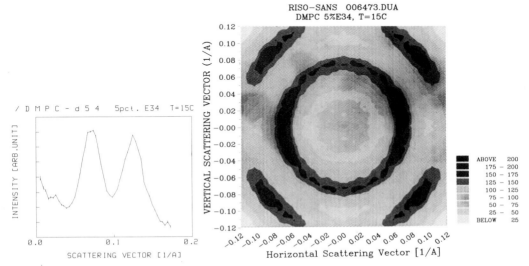

Fig. 4. a) 2D contour plot of the raw scattering data obtained on DMPC incorporated with 5% E34 monomers. b) The radial averaged data of the DMPC 5% E34 sample.

As in the case of cholesterol included in the lamellar phase of DMPC, E34 incorporation shows a Bragg scattering peak related to defect lines or ripples, which amplitude decrease as the temperature is enhanced and vanish as the L_α phase is reached at the main transition (20°C for DMPC). The scattering data are reproducible over time scales of at least days.

When the DMPC-E34 sample is exposed to UV-light at 10°C, i.e. within the P'_β ripple phase, the E34 monomers are expected to polymerize to form chains. Still, however, it has not been possible to observe any changes in the scattering pattern, even after hours. This include both the ripple structure and the scattering at smaller q-values. When the composition successively is heated into the L_α state, however, marked changed are observed at low q-values, as shown in figure 5. After this heat treatment, he scattering pattern reveal Guinier like character with a radius of gyration of the order of 200Å. These characteristics can be easily understood in the frame of the discussions above. When the material was polymerized, all E34-molecules were, naively, lined up in the one-dimensional defect lines. The polymer obtained in that state is therefore to a very high precision a one-dimensional linear polymer. Only after annealing to the fluid state, the polymer entropically relax into a 2D random coil with R_g equal 200Å. This kind of polymerization may not have much biological relevance, though, but unique physical and chemical possibilities certainly appear using this property.

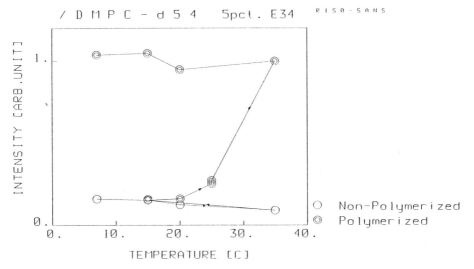

Fig. 5. Small angle scattering of polymerized membranes, as a function of temperature and history.

5. CONCLUSIONS

In conclusion, we have demonstrated that the phase diagram of lamellar bilayer systems including additives to the basic phospholipids, can be characterized by extremely slow kinetic processes which make the material characterization critically dependent on the history of the sample. Many additives, as for example cholesterol, cause phase-separation on the nano-meter range (*microphase separation*) when the material is in the low temperature gel state. Such microphase separation may offer unique possibilities, since it makes it possible to make polymer chains extended ideally within one dimension.

ACKNOWLEDGMENTS

The studies presented here have been done in close collaboration with Walter Pfeiffer (TU, Munich), who prepared the samples. Discussions with Wolfgang Knoll (MPI, Mainz) and Erich Sachmann (TU, Munich) and the supply of polymerizable lipids by H. Ringsdorf (University, Mainz) are greatly acknowledged. The present work was supported by the Danish Natural Science Research Council. The Risø-SANS facility has been established with the support from the Danish and the Swedish Natural Science Research Councils.

REFERENCES

1. P.N.T. Unwin and R.Henderson, *Scientific American* 250:19 (1965).
2. E. Sachmann, *Biophysics* W. Hoppe, W. Lohmann, H. Marki and H. Ziegler, ed., Springer-Verlag (1983).
3. K. Mortensen, W. Pfeiffer, E. Sachmann, and W. Knoll, *Biochim. Biophys. Acta,* 945:221 (1988).
4. J.S. Pedersen, D. Posselt, and K. Mortensen, *J. Appl. Cryst.* 23:321 (1990).
5. D.Ruppel, and E. Sachmann, *J. Phys.* 44:1025 (1983).
6. J.H. Ipsen, G. Karlström, O.G.Mouritsen, H. Wennerström, and M.J.Zuckermann, *Biochim. Biophys. Acta,* 905:162 (1987).
7. J.H. Ipsen, and O.G.Mouritsen, *Private communication.*
8. P. Rock, T.E. Thompson, and T.W. Tillack, *Biochim. Biophys. Acta,* 979:347 (1989).
9. B.M. Pearse, and R.A. Crowther *Ann. Rev. Biophys. Biophys. Chem.* 16:49 (1987).
10. R. Bauer, M. Behan, S. Hansen, G. Jones, K. Mortensen, T. Særmark, *J. Appl. Cryst.,* to be published (1990).

163

MORPHOLOGICAL CHANGES IN PARTIALLY POLYMERIZED VESICLES

M. Mutz and D. Bensimon

Ecole Normale Supérieure, Laboratoire de Physique Statistique
24 rue Lhomond, F-75231 Paris CEDEX 05

1. INTRODUCTION

Phospholipid molecules spontaneously form giant unilamellar vesicles when dispersed in an aqueous solution. These are often used as models of biological cells, e.g. the red blood cell and have technological applications as drug delivery systems. To stabilize these vesicles, polymerizable phospholipids have been developped[1-3].

Polymerizable phospholipids are a class of amphiphilic molecules with the ability to covalently cross-link upon catalysis by a chemical or photochemical process. Among the many different lipids and polymerization mechanisms to have been investigated, the most extensively studied class is that of the diacetylenic phosphatidylcholine lipids which upon UV irradiation polymerize into two dimensional sheets[1].

The first experiments with partially polymerized vesicles have shown that upon cooling they display a reversible folding transition from a state where the membrane is "floppy", fluctuating and rather smooth, to a state where the membrane is rigid and highly convoluted. A theoretical interpretation in terms of a spinglass transition has been proposed[4,5]. The central assumption of this model was the presence of a random local spontaneous curvature term imposed by the embedded polymer nets. The further observation of stable toroidal-vesicles in partially polymerized samples seems to support this assumption[6]. Indeed, toroidal vesicles have been predicted to be stable only for certain negative values of their spontaneous curvature[7].

Continuing our preliminary observations we now report the existence of further interesting shapes and shape transformations in partially polymerized membranes. Besides the already mentioned toroidal vesicles (all well described by the theoretical analysis of Ou-Yang Zhong-can[7]) we have observed two-holed vesicles, a reversible budding transition and a binding-unbinding transition between concentric vesicles.

2. MATERIAL AND METHODS

The polymerizable phospholipids used in our experiments 1,2 bis (10,12-tricosadiynoyl)-sn-glycero-3-phosphocholine ($DC_{8,9}PC$) were purchased from Avanti Polar Lipids as a crystalline powder. The material was used without further purification, though its purity was checked prior to use by silica gel Thin Layer Chromatography (TLC) in a chloroform/methanol/water (65:25:4) solution. Only a single spot was visible with the unpolymerized material when developped in iodine vapour. In view of the sensitivity of the procedure, impurity concentrations larger than 1 % can be excluded. Lipids were kept below $-30°C$ under nitrogen.

A small amount (< 0.1 mg) of the powder was spread on a glass cover slide. A quartz cover slip was put on top with a paraffin film as spacer. This $\approx 100\mu m$ high microchamber was filled by capillarity with bidistilled water, and was sealed with paraffin to prevent evaporation. The sample was then placed in a small temperature controlled oven on a microscope stage. For the observation of the sample an inverted phase contrast microscope (Nikon) was used. A CCD video camera (Sony) backed by an image analysis system (Cyclope) was attached to the microscope. It allowed us to store an image for further processing like contrast enhancement and exact length measurements. Membranes are seen in the focal plane when they are parallel to the optical axis of the microscope.

To measure the degree of polymerization we transferred about $3-30$ mg of the powder to a Hellma quartz cuvette, dissolved it in chloroform and let it evaporate overnight in a vacuum oven protected from light. The lipids were then rehydrated with bidistilled water to a concentration of $1-10$ mg/ml at 50°C for $1-3$ hours. After gently shaking a homogenous milky dispersion was obtained. That solution was then cooled below the melting transition temperature ($T_m = 43°C$)[8] and polymerized by UV irradiation (at 254 nm) for t=0-15 min. The degree of polymerization was checked afterwards by TLC (after evaporation of the water and subsequent redilution in chloroform). Polymerized solution exhibits two spots when developped in iodine vapour: one at the baseline and a further at the same height as the unpolymerized lipids. The relative amount of polymerized molecules was measured by densitometry analysis of the TLC spots. The intensity of the spot corresponding to the monomer was independently calibrated with known solution of the unpolymerized lipids. A linear relationship between concentrations and spot-density was thus established.

3. RESULTS AND DISCUSSION

Above their main transition temperature T_m ($\approx 43°C$)[8] the unpolymerized lipids swelled spontaneously, forming giant unilamellar vesicles and other dilute structures. The membranes display characteristic and well known thermal undulations.

Upon cooling slowly ($\approx 1°$/min) unpolymerized membranes form long ($10-200\,\mu m$) cylindrical tubules of diameter $\approx 1\mu m$ at $T<T_m$, with a supercooling of about 5°C. If cooled rapidly ($\approx 10°$/min) below T_m, the vesicles break up into "shards", i.e. small tubular sections. These structures have been characterised as hollow tubules formed by winding sheets of bilayer[9,10]. Reheating above T_m made all structures resume again their undulating vesicular forms.

The lipids used in our experiments are only polymerizable at $T<T_m$, i.e., in a phase where they spontaneously form tubules. We observed that the effect of polymerization is different depending on the state of the membrane. First we polymerized the membranes after cooling the sample to a certain temperature ($T\approx 25°C$) well below T_m and exposed them to UV radiation (for several seconds up to 30 minutes). If the time of UV irradition exceeds 12 min the rigid structures keep their forms when reheated again above T_m. Presumably, the polymerized net percolated

throughout the structure. However, as long as the time of irradiation is less than about 12 min the rigid structures, when reheated above T_m, reverse to vesicles fluctuating with amplitudes which are slightly reduced with increasing exposure time. When cooling the samples once more below T_m the formation of tubular structures is now prevented. The monomers in the membrane may still have a gel transition at T_m, as indicated by a (reduced) endothermic event in a DSC scan, and by the observation at that temperature of a reversible binding-unbinding transition between concentric vesicles, see Fig. 1. This transition could be triggered by a modification of the elastic properties of the membranes as the monomers order in the gel phase. At a lower temperature ($T \approx 20°C$) all vesicles undergo a spontaneous transition to a folded and rigid structure[4]. That transition is often accompanied by the tearing of the membrane and the subsequent expulsion of the vesicular content.

The formation of the polymer nets was different when we exposed the fresh samples to UV light while cooling through the melting transition. We started with UV irradiation one degree above T_m while cooling the sample to several degrees below. The final exposure time was about 4 min. With this procedure it was possible to freeze the vesicular structures in a polymer network and prevent them from forming tubules or shards. The vesicles treated with this method became completly rigid at all temperatures ($10-60°C$) after UV exposure of about 4 min in contrast to what we observed with the other method, where 12 minutes of UV irradiation were necessary to rigidify the tubular structures irreversibly. We do not understand this difference in polymerization behavior. To check the degree of polymerization with these procedures, we prepared solutions of varying irraditon times for analysis by TLC. If one compares the TLC results of the two methods for an exposure time of 4 min on finds that about 35 % of the monomer are bound in a polymer net in both cases. However, the polymer dispersion obtained when cooling through the transition is higher than when polymerizing the stable low temperature tubular phase.

While investigating the partially polymerized samples we were aware of several giant ring-vesicles (with diameter $> 4 \, \mu$m). As these vesicles were free to float in the aqueous solution they rotated slowly so that one got a view of the different projections of the vesicle in the focal plane. A top (side) view of the same vesicle, where the larger (smaller) generating ring is parallel to the focal plane, is shown in Fig. 2. The side view clearly exhibits the toroidal form of the vesicle and rules out other interpretations. Toroidal-vesicles were stable over days, without changing their mean characteristic form. With the help of the image analysis system we evaluated the width (d) of the ring and the outer diameter (D) of more than 20 ring-vesicles. The mean ratio of the two diameters was found to be: 0.41 ± 0.02. This value has to be compared with $\sqrt{2}-1$ which has been obtained from a theoretical calculation of the stability of toroidal vesicles with nonvanishing spontaneous curvature. The experimental results are shown in Fig. 3. In some rare cases where the ring-vesicles were not axisymmetric, the mean width of the ring was used. From recent numerical calculations of the minimum of the vesicle shape energy Seifert obtained further toroidal shapes for vanishing spontaneous curvature[11]. However, in the case of partially polymerized membranes it seems reasonable to assume a nonzero spontaneous curvature due to the underlying polymer net[4,6].

A further question from the theoretical and experimental sides is the existence of vesicles of higher topological genus. One knows that for permeable membranes (unconstrained volume) of vanishing spontaneous curvature there exist vesicles of arbitrary genus minimizing the curvature energy (Willmore surfaces[12]). However,

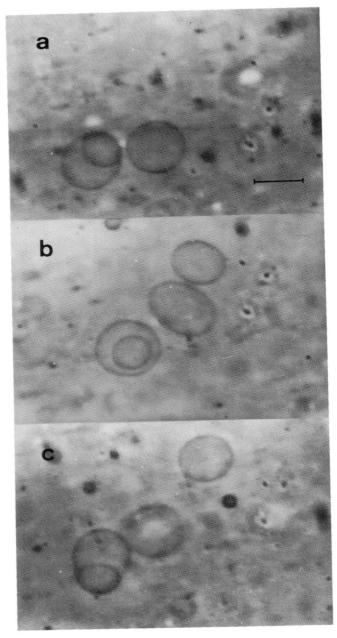

Fig. 1. Reversible binding-unbinding transition. Below T_m at $T= 35°C$ the two concentric vesicle adhere to each other(a), while they unbind (b) when heated above T_m. The vesicles bind (c) again upon cooling below T_m. The bar indicates 10μm.

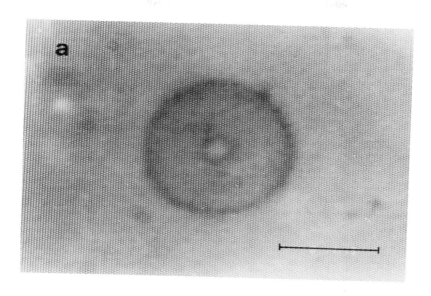

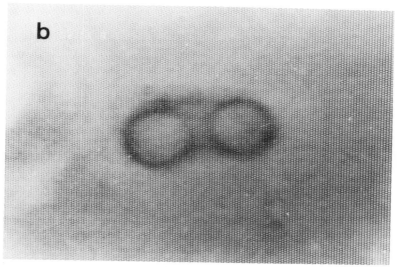

Fig. 2. Two views of a toroidal vesicle of partially polymerized phospholipid membrane at $T = 50°C$, (a) top view, (b) side view. The bar indicates $10\mu m$.

for vesicles of nonzero mean curvature, the question remains theoretically unsolved. Experimentally, however, we have observed a two-holed vesicle, i.e. a vesicle of genus two. As the vesicle rotated freely all projections in the focal plane could be seen. In Fig. 4 three different projections of this vesicle are shown. The side view through a line perpendicular to the line joining the two holes (Fig. 4b) clearly demonstrates the existence of two holes.

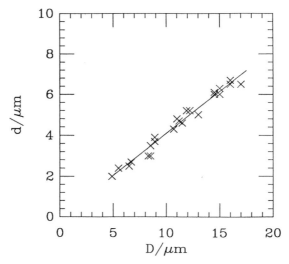

Fig. 3. Results for 24 individual ring-vesicles. The solid line represents the theoretical prediction.

Finally we report the observation of a special shape transformation, known as budding. The creation of a small daughter vesicle from a larger mother one was the subject of recent theoretical calculations and was also observed experimentally in vesicles of unpolymerizable phospholipids[13–16]. We also observed this budding phenomenon in partially polymerized ones but upon cooling not heating as in Ref. 16. A spherical vesicle which shows large fluctuations at T= 50°C (Fig. 5a), i.e. with a large surface to volume ratio, becomes ellipsoid upon cooling to about T= 40°C (Fig. 5b). Further cooling results after passing a dumbbell shape (Fig. 5c) in a pear shaped form at about T= 20°C (Fig. 5d) and leads finally to the formation of a bud at T= 15°C (Fig. 5e). The final form is the double vesicle (Fig. 5f), with a nearly invisible tether connecting both. Reheating above T= 20°C led again to the ellipsoidal form.

To explain this budding phenomenon two theoretical descriptions have been proposed. Both models are based on the minimization of the bending energy, with an assumed asymmetry between the two layers of the membrane. This asymmetry has been expressed by a spontaneous curvature[13–18] or by the assumption of coupled monolayers with no lipid exchange between them [14,18]. Although both models are related by a Legendre transformation they lead to different predictions for the

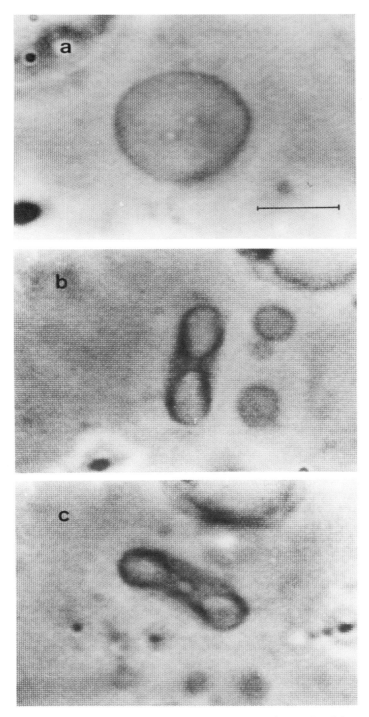

Fig. 4. Three views of a two-holed partially polymerized vesicle, (a) top view, (b) side view through a line perpendicular to the line joining the two holes, (c) side view through a line passing through the two holes. The bar indicates 10μm.

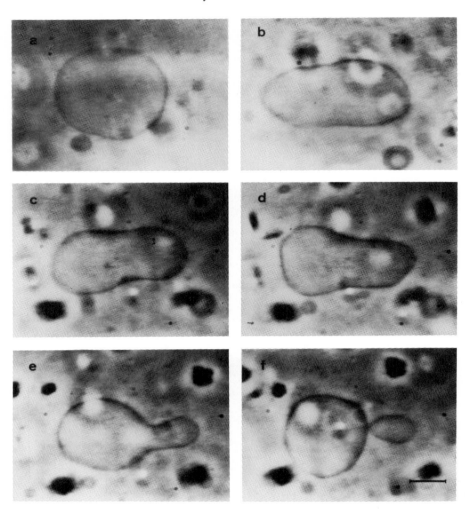

Fig. 5. Budding in a partially polymerized vesicle. Starting from a sphere (a) at T= 50°C the vesicle form becomes ellipsoid (b) at T= 40°C. Further cooling results in a dumbbell shape (c) and then in a pear shaped form (d) at T= 20°C. The vesicle forms a bud (e) which finally leads to a double vesicle (f) at T= 15°C. The two vesicles are still connected by a nearly invisible small tether. The bar indicates 10μm.

budding transition. In the bilayer coupling model temperature dependence is introduced by attributing different thermal expansivities to both monolayers. That leads to a continous shape transformation[14,16] while the same shape transformation in the spontaneous curvature model is discontinous[13,15]. The apparent absence of hysteresis may be taken to suggest that the budding obeys the bilayer coupling theory.

4. CONCLUSION

Partially polymerized vesicles are a new model system, which displays interesting morphological shapes and shape transformations. Based on the general assumption of a heterogeneous bilayer membrane imposed by the embedded polymer nets one can explain the observation of non-zero genus topologies such as toroidal shapes, of the folding transition and the budding phenomenon. As heterogenous (e.g. partially polymerized) vesicles represent biologically relevant models for the cellular membrane they will hopefully stimulate further experimental and theoretical work.

ACKNOWLEDGMENTS

We thank Vincent Croquette for his help and cooperation and the DRET and PROCOPE for financial support.

REFERENCES

1. D. S. Johnson, S. Sanghera, M. Pons and D. Chapman, *Biochim. Biophys. Acta* 602, 57 (1980)
2. E. Sackmann, P. Eggl, C. Fahn, H. Bader, H. Ringsdorf and M. Schollmeier, *Ber. Bunsenges. Phys. Chem.* 89, 1198 (1985).
3. H. Ringsdorf, B. Schlarb and J. Venzemer, *Angew. Chem.* 27, 113 (1988).
4. M. Mutz, D. Bensimon and M. J. Brienne, *ENS preprint* (1990).
5. D. R. Nelson and L. Radzihovsky, *preprint* (1990).
6. M. Mutz, D. Bensimon, *Phys. Rev. A* (in press).
7. Ou-Yang Zhong-can, Phys. Rev. A *41, 4517* (1990).
8. T. G. Burke, P. E. Schoen, C. Davis, R. R. Pice and A. Singh, *Chem. Phys. Lipids* 48, 215 (1988).
9. P. Yager, P. E. Schoen, C. Davis, R. R. Pice and A. Singh, *Biophys. J. 48,* 899 (1985).
10. P. Yager, R. R. Price, J. M. Schnur, P. E. Schoen, A. Singh and D. G. Rhodes, *Chem. Phys. Lipids* 46, 171 (1988).
11. U. Seifert, *Proceeding of the International Workshop on Geometry and Interfaces,* to appear in J. Phys. Colloq. France.
12. For a review on Willmore surfaces, see U. Pinkall and I. Sterling, *Math. Intelligencer* 9, 38 (1987).
13. L. Miao, B. Fourcade, M. Rao, M. Wortis and R. K. P. Zia, *preprint* (1990).
14. U. Seifert, K. Berndl and R. Lipowsky, *preprint* 1990.
15. W. Wiese and W. Helfrich, J. Phys. Cond. Matter, in press.
16. K. Berndl, J. Käs, R. Lipowsky, E. Sackmann and U. Seifert, *Europhys. Lett.* 13, 659 (1990).
17. W. Helfrich, *Z. Naturforschung* 28 c, 693 (1973).
18. H. J. Deuling and W. Helfrich, *J. Phys. (Paris)* 37, 1335 (1976).
19. W. Helfrich, *Z. Naturforschung* 29 c, 510 (1974).

UNBINDING TRANSITIONS OF POLYMERS OR MEMBRANES

IN TWO DIMENSIONS

G. Gompper

Sektion Physik der Ludwig-Maximilians-Universität München
8000 München 2, West Germany

1. INTRODUCTION

Membranes can be thought of as $(d-1)$-dimensional objects fluctuating in d-dimensional space, where usually $d = 3$. The fluctuations of membranes are governed by their bending energy, in contrast to the fluctuations of interfaces, which are controlled by surface tension (see, for example, Ref.1). Continuum models for both fluid[2] and crystalline[3] membranes have recently been proposed. Membranes are believed to be crumpled[4] for all temperatures $T > 0$ in spatial dimension $d < 3$, and to have a crumpling transition at finite temperatures for $d > 3$. The analysis of these models is rather difficult, even for free membranes. In studying unbinding transitions, it is often useful to look at sytems in reduced spatial dimension $d = 2$. In this case, membranes become indistingushable from polymers, 1-dimensional objects in d-dimensional space.[5] Some of the results presented here may therefore be studied experimentally for polymers adsorbed on a flat substrate.

For polymers (or membranes) with a large bending modulus near an attractive wall, a solid-on-solid (SOS) approximation where configurations with overhangs are neglected, may be justified over a large temperature range,[6] even if the free polymer (or membrane) is in its crumpled state. In Section II of this paper the unbinding of such semi-flexible polymers will be considered.[7,8] As the polymer unbinds, its distance from the wall increases, and the SOS-approximation eventually breaks down. This crossover occurs when the distance from the wall becomes large compared to the persistence length[9], ξ_p, which is proportional to the bending modulus. The asymtotic critical behavior of the unbinding transition is therefore dominated by the crumpled state.[10,11] This behavior will be discussed in Sections III and IV. In the crumpled state, all monomers of a real polymer interact with each other, either by electrostatic or van der Waals interactions, or simply via the excluded volume interaction. This makes the analysis of the behavior of real polymers again rather difficult. Therefore, I will first consider a model in Section III in which arbitrary configurations are allowed, but in which different parts of the polymer do not interact. This is usually referred to as a Gaussian (or ideal) polymer.[6] Finally, in Section IV, I will discuss briefly the unbinding transitions of a "true" polymer with excluded volume interactions.

Biologically Inspired Physics, Edited by L. Peliti
Plenum Press, New York, 1991

2. UNBINDING TRANSITIONS OF SEMI-FLEXIBLE POLYMERS

In the solid-on-solid approximation, the configurations of the polymer near an attractive wall can be described by a single valued variable $z(x)$, where x and z are coordinates parallel and perpendicular to the boundary. In the continuum limit, the partition function is defined by the path integral

$$Z_\ell(z, \vartheta | z_0, \vartheta_0) = \int Dz \, \exp\{-\int_0^\ell dx [\frac{\kappa}{2}(\frac{d^2 z}{dx^2})^2 + V(z)]\}. \tag{1}$$

At $x = 0$ and $x = \ell$ the position and slope of the membrane are fixed at the values z, ϑ and z_0, ϑ_0, respectively. The path integral implies the Schrödinger-like equation[7,8]

$$[\frac{\partial}{\partial \ell} - \frac{1}{2\kappa}\frac{\partial^2}{\partial \vartheta^2} + \vartheta \frac{\partial}{\partial z} + V(z)] Z_\ell(z, \vartheta | z_0, \vartheta_0) = 0. \tag{2}$$

The polymer interacts with the boundary via the potential $V(z)$. Potentials of the form

$$V(z) = \begin{cases} -wz^{-p}, & z > a_0; \\ -u, & 0 \le z < a_0; \\ \infty, & z < 0, \end{cases} \tag{3}$$

will be considered. By introducing rescaled variables $\tilde{z} = z/a_0$, $\tilde{\vartheta} = (2\kappa/a_0)^{1/3}\vartheta$, $\tilde{\ell} = (2\kappa a_0^2)^{-1/3}\ell$, and $\tilde{u} = (2\kappa a_0^2)^{1/3}u$, $\tilde{w} = (2\kappa)^{1/3}w$, we can write eq.(2) as

$$[\frac{\partial}{\partial \tilde{\ell}} - \frac{\partial^2}{\partial \tilde{\vartheta}^2} + \tilde{\vartheta} \frac{\partial}{\partial \tilde{z}} + \tilde{V}(\tilde{z})] Z_{\tilde{\ell}}(\tilde{\vartheta}, \tilde{z} | \tilde{\vartheta}_0, \tilde{z}_0) = 0, \tag{4}$$

where the potential is now

$$\tilde{V}(\tilde{z}) = \begin{cases} -\tilde{w}\tilde{z}^{-p}, & \tilde{z} > 1; \\ -\tilde{u}, & 0 \le \tilde{z} < 1; \\ \infty, & \tilde{z} < 0. \end{cases} \tag{5}$$

This shows that the four parameters κ, u, a_0, and w can all be absorbed in two variables $\tilde{u}$ and $\tilde{w}$.

Three different scaling regimes[6,12] can be defined by comparing the asymptotic decay z^{-p} of the potential $V(z)$ and the fluctuation-induced repulsion[6,13] $V_{FL} \sim z^{-\tau}$, with $\tau = -2(d-1)/(d-5)$. The conditions $p > \tau$, $p = \tau$, and $p < \tau$, with $\tau = \frac{2}{3}$ for $d = 2$, correspond to the strong, intermediate, and weak-fluctuation regimes, respectively.

Eq.(4) can be solved[8] for a free polymer, with the results $< \vartheta^2 >= 2\tilde{\ell}$ and $< \tilde{z}^2 >= \frac{2}{3}\tilde{\ell}^3$ for $z_0 = \vartheta_0 = 0$. Below we only consider systems with one end of the polymer pinned close to the wall and suppress the z_0 and the ϑ_0 dependence of Z_ℓ. In order to simplify the notation we will also drop the tilde and replace $\tilde{z}$ by z, etc. The results for the free membrane motivate the scaling ansatz[7]

$$Z_\ell(z, \vartheta) = z^\alpha l^{-\psi} g(z\ell^{-3/2}, \vartheta \ell^{-1/2}) \tag{6}$$

for wall potentials V that decay as $z^{-\tau}$ or faster and are *not* strong enough to *bind* the membrane. In Eq.(6), the exponents α and ψ are defined by the condition $g(0,0) =$const.

The partition function (1) has the Markov property

$$Z_{\ell_1 + \ell_2}(z_2, \vartheta_2 | z_0, \vartheta_0) = \int_0^\infty dz_1 \int_{-\infty}^\infty d\vartheta_1 \; Z_{\ell_2}(z_2, \vartheta_2 | z_1, \vartheta_1) Z_{\ell_1}(z_1, \vartheta_1 | z_0, \vartheta_0). \tag{7}$$

It can be used to relate the exponents α and ψ. Expressing the partition function of a polymer of length 2ℓ with both ends near the wall in terms of the partition function of a membrane of length ℓ with one end close to the wall and making use of Eq.(6), we obtain[8]

$$\psi = 2 + 3\alpha. \tag{8}$$

The exponent α can be determined from the differential equation (4). In the large-ℓ limit with z, ϑ fixed, $z > a_0$, we look for solutions

$$g(z\ell^{-3/2}, \vartheta\ell^{-1/2}) \rightarrow H(\vartheta z^{-1/3}). \tag{9}$$

In the intermediate-fluctuation regime (i.e. for potentials with $p = \frac{2}{3}$), Eqs.(4) and (9) lead to the differential equation

$$H'' + \frac{1}{3}y^2 H' - (\alpha y - w)H = 0, \tag{10}$$

where $y = \vartheta z^{-1/3}$. For $\vartheta \ll -z^{1/3}$ or $y \rightarrow -\infty$, $H(y) \rightarrow const \cdot |y|^{3\alpha}$, and the z-dependence in Eq.(6) cancels out. Due to the bending energy (1), configurations with steep *positive* slopes are energetically suppressed near the wall, so that $H(y)$ decays *exponentially* for $y \rightarrow +\infty$, i.e., $\vartheta \gg z^{-1/3}$. These boundary conditions determine the possible values of α. In the special case $w = 0$ considered in Ref.7, H is a linear combination of Kummer's confluent hypergeometric functions, which implies $\alpha = \frac{1}{6} + n$, with $n = 0, \pm 1, \pm 2, \ldots$ It will be argued below that in fact $n = 0$. For $w \neq 0$ no special function seems available, and we have to rely on a numerical solution of (10). The resulting function $\alpha(w)$, for $p = \frac{2}{3}$, is shown in Fig.1. For all potentials with $p > \frac{2}{3}$, $\alpha = \frac{1}{6}$ and $\psi = \frac{5}{2}$, just as for $w = 0$.

We have also studied numerically a discrete restricted solid-on-solid (RSOS) model.[7,8] The partition function satisfies the recurrence relation

$$Z_{\ell+1}(z, \vartheta) = [(1 - 2q)Z_\ell(z - \vartheta, \vartheta) + qZ_\ell(z - \vartheta + 1, \vartheta - 1)$$
$$+ qZ_\ell(z - \vartheta - 1, \vartheta + 1)] \exp(-V(z)). \tag{11}$$

q is the probability of the polymer to bend at the ℓth monomer. In analogy to eq.(3), potentials of the form

$$V(z) = \begin{cases} -U\delta_{1,z} - Wz^{-p} & \text{for } z > 0 \\ \infty & \text{for } z \leq 0 \end{cases} \tag{12}$$

have been considered. Iterating Eq.(11) with the initial condition $Z_{\ell=1}(1,0) = 1$ we have determined ψ from the ℓ-dependence of $Z_\ell(1,0)$, for $U \equiv 0$. The results for α, calculated using Eq.(8), are shown in Fig.1. The value of α for $w = 0$ corresponds to the $n = 0$ branch of solutions (10). According to Fig.1 the potential parameters of the continuous and discrete models for $q = 0.25$ are related by $w \approx 1.6W$.

The effect of the short-range part of the potential can be easily inferred from the necklace model for wetting.[14,15] As U is varied, there is a unbinding transition

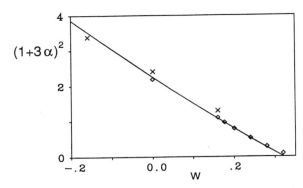

Fig. 1.   Dependence of the exponents α and β, defined in eqs.(6) and (14), with $1 + 3\alpha = -(1 + 3\beta)$, on the potential parameter w. The full line shows the solution to the differential equations (10) and (15). The triangles and circles indicate values of α and of β, respectively, for the RSOS model with $q = 0.25$ and $w = 1.6W$.

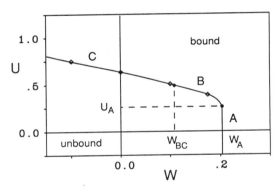

Fig. 2.   Phase diagram for the RSOS model (13) with the potential (2). The phase boundary between the bound and the unbound states has three distinct sub-regimes A, B, and C. The value of U_A is a rough extrapolation.

at a critical value $U = U_c(W)$. For $1 < \psi < 2$, the transition is continuous, with critical exponents determined by ψ. (For example, the exponent $\nu_\|$ of the longitudinal correlation length $\xi_\|$ is given by $\nu_\| = (\psi - 1)^{-1}$.) For $\psi > 2$ the transition is first order (with $\nu_\| = 1$). The results presented in Fig.1 imply the three subregimes[16] A, B, and C of unbinding transitions indicated in Fig.2.

For $w > w_A \approx 0.32114$ or $W > W_A \approx 0.20$, the tail of the potential is strong enough to bind the membrane, and the short-range part of V is irrelevant. As $w - w_A \to 0^+$ with $-\infty < u < u_A$ (subregime A), one expects an essential singularity in the correlation length, in analogy with wetting.[16-18] In the subregime B of continuous transitions, corresponding to $w_{BC} < w < w_A$, with $w_{BC} \approx 0.170245$ or $W_{BC} \approx 0.105$,

$$\xi_\| \sim [u - u_c(w)]^{-\nu_\|}, \quad \text{with} \quad \nu_\| = (1 + 3\alpha)^{-1}. \tag{13}$$

Finally, in subregime C, corresponding to $w < w_{BC}$, the transition is first-order.

Let us take a closer look at subregime C. Since the transition is first order, there is a zero-energy bound state right at the transition.[17] Inserting the scaling ansatz

$$Z_\ell(z, \vartheta) = z^\beta \Gamma(\vartheta z^{-1/3}) \tag{14}$$

for $\ell \to \infty$ into (4), we find

$$\Gamma'' + \frac{1}{3} y^2 \Gamma' - (\beta y - w)\Gamma = 0, \tag{15}$$

where $y = \vartheta z^{-1/3}$ for $z > a_0$. This is the same as differential equation (10) for the scattering states. Since the boundary conditions as $y \to \pm\infty$ are also the same, $\beta(w)$ is determined in exactly the same way as $\alpha(w)$. However, a different branch of solutions now applies. The integral in Eq.(7) only exists for $\beta(w) < -\frac{2}{3}$, which excludes the branches $n \geq 0$. Comparison with the RSOS results leads to the identification $n = -1$ or $\beta = -\frac{5}{6}$. Numerically we find that

$$\beta(w) = -\frac{2}{3} - \alpha(w) < -\frac{2}{3}. \tag{16}$$

We have also calculated the shape function Γ for various values of w. The results obtained from the differential eq.(15) and from the RSOS model (11) are found to be in excellent agreement (see Ref.8).

3. UNBINDING TRANSITIONS OF FLEXIBLE GAUSSIAN POLYMERS

We consider now Gaussian (or ideal) polymers, i.e. polymers which are allowed to crumple, but have no interactions beyond nearest neighbors. The configuration of the chain will be descibed by $(x(s), z(s))$ and $\vartheta(s)$, where x and z are coordinates parallel and perpendicular to the wall, ϑ is the local angle between the polymer and the wall, and s labels the sequence of monomers. $x(s)$ and $z(s)$ are are related to $\vartheta(s)$ by $dx/ds = \cos(\vartheta)$ and $dz/ds = \sin(\vartheta)$. In the continuum limit, the partition function is then defined by the path integral

$$Z_\ell(\vartheta, z, x | \vartheta_0, z_0, x_0) = \int D\vartheta \, \exp\{-\int_0^\ell ds[\frac{\kappa}{2}(\frac{d\vartheta}{ds})^2 + V(z)]\}. \tag{17}$$

Here, κ is the bending modulus, as before. At $s = 0$ and $s = \ell$ the position and angle of the polymer are fixed at the values x, z, ϑ and x_0, z_0, ϑ_0, respectively. The path integral implies the Schrödinger-like equation[11]

$$[\frac{\partial}{\partial \ell} - \frac{1}{2\kappa} \frac{\partial^2}{\partial \vartheta^2} + \sin(\vartheta) \frac{\partial}{\partial z} + \cos(\vartheta) \frac{\partial}{\partial x} + V(z)] Z_\ell(\vartheta, z, x | \vartheta_0, z_0, x_0) = 0. \tag{18}$$

The potentials have again the form (3). If information about the distribution of x is not required, a similar, but somewhat simpler equation can be derived for $Z_\ell(\vartheta, z | \vartheta_0, z_0)$. On length scales less than the persistence length, only angles $\vartheta \ll 1$ contibute to the partition function. The trigonometric functions can then be expanded, and one recovers the semi-flexible limit (2).

For polymers much longer than the persistence length, the effective Hamiltonian[10]

$$H\{x, z\} = \int_0^\ell ds \left[\frac{1}{2} \Sigma \left(\frac{dx}{ds} \right)^2 + \frac{1}{2} \Sigma \left(\frac{dz}{ds} \right)^2 + V_{eff}(z) \right]. \tag{19}$$

can be used to describe the critical behavior of the polymer. Again three different scaling regimes[12] can be defined by comparing the asymptotic decay z^{-p} of the potential $V(z)$ and the fluctuation-induced repulsion $V_{FL} \sim z^{-\tau}$, but now $\tau = -2(d-1)/(d-3)$. The effective Hamiltonian (19) is obtained by integrating out fluctuations on length scales shorter than the persistence length $\xi_p = \kappa$. For a *free* polymer this yields the effective tension $\Sigma = \frac{\Sigma_0}{\kappa}$, with a constant $\Sigma_0 = O(1)$. However, near a wall this integration also leads from the bare potential V to the renormalized potential V_{eff}, which cannot be obtained easily. Only when *all* length scales are much larger than ξ_p is $V_{eff} = V$. With rescaled variables $\hat{z} = z/a_0$, $\hat{x} = x/a_0$, $\hat{\ell} = \ell/(2\Sigma a_0^2)$, and $\hat{u} = 2\Sigma a_0^2 u$, $\hat{w} = 2\Sigma w$, all parameters can then be absorbed in the two variables $\hat{u}$ and $\hat{w}$.

For a numerical study, we consider a discrete version of the model (17), which is defined by the following recurrence relation for the partition function:

$$\begin{aligned}
Z_{\ell+1}(\vartheta, z, x) = &[(1 - 2q) Z_\ell(\vartheta, z - b\sin(\vartheta), x - b\cos(\vartheta)) \\
&+ q Z_\ell(\vartheta - \Delta\vartheta, z - b\sin(\vartheta), x - b\cos(\vartheta)) \\
&+ q Z_\ell(\vartheta + \Delta\vartheta, z - b\sin(\vartheta), x - b\cos(\vartheta))] \\
&\exp[-V(z)].
\end{aligned} \tag{20}$$

The angle ϑ changes in discrete units $0, \pm\Delta\vartheta$. q is the probability of the polymer to have a bend of angle $\Delta\vartheta$ at the ℓth monomer. To implement the recursion numerically, we take discrete integer variables x, z, and a monomer length b, which is considerable larger than unity. If $(z - b\sin(\vartheta))$ or $(x - b\cos(\vartheta))$ do not coincide with a lattice point, the nearest integer is used.

The free polymer is easily solved for the discrete model by transfer matrix methods. For large ℓ and $(\Delta\vartheta) \ll 1$ (so that sums can be replaced by integrals) we obtain the result

$$< x > = 2\kappa(1 - e^{-\ell/2\kappa}), \quad \text{with} \quad \kappa = \frac{b}{2q(\Delta\vartheta)^2}. \tag{21}$$

We consider first the unbinding problem with short range interactions, i.e. $w = 0$ in eq.(3). As u approaches a critical value, u_c, critical unbinding is observed with $< z > \sim (u - u_c)^{-\nu_\perp}$, with $\nu_\perp = 1$. This is exactly the behavior expected from (10). However, for the critical potential strength u_c we find three different scaling regimes, see Fig.3, depending on the relative size of the length scales $\xi_p = \kappa$, a_0 and the effective step size perpendicular to the wall, $b_\perp = q^{-1}b\Delta\vartheta$: (i) the stiff-rod regime, where $a_0 \ll \kappa$ and $b_\perp \gg a_0$, (ii) the semi-flexible regime, where $1 \ll a_0 \ll \kappa$ and

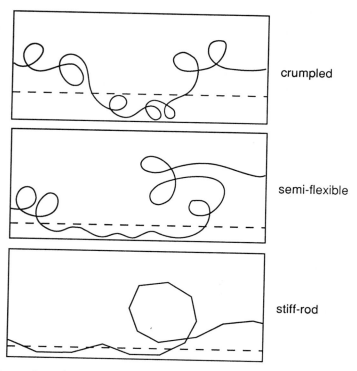

Fig. 3. Schematic polymer configuration near the wall. The dashed line indicates the range of the attractive potential.

$b_\perp \ll a_0$, and (iii) the crumpled regime, where all length scales are large compared to the persistence length.

The behavior in the *stiff-rod regime* can be calculated by the necklace approach of Fisher and Huse[14,15]. For $a_0 = 1$, this approach is exact, and yields[11]

$$\kappa u_c = \frac{b}{(\Delta\vartheta)^2} \quad \text{for} \quad a_0 = 1, \tag{22}$$

where (21) has been used. In the *semi-flexible regime*, the polymer fluctuates *inside* the potential well, but is still essentially parallel to the wall. Therefore, the free energy of the bound state is given by the free energy (per monomer), $f \sim a_0^{-2/3}$, of a semi-flexible polymer between two hard walls with separation a_0.[7] This implies

$$\kappa u_c = const.(\frac{\kappa}{a_0})^{2/3}. \tag{23}$$

Finally, in the *crumpled regime*, the critical strength is given by $u_c \sim (\Sigma a_0^2)^{-1} \sim \kappa/a_0^2$.

For the continuum model, only regimes (ii) and (iii) exist. Therefore, in the continuum limit, u_c has the scaling form[11]

$$u_c = \frac{1}{\kappa}\Omega(\kappa/a_0), \tag{24}$$

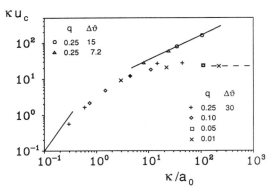

Fig. 4.   Scaled critical potential strength κu_c as a function of the scaled inverse
width $(a_0/\kappa)^{-1}$ of the short-range potential. The scaling function Ω in the
continuum limit (24) is given by the upper envelope of the data points. Its
asymptotic behavior is indicated by the solid lines. The dashed line shows
the stiff-rod result (22) for $\Delta\vartheta = 30°$. The monomer length used is $b = 6$ for
$\Delta\vartheta = 30°$ and $b = 8$ otherwise.

where $\Omega(y) \sim y^{2/3}$ for $y \to \infty$, and $\Omega(y) \sim y^2$ for $y \to 0$. The results obtained from
the discrete model for various values of p, a_0 and $\Delta\vartheta$ are shown in Fig.4. The results
agree very well with (22), (24).

Finally, we want to consider the unbinding transition in the case of long-range
potentials with $p = p_c = 2$. The phase diagram for the crumpled regime has been
calculated before.[16] It has the same topology as the phase diagram of the semi-flexible
polymer. But short range potentials ($w = 0$) are now in subregime B.

The behavior in the regimes (i) and (ii) is more complex, however. In this case,
there is a minimal value, $\psi^* = \psi(w^*)$, of ψ, which can be reached by varying w, with
$3/2 > \psi^* > 1$. For $w > w^*$, the polymer is bound. This implies in particular that
there is no subregime A in this case! In order to investigate this point further, we
introduce a slightly generalized potential

$$
V(z) = \begin{cases}
-w(z + z_0)^{-p}, & z > a_0; \\
-u, & 0 \le z < a_0; \\
\infty, & z < 0.
\end{cases}
\tag{25}
$$

The phase diagram for $a_0 = 1$ and $z_0 = 10$ is shown in Fig.5. For $w/\kappa > 0.91$, the
polymer is always bound, although ψ (calculated with a sufficiently large value of
z_0 in (25)) reaches its 'critical' value $\psi = 1$ only at $w/\kappa = 1.48$. Subregime A is
recovered as z_0 (or a_0) becomes large compared to κ. This is demonstrated in Fig.6
for the case $a_0 = 0$ (i.e. a potential without a square well part). w^* increases with
increasing z_0, and approaches w_A at $w/\kappa \gtrsim 1$. There should be a *finite* value of z_0,
above which subregime A is present in the phase diagram.

It is numerically rather difficult to obtain precise values for $\psi(w)$, but the results

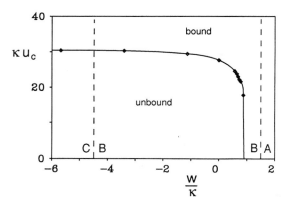

Fig. 5. Phase diagram for the discrete model (20) with the potential (25), with $b = 6$, $\Delta\vartheta = 30°$, $q = 0.25$, $a_0 = 1$, and $z_0 = 10$. The dashed lines indicate the boundaries of between the subregimes A and B, and between B and C, respectively. The solid line is a guide to the eye.

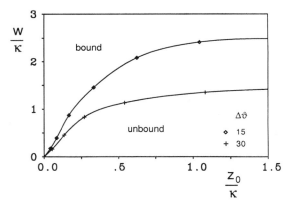

Fig. 6. Phase diagram for the discrete model (20) for $a_0 = 0$ (no square well part). The solid lines are a guide to the eye. The monomer length used is $b = 5$ for $\Delta\vartheta = 30°$ and $b = 8$ for $\Delta\vartheta = 15°$. In both cases $q = 0.25$.

are consistent with $\psi = 1 + \sqrt{\frac{1}{4} - \hat{w}}$, where $\hat{w} = 2\Sigma_0 w/\kappa$. For $b = 6$, $\Delta\vartheta = \pi/6$ and $p = 0.25$, we obtain $w_A/\kappa \approx 1.48$, $w_{BC}/\kappa \approx -4.55$, and $\Sigma_0 = 0.082$. This value of Σ_0 is *not* the same as the value which appears in other properties, like the end-to-end distance of a free polymer.

4. UNBINDING TRANSITIONS OF SELF-AVOIDING POLYMERS

A "true" polymer has interactions between *all* different monomers. If we consider polymers *without* electrostatic or van der Waals interactions for simplicity, there is still the excluded volume interaction, which makes different parts of the polymer repel each other. For such a "true" polymer, subject to a z^{-p} potential, the strong, intermediate, and weak-fluctuation regimes correspond to $p > p_c$, $p = p_c$, and $p < p_c$, where in $d = 2$, $p_c = \frac{4}{3}$. This follows from an independent-blob picture[10] and also from the equivalence with the $O(n)$ model of magnetism[19] in the limit $n \to 0$ and the result $p_c = \nu^{-1}$ for magnetic systems with inhomogeneous coupling constants.[20]

For $p > p_c$ the polymer unbinding transition corresponds to the "special" transition[19,21] of the magnetic system. Therefore, the average distance $< z >$ of the polymer from the wall, and the average surface density, ρ_1 (as well as other quantities) obey scaling laws:[19]

$$< z > = \ell^\nu R_\perp(c\ell^\varphi)$$
$$\rho_1 = \ell^{\varphi-1}\Lambda(c\ell^\varphi) \tag{26}$$

where $c = (u_c - u)$, and $\nu = 3/4$ and $\varphi = 1/2$.[21] In the limit $\ell \to \infty$, one finds

$$\rho_1 \sim \begin{cases} |c|^{1/\varphi-1} & \text{for } c < 0 \\ \ell^{\varphi-1} & \text{for } c = 0 \\ (c\ell)^{-1} & \text{for } c > 0 \end{cases} \tag{27}$$

and

$$< z > \sim \begin{cases} |c|^{-\nu/\varphi} & \text{for } c < 0 \\ \ell^\nu & \text{for } c \geq 0. \end{cases} \tag{28}$$

At $p = p_c$ the surface magnetic exponents are known to be nonuniversal[20], but the magnetic analog of the polymer unbinding transition has not yet been studied in detail. However, we expect the scaling forms (26-28) to hold in the whole subregime B (under the assumption that the topology of the phase diagram does not change). Also, the exponent ν should be unchanged, because it reflects the behavior of the free polymer. It is the exponent φ which becomes non-universal, $\varphi = \varphi(w)$, with $\varphi(0) = 1/2$. As subregime A is approached, $\varphi(w) \to 0$, whereas $\varphi(w) \to 1$ as w tends towards the boundary to subregime C.

We have performed Monte Carlo simulations of self-avoiding polymers modeled by chains of N hard discs of radius r linked together by loose tethers of maximum extension $\ell_0 < 4r$ between centers of adjacent discs. Both ends are pinned at $z = a_0$, but their distance parallel to the wall is allowed to vary. We restrict our interest to the case of small bending energy. Since these simulations are very time consuming, we can report here only some preliminary results. First of all, we have to get an idea about the w-intervall, which corresponds to subregime B, where we can expect the scaling (26-28) to hold. This is complicated by the fact that if the polymer is bound independent of the strength of the square well potential, this may not imply $w > w_A$,

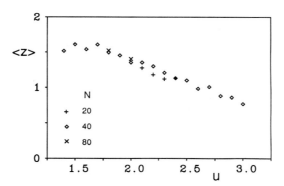

Fig. 7.   Average distance $< z >$ of a self-avoiding polymer with N hard discs as a function of the square well potential with u and $a_0 = 0.1$. The potential tail has the parameters $w = 0.2$ and $z_0 = 0$.

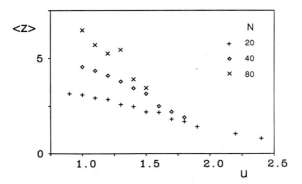

Fig. 8.   Average distance $< z >$ of a self-avoiding polymer with N hard discs as a function of the square well potential with u and $a_0 = 0.5$. The potential tail has the parameters $w = 1.0$ and $z_0 = 2.0$.

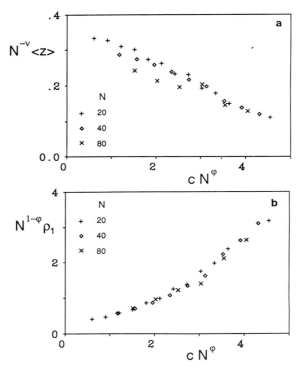

Fig. 9. Scaling plot for the potential of Fig.8 of (a) the average distance $<z>$ and (b) the surface density ρ_1 vs. the scaling argument cN^φ, with $u_c = 0.7$ and $\varphi = 0.37$

as observed for Gaussian polymers. We find indeed that we can obtain values of φ which are significantly different from $\varphi(0) = 1/2$ only for $a_0 \gg r$ or $z_0 \gg r$.

Fig.7 shows $<z>$ as a function of u for various chain lengths N, for $w = 0.2$, $a_0 = 0.1$ and $z_0 = 0$. Since the data points for different N all fall onto the same curve, the polymer is clearly bound for all u. However, for the much larger amplitude $w = 1.0$ of the tail, with $a_0 = 0.5$ and $z_0 = 2.0$, the polymer has an unbinding transition as u varies (see Fig.8). A scaling plot of $<z>$ and ρ_1, shown in Fig.9, reveals that $\varphi \approx 0.37 \pm 0.05$ in this case, still deep in regime B. Nevertheless the data clearly show that $\varphi(w = 1.0)$ differs from $\varphi(w = 0) = 1/2$. This proves that the rich non-universal behavior observed in the intermediate fluctuation regime in other cases prevails if self-avoidance is taken into account.

ACKNOWLEDGMENTS

I thank T. Burkhardt and U. Seifert for enjoyable collaboration.

REFERENCES

1. J. Meunier, D. Langevin, and N. Boccara, *Physics of Amphiphilic Layers,* Springer Proceedings in Physics, Vol.**21**;Springer, Berlin (1987).
2. W. Helfrich, Elastic properties of lipid bilayers: Theory and possible experiments, *Z. Naturforsch.* 28c:693 (1973).
3. D.R. Nelson and L. Peliti, Fluctuation in membranes with crystalline and hexatic order, *J. Phys. (Paris)* 48:1085 (1987).
4. L. Peliti and S. Leibler, Effects of thermal fluctuations on systems with small surface tension, *Phys. Rev. Lett.* 54:1690 (1985).
5. P.-G. deGennes, *Scaling concepts in polymer physics,* Cornell University Press, New York (1979).
6. R. Lipowsky and S. Leibler, Unbinding transitions of interacting membranes, *Phys. Rev. Lett.* 56:2541 (1986); Erratum 59:1983 (1987).
7. A.C. Maggs, D.A. Huse, and S. Leibler, Unbinding transitions of semi-flexible polymers, *Europhys. Lett.* 8:615 (1989).
8. G. Gompper and T.W. Burkhardt, Unbinding transitions of semi-flexible membranes in (1+1) dimensions, *Phys. Rev. A* 40:6124 (1989).
9. P.-G. de Gennes and C. Taupin, Microemulsions and the flexibility of oil/water interfaces, *J. Phys. Chem.* 86:2294 (1982).
10. R. Lipowsky and A. Baumgärtner, Adsorption transitions of polymers and crumpled membranes, *Phys. Rev. A* 40:2078 (1989).
11. G. Gompper and U. Seifert, Unbinding transitions of flexible Gaussian polymers in two dimensions, *(preprint).*
12. R. Lipowsky and M.E. Fisher, Scaling regimes and functional renormalization for wetting transitions, *Phys. Rev. B* 36:2126 (1987).
13. W. Helfrich, Steric interaction of fluid membranes in multilayer systems, *Z. Naturforsch.* 33a:305 (1978).
14. D.A. Huse and M.E. Fisher, Commensurate melting, domain walls, and dislocations, *Phys. Rev. B* 29:239 (1984).
15. M.E. Fisher, Walks, walls, wetting, and melting, *J. Stat. Phys.* 34:667 (1984).
16. R. Lipowsky and T.M. Nieuwenhuizen, Intermediate fluctuation regime for wetting transitions in two dimensions, *J. Phys. A* 21:L89 (1988).
17. R.K.P. Zia, R. Lipowsky, and D.M. Kroll, Quantum unbinding in potentials with $1/r^p$ tails, *Am. J. Phys.* 56:160 (1988).
18. D.M. Kroll and R. Lipowsky, Universality classes for the critical wetting transition in two dimensions, *Phys. Rev. B* 28:5273 (1983).
19. E. Eisenriegler, K. Binder, and K. Kremer, Adsorption of polymer chains at surfaces: Scaling and Monte Carlo analyses, *J. Chem. Phys.* 77:6296 (1982).
20. T.W. Burkhardt, I. Guim, H.J. Hilhorst, and J.M.J. van Leeuwen, Boundary magnetization and spin correlations in inhomogeneous two-dimensional Ising systems, *Phys. Rev. B* 30:1486 (1984).
21. T.W. Burkhardt, E. Eisenriegler, and I. Guim, Conformal theory of energy correlations in the semi-infinite two-dimensional $O(N)$ model, *Nucl. Phys. B* 316:559 (1989).

HEAT-FLUX DRIVEN BIOMEMBRANE TRANSPORT OF MATTER AND CHARGE

F.S. Gaeta, D.G. Mita, E. Ascolese, M.A. Pecorella and P. Russo

Int. Institute of Genetics and Biophysics of C.N.R.
Via G. Marconi, 10 - 80125 Naples

1. RADIATION FORCES ASSOCIATED WITH HEAT PROPAGATION

Heat, flowing through material systems composed of condensed phases produces thermal radiation forces, giving thus rise to various effects, some of them probably relevant to biological membrane transport. We shall use a simple physical argument to prove that the heat flux J_q couples with momentum flux J_p, generating pressure in liquid and liquid-solid interfaces. A theorem due to Boltzmann, applied by Ehrenfest to the adiabatic case[1], shall be our starting point.

Let us consider a system consisting of an isotropic medium made by a great number of particles interacting among them through short-range forces. The system can vibrate at many characteristic frequencies, and, at any temperature above absolute zero, such vibrations spontaneously occur, also in the absence of external excitation. If an amount ΔQ of thermal energy is fed to the system, the Boltzmann theorem states that:

$$\Delta Q = \frac{2}{\tau} \delta \int_0^\tau E_{kyn} \, d\tau = \frac{2}{\tau} \delta \left[\langle E_{kyn} \rangle_\tau \tau \right] , \qquad (1)$$

where $\langle E_{kyn} \rangle_\tau$ is the average kinetic energy associated with vibrations of period τ. Ehrenfest considers the particular case of an adiabatic system, we instead focus our attention on a system at thermal steady-state where the condition $\Delta Q = 0$ follows from the equality of entering and issuing heat fluxes. Such is for instance the case of an isotropic condensed phase - say a cylinder of an homogeneous liquid thermally insulated on the lateral surface; a steady heat flux flowing along its axis (Fig.1), owing to a temperature difference along x. Once thermal effects connected with the initial transient are over, the net exchange of thermal energy of the system with the external world is equal to zero over any time interval, as in the adiabatic case, and from eq. (1) for each value of one then has:

$$\delta \left[\langle E_{kyn} \rangle_\tau \tau \right] = 0 , \quad i.e \quad \langle E_{kyn} \rangle_\tau \tau = \text{constant along x.} \qquad (2)$$

Biologically Inspired Physics, Edited by L. Peliti
Plenum Press, New York, 1991

The difference of course is that here there is a constant rate of entropy production owing to the flux of thermal energy, a circumstance however that does not affect the form of eqs.(1) and (2). Elastic vibrations in the medium satisfy the relation:

$$\langle E_{kyn}\rangle_\tau = \langle E_{pot}\rangle_\tau = \frac{1}{2} U , \tag{3}$$

$\langle E_{pot}\rangle_\tau$ being the average potential and U the total mechanical energy connected to oscillations. From eqs.(2) and (3) then it follows:

$$\frac{1}{\tau} \frac{d\tau}{dx} + \frac{1}{U} \frac{dU}{dx} = 0 , \tag{4}$$

This means that every local change of vibration frequency will be connected with a variation of mechanical energy ΔU. More specifically, the percent variation of the period shall be equal and opposite to that of mechanical energy.

It is well known that thermal energy in the condensed phases mostly consists of very high frequency elastic waves. Part of the spectrum of these phononic thermal excitations has been experimentally investigated in liquids by optical methods. Proceeding from a region of the liquid to another hotter or cooler, the waves undergo frequency changes owing to thermal expansion of the medium; equation (4) allows us to write:

$$dL \equiv Fdx = - dU = U \frac{d\tau}{\tau} = S \left(\frac{U}{S\tau}\right) d\tau, \tag{5}$$

where S is the cylinder cross-section. Thus mechanical work is produced by the system when thermal excitations drift down the temperature gradient. Since we are considering steady-state conditions, every production of work connected with the transient must have disappeared. So the work dL can be due only to energy dissipation in the medium connected with the heat propagation. The temperature gradient affects the elastic properties, which change along x and at the same time it also affects the dynamics of the population of thermal phononic excitations, producing a net drift of the excitations along this same axis. The quantity $(U/S\tau)$ represents mechanical work per unit of surface produced in the medium per period, when heat - in the form of high frequency elastic waves - propagates along x. Obviously $(U/S\tau)$ is proportional to heat flux, i.e. to the equidimensional quantity $J_q = - K \frac{dT}{dx}$, K being the tensor thermal conductivity, reduced here to a simple function of x. We may then write:

$$dL = S \left(\frac{U}{S\tau}\right) d\tau = - S \left(K \frac{dT}{dx} \mathcal{R}\right) \frac{d\tau}{dx} dx , \tag{6}$$

$\mathcal{R}$ being an a-dimensional proportionality constant connecting $(U/S\tau)$ and $K (dT/dx)$ within dx; $\mathcal{R}$ thus is a reflection coefficient due to the non-isothermal condition, i.e. defined by:

$$\mathcal{R} \equiv \left[\frac{d(\rho u)}{\rho u + \rho u + d(\rho u)}\right]^2 , \tag{7}$$

where ρu is the acoustical impedence of the medium. Wave reflection is seen as due to the change of density and sound velocity brought about by temperature change along x. Let's now proceed to consider what occurs when thermal excitations drift from x_1 to x_2 (Fig.1).

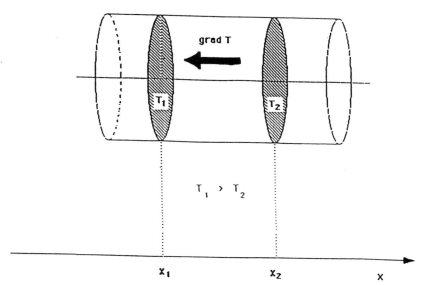

Fig.1   Cylindric portion of an homogeneous liquid thermally insulated on its lateral surface. A temperature gradient is applied along its axis x producing a steady heat flux in the medium.

The work $L_{(x_1,x_2)}$ is given by:

$$L_{(x_1,x_2)} = -S\mathcal{R} \int_{x_1}^{x_2} K \frac{dT}{dx} \frac{d\tau}{dx} dx = -S\mathcal{R} K \frac{dT}{dx} \int_{x_1}^{x_2} \frac{d\tau}{dx} dx , \qquad (8)$$

where $[K(dT/dx)]$, being constant along x can be taken out of the integral. It should be noted that now $\mathcal{R} \equiv [(\rho_2 u_2 - \rho_1 u_1)/(\rho_2 u_2 + \rho_1 u_1)]^2$, if $(\rho u)_{x_1} \equiv \rho_1 u_1$ and $((\rho u)_{x_2} \equiv \rho_2 u_2$. On the other hand, τ is related to the phase velocity by $\tau = 2\pi h_0/u$ where h_0 is the inter-molecular distance in the medium. Thus:

$$\int_{x_1}^{x_2} \frac{d\tau}{dx} dx = 2 \pi h_0 \int_{x_1}^{x_2} \frac{d}{dx} \left(\frac{1}{u}\right) dx = 2 \pi h_0 \left(\frac{1}{u_2} - \frac{1}{u_1}\right). \qquad (9)$$

From (8) and (9) it follows:

$$P_{(x_1,x_2)} \equiv \frac{L_{(x_1,x_2)}}{Sh_0} = -2\pi\mathcal{R}\left(\frac{1}{u_2} - \frac{1}{u_1}\right) K \frac{dT}{dx} = -H\left[\left(\frac{K}{u}\frac{dT}{dx}\right)_{x_2} - \left(\frac{K}{u}\frac{dT}{dx}\right)_{x_1}\right]. \qquad (10)$$

Thus the pressure developed from r_1 to r_2 is proportional to the change of the ratio of heat flux to wave velocity u, i.e. to the variation of momentum flux in the same interval.

2. THERMAL RADIATION FORCES IN BIOLOGICAL SYSTEMS

Transport of matter and electric charge owing to heat flow has been studied in homogeneous non-isothermal solutions[2-4] and thermal radiation forces have been directly measured[5,6]. Most interesting in view of their biological implications however are transport processes occurring in artificial non-isothermal porous membranes[7,11] (effect of thermo-dialysis). Significant analogies observed among the phenomenology of thermodialysis and that of biological membrane transport, induced us to advance the following hypothesis. Cell membranes are generally crossed by fluxes of thermal energy, owing to the different rates of metabolic activity in the two compartments separated by the membrane. Thus, there is a logical basis to ask wether the simultaneous existence of heat and matter transmembrane fluxes involves their mutual interaction. More specifically one is induced to explore the possibility that the heat flux provides the driving force for energy-requiring forms of biological membrane transport. In order to check this hypothesis, experiments were performed with cells of the marine alga *Valonia utricularis,* a convenient organism for these studies, in view of its large size and great sturdiness, that made it a favourite material for electrophysiology. Figure 2 shows the cell's inner structures relevant to our discussion.

Four independent experimental approaches have been attempted todate, each of which yielded evidence consistent with our hypothesis.

Our **first experimental approach** consisted in the determination of the rate of water transport across the cell wall and the two membranes of *Valonia,* the plasmalemma and the tonoplast. This was done by using tritium oxide as a label of external (or internal) medium. The rate of tracer exchange was studied in influx and efflux experiments, using both living and heat killed cells. In the latter, biomembranes had been destroyed by high temperature, transport occurring by diffusion through the unselective, highly porous cell wall. From the observed fluxes (see Figure 3) activation energies for water transport in living and dead cells

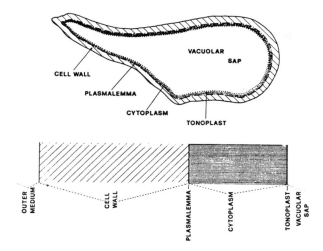

Fig.2   Section through one cell, showing the structures relevant to our investigation.

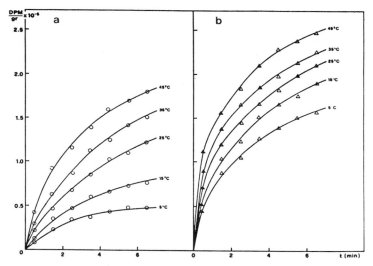

Fig.3   Incorporation of ^3H_2O in DPM/gr (disintegrations per minute per gram of vacuolar fluid) as a function of time in living (a) and heat-killed cells (b) at various temperatures.

can be calculated by means of Arrhenius plots. In dead cells water transport occurs by diffusion (activation energy E_a - 4,3 Kcal/mole); E_a - 6,9 Kcal/mole in living *Valoniae*, a value characteristic of water transport by thermodyalisis[14]. A discussion of these experiments has been given elsewhere[15-16]. Conclusions analogous to those described for water exchange were also reached for transport of sulphate ions[16].

Our **second approach**[17] consisted in the study of the concentration-dependence of the uptake rate of a single permeant, motivated by the fact that its behaviour can be indicative of the mechanism of transport actually exploited. The rate of sulphate influx as a function of concentration in the outer medium was determined in *Valonia* employing [35]S. The rationale for this rests on the circumstance that significant sulphate influxes are observed to occur in various algal cells and in bacteria, and in some species constitute a form of active transport,

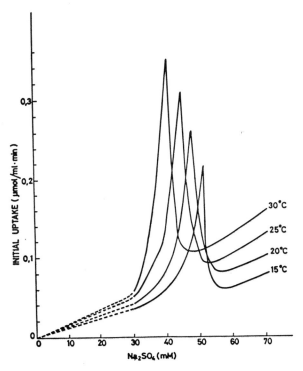

Fig.4   Initial uptake rates of sulphate ion as a function of its concentration in the outer medium. Curve parameter is the temperature of the external medium.

the ions moving against an electrochemical potential gradient[17-18]. In the case of *Salmonella thyphimurium,* a protein component of a system responsible for sulphate uptake has actually been isolated[19]. In Figure 4 the initial uptake rates as a function of sulphate concentration in the outer medium are reported. The pattern found by us is closely reminescent of the concentration-dependence of electrolyte transport by thermodialysis. The sharp peaks superimposed on an otherwise linear concentration-dependence are a well-established characteristic of non-isothermal electrolyte transport, owing to solute-solvent interactions[20]. Thus sulphate uptake in Valonia utricularis is an heat-flux mediated process, and the site of uphill pumping of the permeant can be shown to be at the plasmalemma.

The **third approach**[21] is a direct test of our working hypothesis; it consists in artificially inducing heat fluxes outward or inward-bound across the cell membrane and determining their effects on matter transport. Transmembrane heat fluxes are produced by heating or cooling the cells with their medium at predetermined rates. If V and A are, as usual, cell volume and surface, ρ and C_p the (average) density and specific heat of cellular fluids, and dt is the length of the temperature shift, then the artificial heat flux $J_q(art)$ will be given by:

$$J_q(art) = \frac{V}{A} \rho C_p \frac{dT}{dt} .$$

(11)

Assuming spherical geometry, calling K the thermal conductivity of a layer of radius r and thickness δ, the temperature gradient $\Delta T/\delta$ induced through it will be given by:

$$\frac{\Delta T}{\delta} = \frac{\left(\frac{r}{3} \rho C_p \frac{dT}{dt}\right)}{K} ,$$

(12)

the gradient being directed normal to the cell surface.
At constant temperature, the cell membrane will be crossed by the flux of thermal energy, $J_q(met)$, produced by metabolism inside the cell. During the temperature shifts the artificially produced heat fluxes, $J_q(art)$, are vectorially superimposed to $J_q(met)$, so that the resultant flux, $J_q(res)$, through the cell will be given by:

$$J_q(res) = J_q(met) \pm J_q(art)$$

(13)

During the temperature shifts, increase or decrease of the rate of permeant transport, respectively above and below the values corresponding to the highest and lowest temperatures should occur according to our hypothesis. Vice versa, matter fluxes should only exhibit temperature-dependence and be independent of the rate of temperature change if sulphate is transported enzymatically. In Figure 5, the results of experiments of sulphate uptake in *Valonia* in the presence of temperature shifts are graphically reported. As one can see, the points obtained in runs during which the temperature changes fall well outside the curves relative to influx at the respective initial temperatures. Remembering that $\Delta T/dt$ is proportional to the temperature gradient artificially induced across the plasmalemma one expects that changing the rate of heating (cooling) the effects on transport should be quantitatively modified in accordance; this is indeed the case, in full agreement with theoretical expectations[21].

The **fourth approach** was aimed to ascertain wether transmembrane temperature gradients couple with transport of electric charge in living cells of *Valonia utricularis*. The method is electrophysiological, its rationale being the following: since the temperature gradient modulates transmembrane transport of SO^2_{-4} an ionic permeant, the difference of electric potential should be also affected, membrane potential being directly correlated to ionic fluxes. In these experiments we used aplanospores, i.e. very young cells almost completely constituted of cytoplasm and endowed with a thin cell wall. Aplanospores were obtained by cutting a mature cell and leaving it in sea water. After 5h, some small spherical new cells,

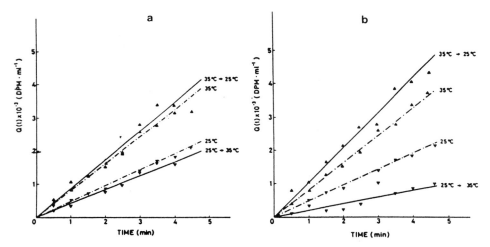

Fig.5   Time-dependence of internal ^{35}S activity in cells at constant temperatures $T_1 = 25^oC$ (∇); and during heating (∇) and cooling (Δ). The rate of temperature change was ($\pm 0.35^oC/min$) in (a) and $\pm 1^oC/min$ (b).

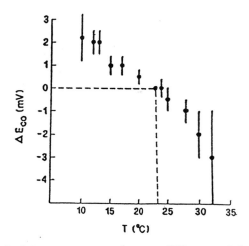

Fig.6   Constant temperature steady-state difference of electric potential $\Delta E_{co}=[E_{co}(T)]_{non-isothermal}\ [E_{co}(T)]_{isothermal}$ as a function of T.

196

deprived of vacuole, could be seen under the microscope, having diameters ranging from 70 and 300 μm. By suitable methods[22] the temperature of the cells was varied linearly in time, at rates up to 10°C/min; equation (12) shows that this corresponds to radial temperature gradients of the order of 1°C/cm. Cycles of heating (cooling) produce heat fluxes antagonist (synergic) to metabolic activity.

The difference of electric potential between cytoplasm and outer medium $[E_{co}(T)]_{isothermal}$ (ie measured at constant temperature), in the range between +10°C and +32°C is reported in Figure 6. The values in the figure represent the difference between the resting potential measured at temperature. T, $[E_{co}(T)]_{isothermal}$ and the one measured at a reference temperature chosen as T_r=23°C. Each point of Fig.6 is the average of 10 measurements with as many aplanospores.

The resting potential at 23°C was $[E_{co}(T_r)] = - 42 \pm 0.5$ mV in these measurents, this value being itself the average on 10 distinct cells; the vertical bars give the respective standard deviations. When the temperature is changed during measurement, $[\Delta E_{co}]$ are found distinctly larger than the $[\Delta E_{co}]$ between the corresponding end temperatures. The changes of p.d. with temperature are reversible, after a suitable time lag, so that the excess effects owing to the temperature shifts disappear after a while, and the $[\Delta E_{co}]$ gradually tend to the steady-state difference between initial and final temperatures.

In Fig. 7 the change with time of $[E_{co}(T)]_{non-isothermal}$ refers to a heating phase, an experiment in which the heat flux induced through the cell membrane is antagonist to the one

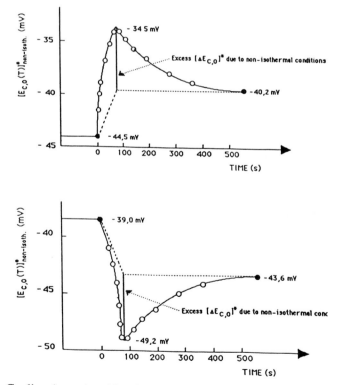

Fig.7   Cooling (upper) and heating (lower) experiments. The behaviour of $[E_{co}(T)]_{non-isothermal}(O)$ in function of time; the full cercles (•) represent $[E_{co}(T)_{isothermal}$ at initial and final temperatures.

produced by metabolism; the effect is reversed in a phase of cooling. From Fig.7 the existence of remarkable effects owing to transmembrane heat flux jumps to the eyes. For a more quantitative appreciation of these effects, the $[\Delta E_{co}]^*$ - that is the excess of variation of p.d. owing to nonisothermal conditions - can be calculated and its dependence on the rate of temperature change experimentally assessed.

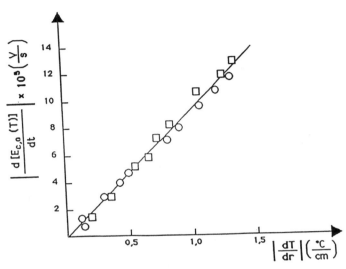

Fig.8   Absolute value of the time rate of membrane potential differences in aplanospores of *Valonia utricularis* as a function of the absolute value of the artificially-induced transmembrane temperature gradient; (O) heating phase, (□) cooling phase.

These results clearly show that a transmembrane flux of thermal energy alters the difference of electric potential across the plasmalemma of aplanospores of *Valonia utricularis*. Furthermore, evidence gathered so far also proves the direct dependence of $[\Delta E_{co}]^*$ on the intensity of the heat flux, as well as the correlation of the sign of $[\Delta E_{co}]^*$ with the sense in which thermal energy crosses the cell membrane. A simple analytical treatment of the data allows to proceed to the measurement of the thermodynamic coupling coefficient connecting charge flux to temperature gradient. Fig.6 exhibits a linear relationship of $[E_{co}(T)]$ to temperature:

$$[E_{co}(T)]_{isothermal} = aT + b, \tag{14}$$

On the other hand, from results such as those described in Fig. 7 the time rates of change of $[E_{co}(T)]_{non\text{-}isothermal}$ can be derived. Values of $\{d[E_{co}(T)]_{non\text{-}isothermal}/dt\}$ have been plotted in Fig.8 as a function of the temperature gradient artificially produced across the plasmalemma. Absolute values of both quantities have been used. A clear-cut linear relationship is evident. The distribution is best approximated by the expression:

$$\left| \frac{d[E_{co}(T)]_{non\text{-}isothermal}}{dt} \right|_{t=0} = a^* \left| \frac{dT}{dr} \right| + b^* , \tag{15}$$

where $a^*=(9.1\pm0.3)\cdot10^{-5}$ (V cm)/(s^{o}C) and $b^*=(0.1\pm0.3)\cdot10^{-5}$ V/s, within the experimental error. On the other hand, calling C the capacitance of the plasmalemma per unit surface one has:

$$I = C \left| \frac{d[E_{co}(T)]_{non\text{-}isothermal}}{dt} \right|_{t=0} . \tag{16}$$

From (15) and (16) follows:

$$I = C\, a^* \left| \frac{dT}{dr} \right| . \tag{17}$$

This equation is the analytical form of the working hypothesis, namely that a temperature gradient produces an ionic current through biomembranes in vivo.
In terms of the field equations of thermodynamics we have:

$$I_{thermo\text{-}electric} = \sum_{I=1}^{n} \alpha_{iq}\, X_q = \frac{1}{T_{AV}} \frac{dT}{dr} \sum_{I=1}^{n} \alpha_{iq} = \frac{C_a}{F} \frac{dT}{dr} . \tag{18}$$

In our experiments, being $I=75_p$A/cm2, equivalent - for an uni-univalent electrolyte - to about $8\cdot10^{-4}$ pmol/cm^2/s, one may conclude that:

$$\sum_{I=1}^{n} \alpha_{iq} \cong 1.7 \times 10^{-3} \frac{p\ mol}{cm\ s} . \tag{19}$$

This proves that the artificially-induced temperature gradient causes transport of electric charge across the plasmalemma of aplanospores of *Valonia utricularis*.
Some closing remarks are necessary at this point. Thermal radiation forces evidently contribute to matter and charge transport across biomebranes. The thermodynamic efficiency $\eta = \Delta T/T_{av}$ of heat-fluxmediated transport however would be very small if transmembrane temperature difference would be due only to the non-localized heat flow resulting from cytoplasmic metabolic activity. However we do not think that the evidence produced here conflicts with the existing experimental proofs of enzymatic involvment in biological membrane transport. There is conflict only with the literal ferryboat carrier hypothesis arbitrarily constructed on this evidence. Since many enzymatic reactions are exothermal, an active site can constitute a point source of thermal energy. If the enzyme structure consists of an anastomizing cavity or channel crossing the entire membrane thickness, with the reaction site at one channel end, then the heat flux due to substrate-enzyme interaction will produce transmembrane transport by thermodialysis. The flux of thermal energy from the point source at the mounth of the channel would then be concentrated within the enclosed aqueous

medium, owing to the higher thermal conductivity of this one relative to the surrounding lipoproteic moiety. The local temperature gradient produced by an enzymatic reaction such as ATP dephosphorilation occurring at one end of a membrane channel some 20 $(\overset{o}{A})^2$ in cross section, with a turnover number of 300 s^{-1} can be readily calculated. If only one-fourth of the energy known to decay to heat in ATP dephosporilation is assumed to flow through the membrane pore, the heat flux in it would be of about 0.5 mcal/cm^2/s; taking for K the thermal conductivity of water, one gets about 0.5°C for the transmembrane temperature gradient in the pore. This of course leads to an improved thermodynamic efficiency. The authors are acutely aware of the limitations of their present conclusions. Their most optimistic hope is that their efforts will promote further interest in the neglected field of heat-flux-mediated transport in biomembranes.

REFERENCES

1. L. Brillouin, Les Tenseurs en Mécanique et en Elasticité, Masson et Cie Ed., Paris (1949).
2. Ch. Soret, Arch. Sci. Phys. Nat. **4**, 209 (1880).
3. Ch. Soret, Ann. Chim. Phys. **22**, 239 (1881).
4. F.S. Gaeta, D.G. Mita, G. Perna and G. Scala, Nuovo Cimento, **30**, 153 (1975).
5. G. Brescia, E. Grossetti and F.S. Gaeta, Il Nuovo Cimento, Ser. Xl, **8B**, 329 (1972).
6. F.S. Gaeta, G. Scala, G. Brescia and A. Di Chiara, J. Poly. Sci., Poly. Phys. Ed., **13**, 177 (1975).
7. F.S. Gaeta, D.G. Mita, G. Perna, "Process of thermal diffusion across porous partitions and relative apparatuses", Patents: Italy, No.928, 656 (1971); U.K., 23, 590 (1972); France, No.260, 497 (1972).
8. F.S. Gaeta, D.G. Mita, J. Membrane Sci., **3**, 191 (1978).
9. F. Bellucci, E. Drioli, F.S. Gaeta, D.G. Mita, N. Pagliuca, F.G. Summa, J. Chem. Soc. Faraday Trans. **2**, 247 (1979).
10. F.S. Gaeta, D.G. Mita, J. Phys. Chem. **83**, 2276 (1979).
11. F. Bellucci, E. Drioli, F.S. Gaeta, D.G. Mita, N. Pagliuca, D. Tomadacis, J. Membrane Sci., **7**, 169 (1980).
12. D.G. Mita, F. Bellucci, M.G. Cutuli and F.S. Gaeta, J. Phys. Chem. **86,** 2975 (1982).
13. N. Pagliuca, D.G. Mita and F.S. Gaeta:, J. Membrane Sci. **14**, 31 (1983).
14. N. Pagliuca, U. Bencivenga, D.G. Mita, G. Perna, F.S. Gaeta, J. Membrane Sci., **33**, 1 (1987).
15. D.G. Mita, M. Bianco, P. Canciglia, A. D'Acunto, C. Minatore, F.S. Gaeta, Gazz. Chim. Ital., **109**, 481, (1979).
16. F.S. Gaeta, D.G. Mita, P. Canciglia, A. D'Acunto, Physical basis of water transport in *Valonia utricularis* in "Membranes, molecules, toxins and cells", K. Block, L. Bolis, D.C. Tosteson, Eds., P.S.G. Publishing Co., Littleton, MA, pp. 271-295, (1981).
17. D.G. Mita, P. Canciglia, A. D'Acunto, F.S. Gaeta, Mol. Physiol., **5**, 9 (1984).
18. J.B. Robinson, J. Exp. Bot., **20**, 201 (1969).
19. J.B. Robinson, J. Exp. Bot., **20**, 212 (1969).
20. A.B. Pardee, J. Biol; Chem., **241**, 5886 (1966).
21. F.S. Gaeta, G. Perna, G. Scala, F. Bellucci, J. Phys. Chem., **86**, 2967 (1982).
22. F.S. Gaeta, P. Canciglia, A. D'Acunto, J. Membrane Sci., **16,** 339 (1983).
23. D.G. Mita, M. Durante, F.S. Gaeta, A. Cotugno, V. Di Maio, C. Taddei-Ferretti, P. Canciglia, Cell Biophysics **16**. 35 (1990).

A LATTICE MODEL FOR STERIC REPULSIVE INTERACTIONS IN PHOSPHOLIPID SYSTEMS

Enrico Scalas[1] and Andrea C. Levi[2]

1. Dipartimento di Fisica, Università di Genova
   Via Dodecaneso 33, I-16146 GENOVA (Italy)
2. Scuola Internazionale Superiore di Studi Avanzati
   Strada Costiera 11, I-34014 TRIESTE (Italy)

1. THE MODEL

We consider a lattice model for phospholipid monolayers and bilayers. The model takes into account the steric repulsive interactions between two phospholipid molecules in an oversimplified way.

A bilayer is assumed to be made up of two non-interacting monolayers. Each monolayer consists of N sites on a two dimensional lattice. In the present work we use, for the sake of simplicity, a square lattice instead of a more realistic triangular one. Every site is occupied by a phospholipid chain which can be in one of five states. One of these states represents the untilted chain, while the other four states represent the chains tilted in one of the four directions of the square lattice. The five states can be pictorially represented as:

- state 1 (the chain is in the rigid all-trans configuration, perpendicular to the monolayer plane);
- ↑ state 2 (the chain is in the excited configuration, tilted "up");
- ↓ state 3 (the chain is in the excited configuration, tilted "down");
- → state 4 (the chain is in the excited configuration, tilted "to the right");
- ← state 5 (the chain is in the excited configuration, tilted "to the left").

The chains can interact only with their four nearest neighbours. In order to take the steric repulsive interactions into account we consider the two-site configurations of Fig. 1 to be energetically unfavoured.

The hamiltonian of the model is:

$$\mathcal{H} = \sum_{ij}\left(\sum_{\alpha\beta} V_{\alpha\beta} L_{ij,\alpha} L_{ij+1,\beta} + \sum_{\alpha\beta} V'_{\alpha\beta} L_{ij,\alpha} L_{i+1j,\beta}\right) \tag{1}$$

the Greek subscripts in (1) run on the chain states while the Latin subscripts run on the lattice sites, $L_{ij,\alpha}$ is equal to 1 if site ij is in state α, and is 0 otherwise; the

Biologically Inspired Physics, Edited by L. Peliti
Plenum Press, New York, 1991

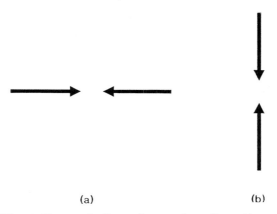

(a)   (b)

Fig. 1. Energetically unfavoured configurations.

interaction matrix V is:

$$V = \begin{pmatrix} -2J & 0 & 0 & 0 & 0 \\ 0 & -J & 0 & 0 & 0 \\ 0 & 0 & -J & 0 & 0 \\ 0 & 0 & 0 & -J & 0 \\ 0 & 0 & 0 & J' & -J \end{pmatrix} \tag{2}$$

and the interaction matrix V' is:

$$V' = \begin{pmatrix} -2J & 0 & 0 & 0 & 0 \\ 0 & -J & 0 & 0 & 0 \\ 0 & J' & -J & 0 & 0 \\ 0 & 0 & 0 & -J & 0 \\ 0 & 0 & 0 & 0 & -J \end{pmatrix}. \tag{3}$$

In (2) and (3) J and J' are positive quantities; periodic boundary conditions are assumed throughout.

As we are interested in the effects of the off-diagonal terms in the interaction matrices, we will consider in the following also a different model with only the four states 2, 3, 4, 5. The interaction matrices of this model are the minors obtained deleting the first row and the first column from (2) and (3). We shall call the five state model A and the four state model B.

2. THE SOLUTION IN ONE DIMENSION

So far we have studied for both model A and model B only the one-dimensional case, which is quite trivial.

Let us first consider model B. The transfer matrix in the one-dimensional case is:

$$T_B = \begin{pmatrix} e^K & 1 & 1 & 1 \\ 1 & e^K & 1 & 1 \\ 1 & 1 & e^K & 1 \\ 1 & 1 & e^{-K'} & e^K \end{pmatrix} \tag{4}$$

where $K = J/kT$ and $K' = J'/kT$.

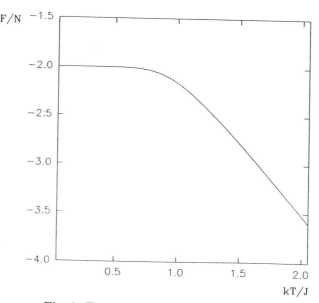

Fig. 2. Free energy per site vs. kT/J.

The largest eigenvalue of the transfer matrix is:

$$\lambda_1 = \exp(-K'/2)[\exp(K'/2) + \exp(K'/2 + K) + \sqrt{1 + 3\exp(K')}] \tag{5}$$

From λ_1 we can calculate the free energy per site and from the free energy we get the other thermodynamic quantities like the entropy and the specific heat. The free energy per site is:

$$F/N = -kT \log(\lambda_1) \tag{6}$$

It is interesting to observe that when $\exp(-K') = 0$ (that is $J' = \infty$) there is little qualitative difference between the solution in one-dimension of the four-state Potts model and that of model B. In fact the greatest eigenvalue of the transfer matrix of the four-state Potts model is:

$$\lambda_p = \exp(K) + 3 \tag{7}$$

while for the model B with infinite repulsion it is:

$$\lambda_\infty = \exp(K) + 1 + \sqrt{3}. \tag{8}$$

The calculation of the greatest eigenvalue in the case of model A is not so easy. In fact the algebraic equations of fifth degree cannot generally be solved by radicals. In the present case the solution by radicals appears to be accidentally possible but extremely complicated and hardly illuminating. Therefore the eigenvalues of the transfer matrix of model A were computed numerically, to obtain the thermodynamic quantities as functions of the temperature.

We present in Fig. 2 the free energy calculated in this way, in Fig. 3 the entropy and in Fig. 4 the specific heat, in the case $J = J'$.

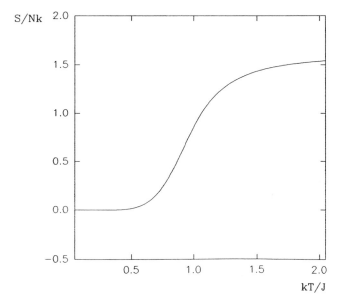

Fig. 3. Adimensional entropy per site vs. kT/J.

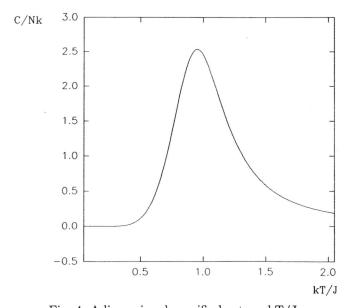

Fig. 4. Adimensional specific heat vs. kT/J.

3. OUTLOOK

We are now studying the two-dimensional versions of models A and B. At $T = 0$ model A is trivial (all chains are in the untilted configuration). For model B, on the other hand, the symmetry is broken, the four ordered states being equivalent. It should be noted, however, that, due to the presence of the J' interaction, at finite temperature islands of "up" and "down" types tend to avoid borders of type b) in Fig. 1 by interposing an island of "right" or "left" type.

ACKNOWLEDGMENTS

It is a pleasure for the authors to acknowledge illuminating discussion with Prof. Sandra Gliozzi. One of the authors (E.S.) was put in conditions to perform this work by a contract of collaboration with INFM (Consorzio Interuniversitario di Fisica della Materia).

DYNEIN-MICROTUBULE INTERACTIONS

Nicolas J. Cordova[1], Ronald D. Vale[2], George F. Oster[3]

1. Department of Applied Mathematics, Weizmann Institute of Science, Rehovot, Israel

2. Department of Pharmacology, University of California School of Medicine, San Francisco, CA 94143

3. Departments of Molecular and Cell Biology, and Entomology University of California, Berkeley, CA 94720

1. INTRODUCTION

Vale and his colleagues (1989) devised a method of attaching single dynein molecules to a substrate and observing the motion of microtubules. In one series of experiments they inhibited the enzymatic activity of the dynein with vanadate. By fitting a Gaussian to this data, they obtained the diffusion constants shown in Table 1. However, according to the theoretical formulas for longitudinal and transverse diffusion of a cylinder these diffusion constants should be much larger. (The longitudinal and transverse diffusion coefficients are given by the following expressions: $D_L = kT\ln(2h/r)/2\pi\eta L$ and $D_T = kT\ln(2h/r)/4\pi\eta L$ (Brennen and Winet, 1977)). This discrepancy suggests two models for the interaction.

TABLE 1 [From Vale, et al., 1989]

$D \times 10^{10}$ [cm2/sec]	MEASURED	COMPUTED
Longitudinal	9.1	67
Transverse	0.08	33

The groove hypothesis: the dynein head rides in an electrostatic groove on the microtubule, so that lateral movement is sterically prevented, while longitudinal motion is hindered by a frictional drag.

Biologically Inspired Physics, Edited by L. Peliti
Plenum Press, New York, 1991

The equilibrium binding hypothesis: dynein heads are retained close to the microtubule by rapid cycles of association and dissociation.

In this note we will construct a simple model for the diffusion of a microtubule on one or a few dynein molecules and show that the second model is correct. Moreover, from the model we will derive an expression for the true diffusion coefficient and the equilibrium binding constant, as well as an assay for the number of dynein molecules attached to the microtubule.

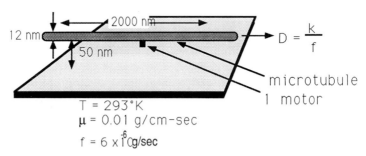

Fig. 1. Diagram of the microtubule gliding assay. Dynein molecules are attached to a coverslip and a microtubule is tracked moving parallel to its long axis. The approximate dimensions for a typical experiment are given in the figure, where f is the friction factor, T the temperature, and μ the fluid viscosity. The analytical expression for the longitudinal diffusion coefficient is $D = kT/f$.

Figure 1 shows a scheme of the experimental setup that was used to track the motion of a single microtubule attached to a single dynein molecule. The dynein arm attaches itself to the microtubule via electrostatic attraction. This was confirmed by noting that D increased when the ionic strength of the medium was increased, which decreases the Debye length, thus attenuating the electrostatic binding, as shown in Figure (2a). Figure (2b) shows a typical histogram of the longitudinal displacement increments for a 2.3μm microtubule, to which has been fit a Gaussian curve, and Figure (2c) shows a plot of the mean squared displacement vs time. The latter plot is quite linear, and if this were a pure diffusion process, the displacement would be given simply by $\langle x^2 \rangle = 2D_L t$, so that the longitudinal diffusion coefficient is given by the slope, $2D_L$.

However, a closer examination of Figure (2b) reveals that the Gaussian fit does not account for the plethora of small steps. This feature of the data allows us to exclude the groove model for the following reason. If the dynein were simply riding in a groove along the microtubule, this would be equivalent to an additional friction

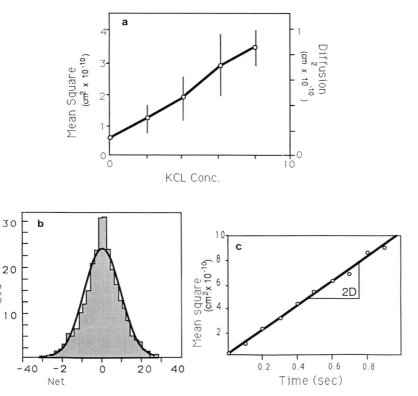

Fig. 2. (a) The mean square longitudinal displacement—and consequently the diffusion coefficient—of a diffusing tubule increases as a function of the salt concentration. Thus the interaction between the dynein head and the tubule is electrostatic. (b) The histogram represents the net displacement distances measured at 2725 successive 0.1 sec time intervals for a $3\mu m$ microtubule undergoing back and forth motion in the presence of $1mM$ ATP and $200\mu M$ Vanadate. The solid line is the best Gaussian fit to the data omitting the excess of small steps (Vale, et al., 1989). (c) The mean square displacement of a tubule as a function of time. The slope gives the apparant diffusion constant according to $\langle x^2 \rangle = 2D_L t$.

drag on the tubule motion, since the diffusion coefficient is inversely proportional to the friction coefficient, f: ($D = \frac{k_b T}{f}$). Thus the probability distribution for the displacements of the microtubule should be just a Gaussian distribution with a somewhat smaller diffusion coefficient, which cannot account for the excess number of displacement values close to zero. Therefore, we can explore the alternative model: that the dynein arm rapidly attaches and detaches from the microtubule, allowing the tubule to diffuse in small steps while it is detached.

2. A DIFFUSION REACTION MODEL

Let us put our coordinate system on the tubule and describe the random walk of the center of mass by the stochastic process shown in Figure (3b). Here α and β are the probabilities of the dynein jumping to the right and left respectively. For thermally driven motion these probabilities are given by Boltzmann factors: $\alpha, \beta \sim e^{-\epsilon/k_b T}$, where ϵ is the energy barrier for intersite jumps. Let X_N be the displacement of the dynein after N time steps, and $p(x,t) = Prob(X_N = x)$, i.e. the probability that at time $t = N\Delta t$ a free dynein is located at point x. Then $p(x,t)$ can be written as the Markov difference equation

$$p(x, t+\tau) = \alpha p(x-\Delta, t) + \beta p(x+\Delta, t) + k_d p^\star - k_b p. \tag{1}$$

That is, the probability of a free dynein being at location x at time $t + \Delta t$ is the probability that it was at $x - \Delta$ at time t multiplied by the probability that it jumped to the right, plus the probability that it was at $x + \Delta$ at time t multiplied by the probability that it jumped to the left, plus the probability of a bound dynein, $p^\star$, at x coming off the tubule, minus the probability of the free dynein at x binding to the microtubule at x. A Markov model of this sort assumes that the dynein spends most of its time bound to the tubule, and occasionally makes an instantaneous jump of distance Δ to an adjacent site. To be more realistic, let us allow the possibility of finite speed motions by making space and time continuous. We do this in the obvious way by expanding $p(x,t)$ in a Taylor series about x and t:

$$p(x \pm \Delta, t) = p(x,t) \pm \Delta \frac{\partial p(x,t)}{\partial x} + \dots \tag{2}$$

$$p(x, t+\tau) = p(x, \tau) + \tau \frac{\partial p(x,t)}{\partial t} + \dots \tag{3}$$

Substituting this into the jump equation above yields the pair of diffusion reaction equations

$$\frac{\partial p}{\partial t} = D \frac{\partial^2 p}{\partial x^2} - v \frac{\partial p}{\partial x} + (k_d p^\star - k_a p), \tag{4}$$

$$\frac{\partial p^\star}{\partial t} = -(k_d p^\star - k_a p), \tag{5}$$

where the diffusion coefficient D, and the drift velocity v, are given by

$$D = (\alpha + \beta)\frac{\Delta^2}{2\tau} \qquad \text{and} \qquad v = (\alpha - \beta)\frac{\Delta}{\tau}. \tag{6}$$

If the tubule motion is resisted by a force F, then the work associated with each jump is $F\Delta$, and the drift velocity is given by the asymmetric expression

$$v = \frac{\Delta}{\tau}e^{-\epsilon/k_b T}(1 - e^{-F\Delta/k_b T}), \tag{7}$$

where τ is the mean time between jumps. This is just the expression for the Feynman thermal ratchet (Feynman et al, 1963; Vale and Oosawa, 1990). Of course, for the diffusing microtubule there seems to be no molecular mechanism for erecting asymmetric energy barriers so that $\alpha = \beta$ and $v = 0$. Consequently we have neglected the drift term when fitting the vanadate data, and we use $D = k_b T / f$ as the diffusion coefficient of the tubule.

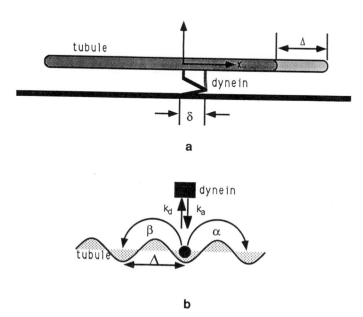

Fig. 3. (a) Microtubule bound to a dynein. (b) The interaction is viewed as a random walk process coupled to a first order reversible adsorption desorption reaction. Here, α and β denote the probability of jumping to the right and to the left, respectively.

The initial value problem describing the microtubule-dynein system is

$$\frac{\partial p}{\partial t} = D \frac{\partial^2 p}{\partial x^2} + (k_d p^\star - k_a p), \qquad p(x, t = 0) = \delta(x), \qquad p^\star(t = 0) = 0, \qquad (8)$$

$$\frac{\partial p^\star}{\partial t} = -(k_d p^\star - k_a p), \qquad \frac{\partial p}{\partial t}(x = L, t) = 0. \qquad (9)$$

Although these equations are linear they cannot be solved analytically, and so we employed a Crank-Nicholson scheme to solve them numerically, and fit the solution to the data using a least square criterion. This assumes that the measurement errors are independent and normally distributed with constant standard deviation. The results are shown in Figure (4b).

In the case where the microtubule interacts with one dynein molecule, for $t \gg k_d^{-1}$, the probability $p(x,t)$ is governed by a diffusion equation, with an instantaneous point source located at $x = 0$ at time $t = 0$. Following Goldstein, et al. (1975), we take the Laplace transform of the reaction equation to obtain

$$s\bar{p}^\star = k_a\bar{p} - k_d\bar{p}^\star. \tag{10}$$

where $\bar{p}$ and $\bar{p}^\star$ are the Laplace transform of p and $p^\star$ respectively. This equation is equivalent to

$$\bar{p}^\star = \frac{k_a}{k_d + s}\bar{p}. \tag{11}$$

For $t \gg k_d^{-1}$, we can expand the Laplace transforms in powers of s/k_d. If we keep only first order corrections we obtain

$$\bar{p}^\star = K(1 - (s/k_d) + O(s/k_d)^2)\bar{p}, \tag{12}$$

where $K = k_a/k_d$. This can be inverted to yield

$$p^\star(x,t) = Kp(x,t) - \frac{K}{k_d}\frac{\partial p}{\partial t}(x,t) - \frac{K}{k_d}\delta(t)\delta(x). \tag{13}$$

Substituting back into the equation for $p(x,t)$ we finally obtain

$$\frac{\partial p}{\partial t}(1 + K) = D\frac{\partial^2 p}{\partial x^2} - v\frac{\partial p}{\partial x} - K\delta(t)\delta(x). \tag{14}$$

Now we can compute the expected value $\langle x \rangle$ of the random variable x. Multiplying (14) by x, integrating from $-\infty$ to $+\infty$ and using integration by parts, we find that

$$\frac{d}{dt}\langle x \rangle = v/(1 + K). \tag{15}$$

Since $\langle x \rangle = 0$ at $t = 0$, we obtain

$$\langle x \rangle = vt/(1 + K). \tag{16}$$

In a similar fashion, the second moment of x is found to be $\langle x^2 \rangle = (v/(1 + K))^2 t^2 + Dt/(1 + K)$, so that the variance is given by

$$\langle x^2 \rangle - \langle x \rangle^2 = Dt/(1 + K). \tag{17}$$

Thus for very long times compared to k_d^{-1} the probability distribution of microtubule displacements is described by the effective diffusion coefficient

$$D^\star = D/(1 + K). \tag{18}$$

This formula can be used to determined the equilibrium rate constant of the dynein-microtubule complex. Using $D = 67 \times 10^{-10} cm^2/s$ and $D^\star = 8.5 \times 10^{-10} cm^2/s$ (Vale et al, 1989), we obtain $K = 6.88$.

We can repeat the same exercise in the case where a microtubule interacts with N dynein molecules. We will assume that the binding of the dynein molecules are

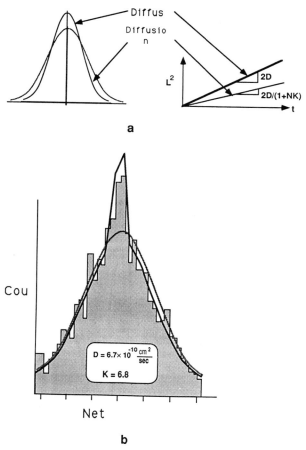

Fig. 4. (a) The diffusion-reaction model gives a non-Gaussian displacement curve that is narrower and more peaked than the corresponding pure diffusion process. A plot of the mean square displacement vs time gives a straight line whose slope is $2D/(1 + NK)$, rather than simply 2D. N is the number of attached motors and K is the binding constant, $K = k_{on}/k_{off}$. Thus the slope of the line is quantized by the number of motors, N, and so can be used as an assay for the number of attached motors. (b) The model fitted to the displacement data accounts for the excess of small displacements, and gives a value for the diffusion coefficient, $D = 6.7 \times 10^{-10}\,\mathrm{cm}^2/\mathrm{sec}$ and $K = \frac{k_{on}}{k_{off}} = 4.74/0.69$.

independent of each other. For simplicity we will assume $N = 2$; the generalization to $N > 2$ is straightforward. The corresponding initial value problem is:

$$\frac{\partial p}{\partial t} = D\frac{\partial p^2}{\partial x^2} - v\frac{\partial p}{\partial x} + k_a p_1^\star - k_a p + k_d p_2^\star - k_a p, \qquad (19)$$

$$\frac{\partial p_1^\star}{\partial t} = k_a p - k_d p_1^\star, \qquad (20)$$

$$\frac{\partial p_2^\star}{\partial t} = k_a p - k_d p_2^\star, \qquad (21)$$

where $p_1^\star$ and $p_2^\star$ are the probability of binding each dynein molecule respectively. For $t \gg k_d^{-1}$ the diffusion equation reduces to

$$\frac{\partial p}{\partial t}(1 + 2K) = D\frac{\partial^2 p}{\partial x^2} - v\frac{\partial p}{\partial x} - 2K\delta(x)\delta(t). \qquad (22)$$

Therefore if a microtubule interacts with N dynein molecules the model predicts that after a long time, the microtubule will diffuse with the diffusion constant

$$D^\star = D/(1 + NK). \qquad (23)$$

If we know $D, D^\star$ and K we can use equation (23) to infer how many dynein molecules interact with the microtubule (c.f. Figure 4a).

If we fit the data with the above equations we can deduce the jump probabilities α and β. Therefore, it would be important to obtain a long record of displacements so that one could measure not only v, but D as well. Together, they allow calculation of the energy barriers, providing we have previously obtained the values for the binding and dissociation constants, k_b and k_d, as described above.

3. CONCLUSIONS

We have presented a simple diffusion-reaction model to describe dynein-microtubule interactions. We were able to reproduce the abnormal amount of small displacements reported by Vale, et al (1989) by assuming that the dynein-microtubule complex follows rapid cycles of association and dissociation. The asymptotic micro-tubule diffusion coefficient was found to be $D^\star = D/(1 + NK)$, where K is the equilibrium constant for dynein-microtubule interaction and N is the number of dynein molecules involved. Using Vale's results we estimated $K = 6.88$.

ACKNOWLEDGMENTS

GO was supported by NSF Grant No. MCS-8110557. RV was supported by NIH Grant No. GM38499. NC was partially supported by US-Israel Binational Science Foundation Grant No. 3777. The authors would like to thank Byron Goldstein for calling our attention to the calculation cited in the text.

REFERENCES

1. Feynman, R. P., Leighton, R. B., Sands, M. (1963). *The Feynman Lectures on Physics*. Addison-Wesley, Reading, Mass., pp. 46-1-46-5.

2. Goldstein, B., Delisi, C., Abate, J. (1975). *Immunodiffusion in gels containing erythrocyte antigen. I. Theory for diffusion of antiserum from a circular well.* J. Theor. Biol. 52, 317-334.
3. Vale, R. D., Soll, D. M., Gibbons, I. R. (1989). *One dimensional diffusion of microtubules bound to flagellar dynein.* Cell. 59, 915-925.
4. Vale, R.D. and Oosawa, F. (1990). *Protein motors and Maxwell's demons: does mechanochemical transduction involve a thermal ratchet?* Adv. Biophys. 26, 97-134.

CELL MOVEMENT AND AUTOMATIC CONTROL

Hans Gruler

Department of Biophysics, University of Ulm
D 7900 Ulm, Germany

1. INTRODUCTION

An important concept for the understanding of biological phenomena is given by cybernetics, also known as the theory of automatic control [1]. It is shown that chemotaxis, galvanotaxis, galvanotropism, contact guidance, etc., are functions of cells having a goal-seeking system which is an automatic controller having a closed-loop feedback system. The model is verified by means of galvanotaxis, chemotaxis and contact guidance data of granulocytes [2,3,4].

The first defense line of mammalian immune systems against the invasion of microorganisms are polymorphonuclear leukocytes (=granulocytes) [5]. These cells are attracted to sites of inflammation to destroy by phagocytosis microorganisms and invaded cells. At least two different mechanisms exist which cause granulocytes to find these microorganisms and those invading cells: (i) The signal can be chemical in nature: the resulting directed movement is then called chemotaxis. The microorganisms or the invaded cells have a special "smell" and thus granulocytes find their way: The initial part of the peptides responsible for chemotaxis is a formylmethionyl moiety. These peptides allow the granulocytes to distinguish between prokaryote and eukaryote cells. Interestingly, horse granulocytes do not have a receptor for these peptides. (ii) The signal may also be electrical in nature: the directed migration is then called galvanotaxis. For example, when a cell is lysed, ions inside and outside the cell diffuse to reestablish a concentration equilibrium. The different diffusion constants of the ions involved result in the separation of small and large ions, and a diffusion potential is created. The cellular membrane of an invaded cell is altered and thus the cell acts as a battery and granulocytes find their way to the source [6].

The directional movement of cells during embryogenesis, wound healing, and tumor invasion has been well documented [7]. In particular, there are many instances when embryonic cells migrate as individual cells during embryogenesis and consistently follow a precise pathway toward their final destination (e.g. neural crest, precardiac mesenchyme, to name a few). The extracellular guiding signal can be chemical in nature: the directed growth is then called chemotropism. (ii) The signal can also be electrical in nature and the corresponding growth is called galvanotropism. (iii) Cells can also guided by the topology around the cells. This directed movement (or growth) is called contact guidance.

Biologically Inspired Physics, Edited by L. Peliti
Plenum Press, New York, 1991

Embryonic electric fields can be generated by ion current driven into the embryo by the epithelium which is well known to pump in positive ions. Such embryonic currents have already been measured exiting the primitive streak of chick gastrulae and the blastpore of frog neurulae in addition to emanating from the cut surface of wounds and regenerating limbs, where morphogenesis is also occuring. A related phenomenon of embryonic cell galvanotropism has also been observed in developing neurons [8].

Phenomena like the directed movement have different levels of understanding. For example the macroscopic and the microscopic level. Here the macroscopic level is discussed where consequently no detailed knowledge of the biochemistry is necessary. One is just interested of the principles how a cell works. It will be shown that (i) the chemokinesis can be explained by a steering device, and (ii) the direced movement, directed growth and contact guidance by an automatic controller.

2. STEERER

The cell as a steering device recieves an external signal. By means of the signal transduction/response system of the cell a product, p, corresponding to the received signal, S, is released ([9,10]). But a steerer has no device which controls whether the product really corresponds to the received signal or not. The rate equation for the steerer is

$$\frac{dp}{dt} = k_t \cdot S - k_d \cdot p + \Gamma(t) \tag{1}$$

The increase of the product, p, is assumed to be proportional to the deterministic primary cellular signal, S, where the transduction coefficient k_t characterizes the signal transduction/response system. The second term describes the memory of the steerer where the inverse of the decay coefficient, k_d, is the memory time of the steerer. The stochastic signal in the sterring device is described by the third term where a δ-correlated white noise with the strength q is assumed.

$$<\Gamma(t)> = 0 \qquad \text{and} \qquad <\Gamma(t) \cdot \Gamma(t')> = q \cdot \delta(t - t') \tag{2}$$

In the two following sections it is shown that the chemokinetic response can be predicted by a proportional steerer.

2.1 Chemokinetic Dose-Response Curve

The number of loaded receptors is regarded as the primary cellular signal in chemokinesis. Consequently, the concentration dependence of the track velocity, v_c, is given by the equilibrium binding constant, K_R, of the chemokinetic molecule to the receptor. One obtains from eq.(1)

$$v_c = A \cdot R_0 \cdot \frac{k_t}{k_d} \cdot \frac{[c]}{[c] + K_R} \tag{3}$$

218

A is a constant for a proportional steerer. R_0 is the total number of receptors exposed to the cell surface, and [c] is the concentration of the chemokinetic molecule. The measured chemokinetic dose-response curve has the predicted concentration dependence as shown in Fig. la.

2.2 Track Velocity Distribution Density

The deterministic signal in eq.1 determines the chemokinetic dose-response curve while the noise in the signal transduction/response system determines the distribution of the track velocity around the expected velocity, v_c^{det}. (Granulocytes seem to be a homogeneous

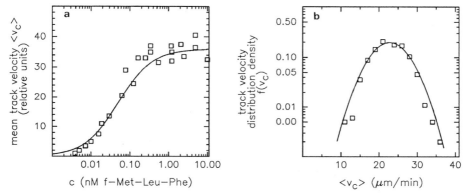

Figure 1. a: Chemokinesis of granulocytes (data from [11]). The solid line is the theoretical prediction (eq.3) where $K_R(=0.05$ nM) and $A \cdot R_0 \cdot k_t/k_d(=$const) are fitting parameters. The direct measured high affinity site of the receptor is between 0.05 to 0.5 nM. b: Track velocity distribution density of migrating granulocytes (data from [12]). The solid line is the theoretical prediction (eq.4) where v_c^{det} (= 22.8μm/min) and σ(= 4.5μm/min) are fitting parameters.

population with respect to their chemokinesis since there are no significant differences between the distribution obtained from a single cell observed over a long period of time and the distribution obtained from multiple cells [12].) The track velocity distribution density, $f(v_c)$, predicted from the stochastic differential equation (eq.1), can be obtained from the corresponding Fokker-Planck equation. One obtains in the case of a white noise source (eqs.2)[13]

$$f(v_c) = f_0 \, e^{-(v_c - v_c^{det})^2/2\sigma^2} \quad \text{with} \quad v_c^{det} = \frac{k_t}{k_d} S \quad \text{and} \quad 2\sigma^2 = \frac{q}{k_d} \qquad (4)$$

where f_0 is a calibration constant. The maximum and the width of the gaussian distribution are described by the deterministic and stochastic signal. The measured track velocity distribution density is gaussian (Fig. 1b) as predicted by the model.

3. AUTOMATIC CONTROLLER

One important function of cells is their capacity for directed movement or growth which is quantified by the mean displacement, $< x >$, parallel to the polar field. The translational movement in a polar field requires two components of the cellular response: the track velocity, v_c, and the direction of migration, Φ. One might well expect that these two parameters would depend on each other. But they are independent of each other since (i) the temporal variations of $v_c(t)$ and $\Phi(t)$ are independent of each other, (ii) the mean track velocity is independent of the angle of migration, and (iii) in galvanotaxis the mean track velocity is independent of the electric field strength. This holds at least for human granulocytes and monocytes [6,12], somitic fibroblasts [14], and neural crest cells [15]. The speed as one variable was already discussed in the previous section. In the following ones, the second variable will be discussed within the context of an automatic controller.

The basic elements of an automatic controller are: first, an element which measures the output of the biological system; second, a means of comparing that output with the desired one; third, a means of feeding back this information into the input in such a way as to minimize the deviation of the output from the desired level.

This means in the case of galvanotaxis the cell must have the ability to measure its orientation with respect to the applied electric field. This measured orientation is compared with the desired one - e.g. to be parallel to the electric field in the case of neural crest cells or to be antiparallel in the case of granulocytes. The created intracellular signal is such that the cell rotates to approach the desired orientation.

A similar behaviour is expected in the case of chemotaxis: The cell must have the ability to measure its orientation with respect to the concentration gradient. And then an intracellular signal is created such that the cell rotates towards increasing concentration .

If the cellular device for contact guidance is an automatic controller then the cell must have the ability to measure its orientation in respect to the direction of surface undulations. The created intracellular signal is such that the cell rotates to approach the desired orientation.

The rate equation of the angle of migration is in the case of an automatic controller composed of two terms: (i) The deterministic torque $\Gamma_D(Signal, \Phi)$, tries to render the movement in the desired direction and (ii) the stochastic torque, $\Gamma_N(t)$.

$$\frac{d\Phi}{dt} = - \Gamma_D(Signal, \Phi) + \Gamma_N(t) \tag{5}$$

The steady state angle distribution density, $f(\Phi)$, predicted from the stochastic differential equation (eq. 5), can be obtained from the corresponding Fokker-Planck equation. One obtains in the case of a white noise source (eqs.2 with the strength Q)

$$f(\Phi) = e^{V(\Phi)} \quad \text{with} \quad V(\Phi) = \frac{2}{Q} \int_0^\Phi \Gamma_D(Signal, \Phi') \, d\Phi' \tag{6}$$

The unknown deterministic torque, $\Gamma_D(Signal, \Phi)$, can approximately determined by a Fourier Series

$$\Gamma_D(Signal, \Phi) = c_1 \cdot \sin \Phi + c_2 \cdot \sin 2\Phi + \ldots \tag{7}$$

where in addition the symmetry of the cellular environment is used to reduce the number of terms: (i) the physical state is unchanged if the coordinate system is rotated by 2π and (ii) the torque changes sign if the coordinate system is reflected at a mirror containing the symmetry axis like the electric field vector. The first term on the right side describes polar phenomena like galvanotaxis, chemotaxis, galvanotropism, etc., and the second term is responsible for apolar phenomena like contact guidance, bidirectional movement or growth, etc.. The coefficients, c_1, c_2, etc. are functions of the strength of the applied signal.

In the next section, galvanotaxis is discussed in detail. Then in further sections, chemotaxis, contact guidance and bidirectional orientation are considered.

3. 1 Galvanotaxis

Cells like granulocytes have the ability to recognize electric field and direct their angle of migration with respect to the received signal. The distribution density, $f(\Phi)$, of the angle of migration is a bell-shaped curve around the expected direction due to the noise in the signal transduction/response system. One expects a straight line (eqs. 6 and 7) if the logarithm of $f(\Phi)$ is plotted versus $\cos \Phi$. A typical result is shown in Fig. 2a. The straight line proves that the only important torque in directed movement of granulocytes is $c_1 \cdot \sin \Phi$. (The slope of the shown line is given by $a_1 = \dfrac{2c_1}{Q}$).

The automatic controller works linearly if the deterministic torque is proportional to the applied electric field strength, E. This is shown by different ways: (i) $c_1 \propto E$ (c_1 can be determined from the angle autocorrelation function [3]); and (ii) $a_1 \propto E$. A plot of the experimentally obtained data (a_1 vs E) yields a straight line for granulocytes where the galvanotaxis coefficient, K_G, describes the cellular sensitivity. The linear response is also shown by means of the galvanotaxis dose-response curve (=average of $\cos \Phi$ vs E).

$$< \cos \Phi > = \frac{I_1(a_1)}{I_0(a_1)} \qquad \text{with} \qquad a_1 = K_G \cdot E \tag{8}$$

The theoretical prediction is the ratio of two hyperbolic Bessel functions $I_1(a_1)$ and $I_0(a_1)$. The galvanotaxis dose-response curve as shown in Fig. 2b, proves the linear response of granulocytes. This result is not specific for granulocytes. Similar results were obtained for galvanotaxis of spermatozoids [16] and for the directed growth of spores in an electric field [17].

3.2 Chemotaxis

Cells like granulocytes have the ability to measure concentration differences and guide their direction of movement according to the measured values. The chemotaxis can be treated in the same way as galvanotaxis: The electric field, E, has to be replaced by the gradient of the chemical activity of the chemotactic molecule [18]. In the case of a proportional controller, the torque should be proportional to grad ln[c] as it is actually found (Fig. 3a). But there is a problem since the constant of proportionality is a function of the mean concentration of the chemotactic molecule. This experimental fact can be explained in the following way [18]: (i) The cell measures the concentration at two separate parts of the membrane and (ii) the deterministic torque is proportional to the difference between these cellular signals (=difference in the number of loaded receptors). The predicted concentration dependence of the chemotaxis coefficient is then

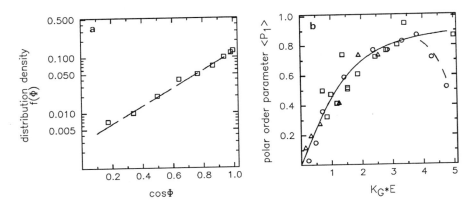

Figure 2. Galvanotaxis of granulocytes (data from [2]). a: Logarithm of the angle distribution density as a function of cos Φ. b: Galvanotaxis dose-response curve. The solid line is a theoretical prediction (equ.8) with K_G^{-1} as fitting parameter. o granulocytes $K_G^{-1} = -0.2\text{V/mm}$ [2], $\Diamond$ braken spermatozoids $K_G^{-1} = 0.024\text{V/mm}$ [18,19], Δ galvanotropism of fungal hyphae (*Neurospora crassa*) $K_G^{-1} = -2\text{V/mm}$ [17].

$$K_{CT} = K_{CT}^0 \cdot \frac{[c] \cdot K_R}{([c] + K_R)^2} \qquad (9)$$

In Fig. 3b the predicted and the measured chemotaxis coefficients are shown. The hypothesis of a spatial recognition system in connection with a proportional controller is consistent with the data.

3.3 Contact Guidance

A random walk is obtained if cells like granulocytes migrate on a flat surface and the chemical environment is homogeneous. The cell migrates due to the chemical stimulation but the mean displacement in the x and y direction are zero since the probability for moving in the + or − direction are equal probable. But the mean squared displacement are unequal zero and the mean squared displacement *versus* time can be described by the Langevin equation where the diffusion coefficient, D, is one fitting parameter [21].

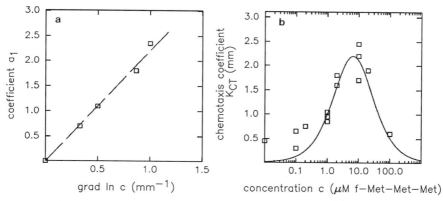

Figure 3. Chemotaxis of granulocytes. a: The coefficient a_1 as a function of grad ln[c] at a constant mean concentration [c] is shown for the chemotactic tripeptide f-Met-Met-Met (=formylmethionylmethionylmethionyl) (data from [20]). b: Chemotaxis coefficient of granulocytes as a function of the concentration of the chemotactic tripeptide (f-Met-Met-Met) (data from [20]). The solid line (eq. 9) is a theoretical prediction with the fitting parameters K_R (=6.6 μM) and K_{CT}^0 (= 9 mm). The direct measured binding constant was several μM.

Cells like granulocytes have the ability to recognize surface undulations (e.g. parallel lines) and consequently their movement is influenced by the undulated surface. Only the very simple case is regarded here where the periodic length of the undulations is smaller as the size of the cell. This situation is referred as *nematic* contact guidance ([22]): the cellular movement is angle dependent but the mean cell area density is a constant. Whether the nematic contact guidance is described by a steerer or by an automatic controller may be answered by a detailed analysis of the cellular motion. The following results were obtained [22]: (i) The diffusion coefficients determined from the mean squared displacement measured parallel and perpendicular to the lines are different. The diffusion coefficients are angle dependent described by an ellipse (Fig. 4a).

$$D(\Phi) = D_x \cos^2 \Phi + D_y \sin^2 \Phi \qquad\qquad\qquad (10)$$

(ii) The track velocity is angular dependent and also described by an ellipse. (iii) The diffusion coefficient is proportional to the mean squared velocity.

The angle dependence of the ratio of diffusion coefficient, $D(\Phi)$, and mean squared track velocity, $<v_c^2>$, is of basic interest: (i) The cell is simply described by a steerer device if the ratio of $D(\Phi)$ and v_c^2 is a constant independent of the angle Φ. (ii) However, if an automatic controller is involved then the direction of migration is also influenced by the undulated surface. In this case the ratio of $D(\Phi)$ and v_c^2 is angular dependent. Granulocytes

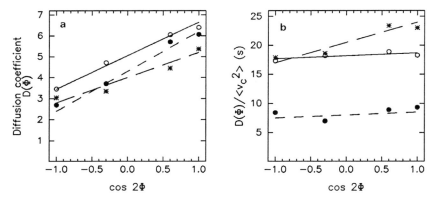

Figure 4. Contact guidance of granulocytes [22]. a: The angular dependence of the diffusion coefficients on different substrates (• machine drilled aluminum surface (p=11 μm), o optical grid (p=5 μm), and * stretched polyethylene foil (undulated surface p=0.5-1.0 μm)). b: The angle dependence of the ratio of the diffusion coefficient and the mean square track velocity as a function of cos 2Φ are shown.

moving on an optical grid or an undulated aluminium surface respond as they had only a steerer. But granulocytes exposed to a stretched polyethylene foil respond as had an automatic controller (Fig.4b). This indicates that in general the nematic contact guidance should be regarded as a cellular response which is based on an automatic controller.

The question how the coefficient c_2 of eq. 7 depends on the strength of the undulation is not yet explored sytematically. But some data are available for another types of contact guidance. Fibroblasts on a cylinder orient their long axis in the mean parallel to the axis of the cylinder. The logarithm of the angle distribution function versus cos 2Φ yields a straight line with the slope a_2 (analogous to the directed movement). The slope is small for large R and large for small R. One finds

$$a_2 = (K_{CG} \cdot \frac{1}{R})^2 \qquad\qquad (11)$$

The value of the conduct guidance coefficient, K_{CG}, is 87 μm for fibroblasts on glass fibers [18,23]. The quadratic dependence has not yet been confirmed.

3.4 Apolar Orientation and Bidirectional Growth

Cells like granulocytes, fibroblasts, neural crest cells etc. have the ability to measure electric fields and guide their direction of migration. Granulocytes are mainly elongated in the direction of migration. But other cells like fibroblasts, neural crest cells, etc., in addition have the ability to orient their long axis in respect to the applied electric field: They orient

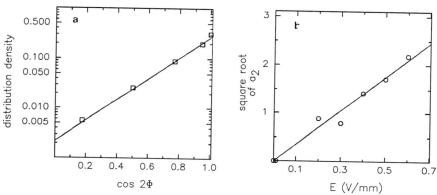

Figure 5. Apolar orientation of somitic fibroblasts in an electric field. a: The logarithm of the orientation angle distribution function *versus* cos 2Φ (data from [24]). b: The square root of the coefficient a_2 *versus* electric field strength E data from [24]).

their long axis in the mean perpendicular to the applied electric field. The logarithm of the angle distribution function *versus* cos 2Φ yields a straight line (Fig. 5a) as expected from eq. 7. The coefficient a_2, obtained from the slope of the straight line, is a function of the applied electric field. It is found that the apolar orientation is described by a square dependence (Fig. 5):

$$a_2 = (K_{BO} \cdot E)^2 \qquad\qquad (12)$$

K_{BO}=i3.5 mm/V for somitic fibroblasts [24]. A similar situation holds for hyphae growing in an electric field with K_{BO}=i0.4 mm/V for *Neurospora crassa* [17]. Obviously, one has a field induced effect: the electric field alters somehow the cell e.g. the membrane potential difference, and then the field interacts with the altered cell. The controller device for apolar orientation or for bidirectional growth is not linear. There is a quadratic relationship.

4. SUBUNITS OF THE AUTOMATIC CONTROLLER

Mobile fragments (cytokineplasts) can be formed out of the pseudopod of granulocytes. The cytokineplasts still have the ability of directed movement ([21,25]). Thus the machinery which creates the cell locomotion contains only a few elements: the plasma membrane, the microfilaments, and an unstructured cytoplasma as seen by light microscopy. The microtubules are not essential for directed movement.

The first step in chemotaxis is the binding of the signal molecule (=chemoattractant) to a specific membrane-bound receptor which activates a G-protein. It is important to note that this G-protein is very likely the essential protein on which the electric field acts in the case of galvanotaxis [6]. The G-proteins activate the phospholipase C which hydrolyzes the membrane phospholipid phosphatidyl inositol-4,5-biphosphate to form two second messengers. The lipid portion of the phospholipid 1,2-diacylglycerol remains in the membrane and activates a further membrane-attached enzyme, the protein kinase C. The other product of the hydrolysis reaction, inositoltriphosphate, is released into the cytoplasma where it causes the release of calcium from internal stores. The calcium ions are involved in triggering the opening of channels for other ions (Na^+, K^+, and H^+ are further essential ions used in directed movement) [26]. The actin-myosin polymerization takes place after these events and the cell migrates in a new direction.

A part of the signal transduction/response system is described above. The signal chain as a whole works linearly as we have demonstrated for the directed movement. It is still unknown how the cell "makes its decision". But we hope to answer this question in the near future.

5. CONCLUSION

Usage of the concepts of steering and automatic control applied to cell movement and to cell growth, is very useful as it was shown with a few examples. The descripton of the proportional controller responsible for galvanotaxis, chemotaxis, etc., is closely related to the descripton of a permanent electric dipol in an electric field. Both effects are linear with respect to the polar field. While the description of the apolar orientation and bidirectional growth is analogous to the description of the induced electric dipole in an electric field. Again both effects are quadratic with respect to the applied polar field.

ACKNOWLEDGEMENTS

This work was supported by "Fond der chemischen Industrie" and by a NATO travel grant.

REFERENCES

1. N. Wiener, Cybernetics: or Control and Communication in Animal and the Machine, M.I.T. Press, Cambridge (1961).
2. K. Franke and H. Gruler, Galvanotaxis of human granulocytes: Electric field jump studies, Eur.Biophys.J. 18: 335 (1990).
3. H. Gruler and K. Franke, Automatic Control and Directed Cell Movement, Z.Naturforsch. 45c: 1241 (1990).

4. H. Gruler, Chemokinesis, chemotaxis and galvanotaxis, in: Biological Motion, W. Alt and G. Hoffmann, Eds. in: Lecture Notes in Biomathematics, Springer Verlag, Berlin, Heidelberg, New York (1990).

5. P.C.Wilkinson, Chemotaxis and Inflammation, Churchill, London (1974).

6. B. Rapp, A. de Boisfleury-Chevance, and H. Gruler, Galvanotaxis of human granulocytes. Dose-response curve. Eur. Biophys. J. 16: 313 (1988).

7. J.P.Trinkaus, Cells into Organs, Prentice-Hall Inc., Englewood Cliffs (1984).

8. R. Nuccitelli, Transcellular ion currents: Signals and effectors of cell polarity, Modern Cell Biology, 2: 451(1983).

9. R.T. Tranquillo and D.A. Lauffenburger, Stochastic model for leukocyte chemosensory movement, J.Math.Biol. 25: 229 (1987).

10. R.T. Tranquillo, D.A. Lauffenburger, and S.H. Zigmond, A Stochastic model for Leukocyte Random Mobility and Chemotaxis Based on Receptorbinding Fluctuations, J.Cell Biol. 106: 303 (1988).

11. E.L. Becker, H.J. Showell, P.H. Naccache, and R. Sha'afi, Enzymes in Granulocyte Movement: Preliminary Evidence for the Involvement of Na^+, K^+ AT-Pase, in Leukocyte Chemotaxis, J.I. Gallin and P.G. Quie, eds. Raven Press, New York (1978).

12. H. Gruler, Cell Movement Analysis in a Necrotactic Assay, Blood Cells 10: 107 (1984).

13. H. Risken, The Fokker-Planck Equation, Springer Verlag, Heidelberg (1985).

14. H. Gruler and R. Nuccitelli, New insights into galvanotaxis and other directed cell movements an analysis of the translocation distribution function, in: Ionic Currents in Development, R. Nuccitelli, ed. A.R. Liss, New York (1986).

15. H. Gruler and R. Nuccitelli, Neural Crest Cell Galvanotaxis: New Data and Novel Approach to the Analysis of Both Galvanotaxis and Chemotaxis, Cell Motility and Cytoskeleton, 19: (1991).

16. H. Gruler, Biophysics of Leukocytes: Neutrophil Chemotaxis, Characteristics and Mechanisms, in: The Cellular Biochemistry and Physiology of Neutrophil, M.B. Hallett, ed., CRC- Press UNISCIENCE, (1989).

17. H. Gruler and N.A.R. Gow, Directed Growth of Fungal Hyphae in an Electric Field, Z.Naturforsch. 45c: 306 (1990).

18. H. Gruler, Cell Movement and Symmetry of the Cellular Environment, Z.Naturforsch. 43c: 754 (1988).

19. C.J. Brokaw, Chemotaxis of Bracken Spermatozoids, J.Exp.Biol. 35: 197 (1958).

20. S.H. Zigmond, Ability of Polymorphonuclear Leukocytes to orient in Gradients of Chemotactic Factors, J.Cell Biol. 75: 606 (1977).

21. H.Gruler and A. de Boisfleury-Chevance, Chemokinesis and Necrotaxis of Human Granulocytes: the Important Cellular Organelles, Z.Naturforsch. 42c: 1126 (1987).

22. T. Matthes and H. Gruler,Analysis of cell locomotion. Contact guidance of human polymorphonuclear leukocytes, Eur.Biophys.J. 15: 343 (1988).

23. M. Abercrombie, The Crawling Movernent of Metazoan Cells, in: Cell Behaviour, R. Bellairs, A.Curtis, and G.Dunn, eds., Cambridge University Press, Cambridge (1982).

24. C.A.Erickson and R.Nuccitelli, Embryonic Fibroblast Motility and Orientation Can Be Influenced by Physiological Electric Fields, J.Cell Biol. 98: 296 (1984).

25. S.E. Malawista and A. de Boisfleury-Chevance, The cytokineplast: purified, stable and functional motile machinery from human blood polymorphonuclear leukocytes, J.Cell Biol. 95: 960 (1982).

26. E.L. Becker, Y. Kanaho, and J.C. Kermode, Nature and Functioning of the Pertrussis Toxin-Sensitive G Protein of Neutrophils, Biomedicine and Pharmacotherapy, 41: 289 (1987).

DIELECTRIC STUDY OF THE HYDRATION PROCESS

IN BIOLOGICAL MATERIALS

A. Anagnostopoulou-Konsta, L. Apekis, C. Christodoulides,
D. Daoukaki, and P. Pissis

Department of Physics, National Technical University of Athens
Zografou Campus, 15773, and E.G. Sideris, Biology Institute
NCSR "Democritos" 15310, GREECE

1. INTRODUCTION

The sorption of water vapour by biological macromolecules is generally assumed to involve the binding of H_2O molecules to specific hydrophilic sites at lower relative humidities, followed by condensation of multi-molecular adsorption as the humidity increases. Several methods have been applied for the investigation and detailed study of the structure, mobility, extent and modes of binding of water molecules in various systems. Among them the most commonly used are IR and Raman spectroscopy (Luck, 1985), differential scanning calorimetry (Berlin et al., 1970), NMR spectroscopy (Kuntz and Kautzmann, 1974, Mathur de Vré, 1979), neutron scattering (Lehmann, 1984), sorption and desorption methods (Pethig, 1979) and dielectric methods (Bone and Pethig, 1982, Pethig and Kell, 1987, Grant et al. 1978, Kent and Meyer 1984). All of them yield some insight into the problem. One common feature observed in nearly all cases is that the relaxation times for reorientation and the diffusion constants of water molecules sorbed in various biological systems are much lower than the values observed for free water, while the enthalpy of vaporisation of the water sorbed is by about 100 cal g^{-1} higher than the value of liquid water (Berlin et al., 1970, Pethig, 1979, Grant et al., 1978). This behaviour suggests that the water molecules contributing to the first hydration layer exhibit restricted motion due to a significant decrease in the translational and rotational modes of motion caused by macromolecular-water interaction. Moreover, the dynamics of the material itself (relaxation and conductivity mechanisms) is strongly influenced by the presence of sorbed water.

In this paper we report on results of systematic dielectric studies of the hydration properties of a variety of water-containing organic systems by means of the thermally stimulated depolarisation current (TSDC) method in the temperature range 77-300 K. These results are part of a more general program of study of the state

Biologically Inspired Physics, Edited by L. Peliti
Plenum Press, New York, 1991

of water in biological systems in the form of aqueous solutions and hydrated solid samples, in order to elucidate the effect of water of hydration and to investigate the microscopic mechanisms responsible for the characteristics observed.

2. EXPERIMENTAL

2.1. *The Method*

The principal of the TSDC technique is the following. The sample is polarized by applying an electric field E_P at some suitable polarization temperature T_P for a time t_P large compared with the relaxation time at T_P. With the electric field still applied, the sample is cooled down to a temperature T_0 sufficiently low to prevent depolarization by thermal energy. The field is then switched off and the sample is heated at a linear heating rate b, while the resulting depolarization current, as the dipoles relax, is detected by an electrometer. Thus for each polarization mechanism an inherent current peak can be detected. In the case of a single relaxation process obeying the Arrhenius equation, $\tau = \tau_0 \exp(W/kT)$, the depolarization current density $J(T)$ is given by the equation

$$J(T) = \frac{P_0}{\tau_0} \exp\left(\frac{-W}{kT}\right) \exp\left(-\frac{1}{b\tau_0} \int_{T_0}^{T} exp\left(\frac{-W}{kT'}\right) dT'\right)$$

where τ is the relaxation time, W the activation energy of the relaxation, τ_0 the pre-exponential factor, T the absolute temperature, k the Boltzmann constant and P_0 the initial polarisation. By analysis of the shape of this curve the activation energy, W, the pre-exponential factor, τ_0, and the contribution of the peak to the static permittivity, $\Delta\epsilon$, may be evaluated (Bucci et al., 1966, Vanderscueren and Gasiot, 1979, Christodoulides, 1985).

Characteristic features of the TSDC method that make it especially suited to this aim are: its high resolving power, its sensitivity (dipole concentrations as low as 0.1 ppm can be measured) and the special experimental techniques it offers, by means of which it is possible to resolve relaxation processes, arising from different sets of dipoles with slightly different relaxation times, into their simpler components (Bucci et al., 1966, van Turnhout, 1980, Pissis et al., 1983a). Moreover, a TSDC spectrum recorded between 77 and 300 K comprises several distinct mechanisms which in the conventional AC-dielectric methods would correspond to a frequency range from several GHz to a few Hz, it covers thus the whole region which normally has to be spanned in order to observe all the dielectric dispersions related to the water of hydration.

A standard equipment described elsewhere (Pissis et al., 1983a) was used for TSDC measurements in the temperature range 77-300 K. If not otherwise stated the experimental conditions used were the following: Polarizing field 3 kV cm^{-1}, polarization time 5 min, cooling rate 10 K min^{-1} and heating rate between 3 and 4 K min^{-1}. Typical experimental errors were 0.001 for the water content h, 1 K for the temperature of current maximum (peak temperature) T_M, 0.001 eV for the activation energy W, a factor of about 2 in τ_0 and 10% for the contribution of a relaxation mechanism to the static permittivity.

2.2. *The Samples*

Where possible, water solutions and hydrated solid samples of the same material were used. In the case of solutions the samples were grown in the measuring capacitor by injecting 25 μl of the liquid between the capacitor plates. The concentration of the solutions was determined by weighing. Solid samples of saccharides and proteins were made by compressing powder or fibres to cylindrical pellets of typically 13 mm diameter and 1 mm thickness. They were hydrated over saturated salt solutions in sealed jars and allowed to adsorb (or desorb) moisture until equilibrium was obtained. Their water content h, defined as the ratio of the mass of sorbed water to the mass of dry sample, was determined by weighing. Drying in vacuo (5×10^{-2} Torr) at typically 110°C or over P_2O_5 at room temperature, until no further weight changes were observed, was adopted as a method for the determination of the dry weight.

Plant tissue samples were taken by transversal cuts of fresh leaves, stems and wood (typically 1.5 mm thickness). Measurements were carried out at different water contents by allowing the fresh samples to dry in air. Wood was hydrated and dehydrated as above.

Several kinds of contacts were used: silver paste, direct contact of the sample with the brass electrodes of the measuring capacitor, thin insulating (teflon, mica) or conducting (Au, Al, Cu) foils between the sample and the brass electrodes.

3. RESULTS AND DISCUSSION

In order to use the TSDC method for the investigation of the hydration properties of materials at subzero temperatures, the good knowledge of the dielectric behaviour of polycrystalline ice is essential. Fig. 1 shows a TSDC plot of polycrystalline ice (curve a), which consists of two main peaks at about 120 and 220 K. Systematic investigations (Pissis et al., 1987, Apekis et al., 1983, 1987) have shown that the main part of the low temperature peak can be described by a continuous distribution of relaxation times, with mean W and τ_0 values of 0.31 eV and 5×10^{-12}s respectively. It is attributed to the reorientation of the water molecule dipoles through the motion of extrinsic D and L defects, generated mainly by physical imperfec tions. The high-temperature peak with mean W and τ_0 values of 0.42 eV and 5×10^{-8} s respectively is attributed to the thermal release of space charges generated inside the crystallites and trapped at the boundaries between them.

In the following, results obtained with some organic and biological materials are going to be discussed.

3.1. *Methylcellulose*

Curves b and c in Fig. 1 show TSDC plots obtained with a methylcellulose solution and a hydrated solid sample respectively (Pissis and Daoukaki, 1988). The solution plot shows a double low-temperature peak and a high-temperature peak at about 220 K. A low-temperature peak of the same type has been recorded with all water-solutions of saccharides (Daoukaki et al. 1984, Pissis et al., 1983b) as well as with model systems consisting of oil-in-water emulsions (Pissis et al., 1982). Experimental analysis of this peak by means of the special techniques offered by the TSDC method has shown that it can, in fact, be resolved into two separate components, the first of which is due to relaxation of free water molecules and the second to relaxation of water molecules modified by the presence of the solute, as shown by a systematic investigation of the above systems as a function of the solute concentration. The second component is characterized by larger T_M and W values than the first one, suggesting restricted motion of the water dipoles modified by the solute. The T_M and W values of the high-temperature peak are, within the limits of experimental errors, equal to those of the high-temperature peak in polycrystalline ice, indicating a similar origin of the peak in both cases.

TSDC plots recorded with hydrated solid samples of the same material (Fig. 1, curve c) show a broad low-temperature peak and a complex high-remperature band. The contribution of the low-temperature peak to the static permittivity, $\Delta\epsilon$, increases linearly with increasing water content. A long series of investigations on other hydrated systems (Anagnostopoulou- Konsta and Pissis, 1987, Pissis, 1989) has shown that this peak originates from reorientation of loosely bound water molecules. It appears only for water contents higher than a critical value, h_c, equal to the fraction of water molecules tightly (irrotationally) bound at primary hydration sites. Comparison between plots (b) and (c) of Fig. 1 has shown that the dispersion due to loosely bound water molecules in hydrated solid samples is described by the same values of T_M and W as the dispersion due to the modified water in the solutions of the same material, indicating that these two fractions of water molecules behave, at least dielectrically, in the same way.

The high-temperature band, which is recorded with all hydrated solid systems, is in general complex consisting of several components. The use of different electrode configurations and special analysis techniques help to distinguish between components related to dipolar reorientation, dc-conductivity, electrode effects, etc. (Pissis and Anagnostopoulou-Konsta 1985, 1988). The different components of the high-temperature band shift to lower temperatures with increasing water content h, while their magnitude increases, a fact which is going to be explained later.

3.2. Cellulose

Fig. 2 shows TSDC plots recorded with native cellulose samples with four different water contents. Here again the low-temperature peak is due to loosely bound water molecules relaxation, while it can be shown that the high-temperature band is related to water-assisted relaxation of dipolar groups other than water (Pissis, 1985, Pissis and Anagnostopoulou-Konsta, 1985). The variation of the contribution of the low-temperature peak to the static permittivity, $\Delta\epsilon$, with water content is shown in Fig. 4, curve a. The amount of tightly bound water is seen to be equal to about 8%, the h-value corresponding to the intersection of the two linear parts of the graph. It is a general belief that hydration in native cellulose takes mainly place in its disordered or amorphous segments. In order to investigate this point we measured TSDC spectra of microcrystalline cellulose (MCC), i.e. cellulose from which all amorphous parts have been eliminated by hydrolysis. From Fig. 3, as well as from Fig. 4, curve b, it may be seen that, in fact, the amount of tightly bound water in MCC, if any, is much less than 8%. Moreover, the maximum value of sorbed water content, which was larger than 25% for native cellulose, is only of the order of 17% for MCC (Apekis 1988). Both facts support the above assumption concerning cellulose hydration.

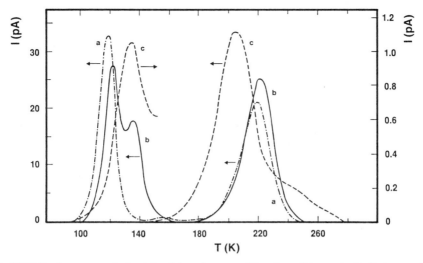

Fig. 1. TSDC plots recorded with: (a) polycrystalline ice; (b) methylcellulose solution with $c=0.075$ g/l; and (c) methylcellulose solid sample with $h=27$.

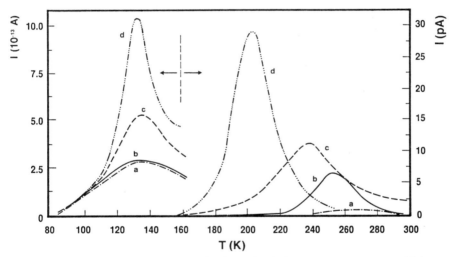

Fig. 2. TSDC plots of native cellulose hydrated solid samples with $h=1.1\%$ (a); 5.6% (b); 9.8% (c); and 19.5% (d).

3.3. Wood

TSDC recorded with wood samples (beach, oak and pine wood, Anagnostopoulou-Konsta and Pissis, 1989) show a behaviour very similar to the one displayed by native cellulose. Fig. 5 (curves d and e) shows the variation of $\Delta\epsilon$ of the low-temperature peak as a function of hydration for two different contact configurations. It is seen that bound water content lies between 7 and 9%. The same figure

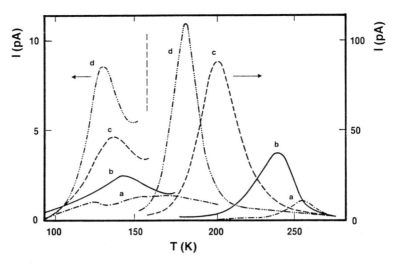

Fig. 3. TSDC plots of microcrystalline cellulose hydrated samples with $h=0.6\%$ (a); 3.7% (b); 8.5% (c); and 13.2% (d).

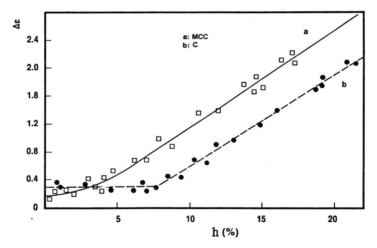

Fig. 4. Contribution of the low-temperature TSDC peak to the static permittivity, $\Delta\epsilon$, vs water content, h, in (a) microcrystalline cellulose and (b) native cellulose.

(curves a, b, c) shows the variation of the elastic parameters of wood with moisture content (mc), as reported by other authors (Kollmann and Côté, 1968, Gerhards, 1982). It is shown that all elastic characteristics decrease linearly with increasing mc between about 8% and the fiber saturation point. On the contrary, in the oven-dry to 8% mc-range all the above properties either remain practically constant or are characterized by a maximum between 4 and 8%.

According to our findings, this maximum of elastic properties may be interpreted by the fact that, up to an approximately 8% mc-value, all water molecules are tightly bound to polar groups of the wood constituants (mainly cellulose). These wa-

234

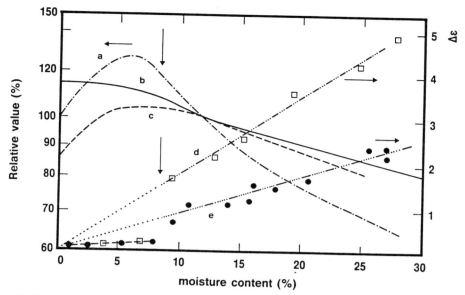

Fig. 5. Effect of moisture content on: (a) tensile strength; (b) modulus of elasticity; and (c) shear modulus of wood. 100% corresponds to 12% *mc*. Contribution of the low-temperature peak to the static permittivity, $\Delta\epsilon$, of Fagus Sylvatica (beechwood) samples with: (d) brass and (e) Ag contacts, as a function of moisture content.

ter molecules in the first hydration sites are bound by two or more hydrogen bonds (Bone and Pethig, 1985), which connect polar groups either of the same macromolecule or of adjacent fibrils, increasing thus the cohesion and stiffness of the material as water is added into the wood structure. When the number of primary hydration sites is consumed, additional water molecules bind more weakly between the micelles increasing the distance between them. Consequently all elastic properties start degrading with increasing moisture content and, at the same time, the water starts acting as a "plasticiser" by augmenting the free volume, facilitating the molecular motion and thus displacing the high-temperature band, caused by water-assisted mechanisms, towards the lower temperatures (Figs. 3 and 4).

3.4. *Plant stems and leaves*

TSDC peak recorded with plant leaves and stems (Pissis et al., 1987a, 1987b) show complete absence of dielectrically free water even for water contents as high as 60% and different organization of water in different parts of the same plant. In fact, water in stems seems to be less tightly bound than in leaves but more tightly bound than in the woody parts of the plant. The amount of tightly bound water, on the other hand, decreases in the order: leaves, stems, wood. All the above findings may reflect a general property of living systems regarding their ability to organize water in their structure.

3.5. DNA

Fig. 6 shows TSDC spectra recorded with solid DNA samples with water content between 6% and 90%, while spectra measured on DNA solutions in 10^{-3} Tris buffer, for concentrations (c) between 0% and 2% w/w, are displayed in Figures 7 and 8 (Anagnostopou- lou-Konsta et al., 1988, 1990) . From measurements on solid samples (Fig. 6) it is shown that the proportion of irrotationally bound water in DNA is of the order of 50%, while loosely bound water corresponds to about 80% of the dry sample weight. In fact a peak at around 120 K, showing the presence of bulk water, appears for water contents higher than this critical h-value. The irrotationally bound water content corresponds to 18-22 water molecules per nucleotide, a value which is in rather good agreement with results reported by other authors (Cross and Pethig, 1983, Clementi, 1983). These tightly bound water molecules exhibit properties significantly different from those of bulk water and seem to remain unfrozen even at temperatures as low as 150 K.

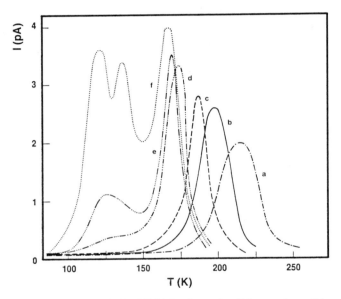

Fig. 6. TSDC plots recorded with Na-DNA hydrated solid samples with water content, h, equal to: (a) 6.7%; (b) 11.6%; (c) 24.0%; (d) 44.0%; (e) 74.2%; and (f) 90.0%.

The behaviour of the low-temperature double peak in DNA solutions (Figs. 7 and 8) as a function of DNA concentration may be explained in the following way. Before addition of DNA (curve a) peak II originates from water molecules modified by the presence of the buffer in the solution. Addition of DNA results to DNA-buffer binding, releasing water molecules which were solvating the buffer, in accordance with Bonincontro et al. (1987) and Marky et al. (1983). Peak I (bulk ice) would thus increase at the expense of peak II (modified ice) up to a certain DNA concentration corresponding to a maximum of free-water fraction (curves b, c, d). As DNA concentration increases further the free water fraction starts decreasing again, since water molecules now hydrate unoccupied sites of added DNA molecules, causing the depression of peak I and a corresponding rise of peak II

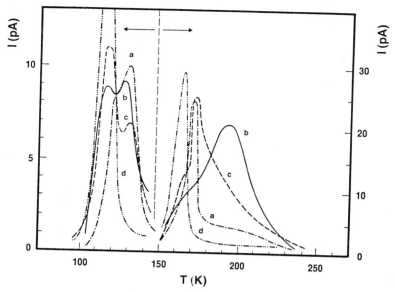

Fig. 7. TSDC plots recorded with Na-DNA solutions in 10^{-3} Tris buffer with concentrations: (a) 0%; (b) 0.12%; (c) 0.25% and (d) 0.5%.

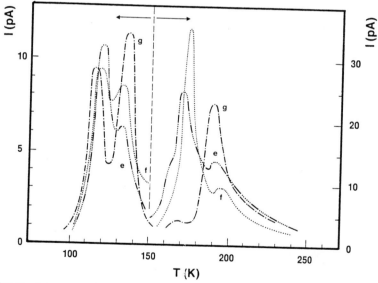

Fig. 8. TSDC plots recorded with Na-DNA solutions in 10^{-3} Tris buffer with concentrations: (e) 0.7%; (f) 1.0%; and (g) 2.0% w/w.

4. CONCLUSIONS

The use of thermally stimulated depolarization current (TSDC) techniques has proved very useful in investigating the state of water of hydration in organic and biological macromolecular systems, both in aqueous solutions and in hydrated solid samples. TSDC measurements on hydrated solid samples show that water adsorbed up to a certain critical level is irrotationally bound to the macromolecule polar bonding sites. For hydration levels higher than this value there is a sudden discontinuity in the TSDC plots, indicating an increase in the motional freedom of the bound water molecules and of the macromolecular structure and conductivity. In certain cases the results permit to speculate about the possible hydration sites and the relevant hydration mechanisms. The influence of water on the molecular mobility and the physicochemical properties of the system under investigation has also been studied in various cases.

ACKNOWLEDGMENTS

The authors wish to express their gratitude to professor E.H. Grant, Dr. C. Gabriel, and Dr. R. Tata, of King's College, London, for very valuable help and discussion. They also acknowledge the assistance of Mr. G. Loukakis in DNA samples preparation.

REFERENCES

1. Anagnostopoulou-Konsta, A., Daoukaki-Diamanti, D., Pissis, P., and Sideris, E.G., 1988, Dielectric study of the interaction of DNA and water, in *Proceedings of the 6th International Symposium on Electrets (ISE 6)*, D.K. Das-Gupta and A.W. Pattullo, ed., IEEE, New York, 271-275.
2. Anagnostopoulou-Konsta, A., Daoukaki-Diamanti, D., Pissis, P., Loukakis, G., and Sideris, E.G., 1990, Dielectric study of DNA-water systems by the thermally stimulated currents method, in *Proceedings of International Discussion Meeting on Relaxations in Complex Systems*, Crete 1990 (in press).
3. Anagnostopoulou-Konsta, A. and Pissis, P., 1987, A study of casein hydration by the thermally stimulated depolarization currents method, *J. Phys. D: Appl. Phys.* 20, 1168-1174.
4. Anagnostopoulou-Konsta, A. and Pissis, P., 1989, Dielectric study of the hydration process in wood, *Holzforschung* 43, 363-369.
5. Apekis, L., 1988, Dielectric study of dry and hydrated microcrystalline cellulose, in *Proceedings of the 6th International Symposium on Electrets*, D.K. Das-Gupta and A.W.Pattullo, ed., IEEE, New York, 281-285.
6. Apekis, L., Pissis, P., and Boudouris, G., 1983, Depolarization thermocur rents in ice Ih at low temperature depending on the electrode material. *Polarization mechanism, Nuovo Cimento* 2D, 932-946.
7. Apekis, L. and Pissis, P., 1987, Study of the multiplicity of dielectric relaxation times in ice at low temperatures, in *Proceedings of the VIIth Symposium on the Physics and Chemistry of Ice*, J. de Physique C1, 127-133.
8. Berlin, E., Kliman, P.G., and Pallansch, M.J., 1970, Changes in state of water in proteinaceous systems, *J. Colloid Interface Sci.*, 34, 488-494.
9. Bonincontro, A., Caneva, R., and Pedone, F., 1987, Hydration properties of DNA-lysine gels by microwave dielectric measurements as a function of temperature, *Eur. Biophys. J.*, 15, 59-63.
10. Bone, S. and Pethig, R., 1982, Dielectric studies of protein hydration and hydration-induced flexibility, *J. Mol. Biol.*, 181, 323-326.

11. Bucci, C., Fieschi, R., and Guidi, G. 1966, Ionic thermocurrents in dielectrics, *Phys. Rev.*, 148, 816-823.

12. Christodoulides, C., 1985, Determination of activation energies by using the widths of peaks of thermoluminescence and thermally stimulated depolarization currents, *J. Phys. D: Appl. Phys.*, 18, 1501-1510.

13. Clementi, E, 1983, Structure of water and ions for DNA, in *Structure and Dynamics: Nucleic Acids and Proteins*,
E. Clementi and R.H. Sarma, ed., Adenine Press, New York, 321-364.

14. Cross, T.E. and Pethig, R. 1983, *Int. J. Quantum Chemistry: Quantum Biology Symposium* 10, 143-152.

15. Daoukaki-Diamanti, D., Pissis, P., and Boudouris, G., 1984, Depolarization thermocurrents in frozen aqueous solutions of mono- and disaccharides, *Chem. Phys.* 91, 315-325.

16. Foster, K., Stuchly, M.A., Kraszewski, A. and Stuchly, S.S., 1984, Microwave dielectric absorption of DNA in aqueous solution, *Biopolymers*, 23, 593-599.

17. Gerhards, C.C., 1982, Effect of moisture content and temperature on the mechanical properties of wood: an analysis of immediate effects, *Wood Fiber Sci.* 14, 4-36.

18. Grant E.H., Sheppard R.J., and South, G.P., 1978, *Dielectrical behaviour of biological molecules in solution*, Clarendon Press, Oxford.

19. Kent, M. and Meyer, W., 1984, Complex permittivity spectra of protein powders as a function of temperature and hydration, *J. Phys. D*, 17, 1687-1698.

20. Kollmann, F.F. and Côté, W.A. Jr, 1968, *Principles of Wood Science and Technology, I, Solid Wood*,
Springer Verlag, 292-419.

21. Kuntz, I.D. and Kauzmann, W., 1974, Hydration of proteins and polypeptides, *Adv. Protein Chem.* 28, 239-345.

22. Lehmann, M.S., 1984, Probing the protein-bound water with other small molecules using neutron small-angle scattering, *J. Pkysique Colloque*, C7, 235-239.

23. Luck, W.A.P., 1985, Spectroscopic attempts to determine the structure of water and of polymer hydration phenomena, *Optica Pura Appl.*, 18, 71-82.

24. Marky L.A., Snyder, G.S., and Breslauer, K.J., 1983, Calorimetric and spectroscopic investigation of drug-DNA interactions, *Nucleic Acid Research*, 11, 5701-15.

25. Mathur-de-Vré, R., 1979, The NMR studies of water in biological systems, *Prog. Biophys. J.*, 50, 213-219.

26. Pethig, R., 1979, Dielectric and Electronic Properties of Biological Materials, Wiley, Chichester.

27. Pethig, R. and Kell, D.B., 1987, The passive electrical properties of biological systems: their significance in physiology, biophysics and biotechnology, *Phys. Med. Biol.*, 32, 933-970.

28. Pissis, P., 1989, Dielectric studies of protein hydration, *J. Mol. Liq.* 41, 271-289.

29. Pissis, P., 1985, A study of sorbed water on cellulose by the thermally stimulated depolarization technique, *J. Phys. D: Appl. Phys.*, 15, 1897- 1908.

30. Pissis, P. and Anagnostopoulou-Konsta, A., 1985, Depolarization thermocur rents in hydrated cellulose, in *Proceedings of the 5th International Symposium on Electrets*, G.M. Sessler and R. Gerhardt-Mulhaupt, ed., IEEE, New York 842-847.

31. Pissis, P. and Anagnostopoulou-Konsta, A., 1988, Thermally stimulated depolarization currents in hydrated casein solid samples, *Progr. Colloid Polym. Sci.*, 78, 116-118.

32. Pissis, P., Apekis, L., Christodoulides, C., and Boudouris, G., 1982, Depolarization thermocurrents in oil-in-water emulsions at subzero temperatures, *J. Phys. D: Appl. Phys.*, 15, 2513-2522.

33. Pissis, P., Apekis, L., Christodoulides, C., and Boudouris, G., 1983a, Dielectric study of dispersed ice microcrystals by the depolarization thermocurrent technique, *J. Phys. Chem.*, 87, 4034-4037.

34. Pissis, P., Diamanti, D., and Boudouris, G., 1983b, Depolarization thermocurrents in frozen aqueous solutions of glucose, *J. Phys. D: Appl. Phys.*, 16, 1311-1322.

35. Pissis, P., Anagnostopoulou-Konsta, A, and Apekis, L., 1987a, Binding modes of water in plant leaves: a dielectric study, *Europhysics Letters*, 3, 119- 125.

36. Pissis, P., Anagnostopoulou-Konsta, A., and Apekis, L., 1987b, A dielectric study of the state of water in plant stems, *J. Exp. Botany*, 38, 1528-1540.

37. Pissis, P. and Daoukaki-Diamanti, D., 1988, Dielectric study of aqueous solutions and solid samples of methylcellulose, *Progr. Colloid Polym. Sci.*, 78, 27-29.

38. van Turnhout, J., 1980, Thermally stimulated discharge of electrets, in *Topics in Applied Physics, Vol. 33: Electrets*, G.M. Sessler, ed., Springer, Berlin, 81-215.

39. Vanderschueren, J. and Gasiot, J., 1979, Field-induced thermally stimulated currents", in *Topics in Applied Physics, Vol. 37: Thermally Stimulated Relaxations in Solids*, P. Braunlich, ed., Springer, Berlin, 135-223.

PROTONIC CONDUCTIVITY IN BIOMATERIALS

IN THE FRAME OF PERCOLATION MODEL

Giorgio Careri and Andrea Giansanti

Dipartimento di Fisica, Università di Roma I
P.le A. Moro 2, I-00185 Roma (Italy)

1. INTRODUCTION

In the last few years we have studied the dielectric properties of nearly dry biological materials as a function of their water content. Details of the experimental apparatus have been given[1]. Let us only mention here that our technique combines a digital balance to monitor sample's weight with an a.c. bridge to record radiofrequency dielectric properties of a composite capacitor, designed to avoid metal contacts and to ensure uniform evaporation from the sample. The investigation of lysozyme powders was quite fruitful because we were able to assess the onset of a protonic percolative conductivity at a critical hydration threshold. This threshold was found to be coincident with the critical hydration for the onset of enzymatic activity[3-7]. This study was extended to other more complex biomaterials, of different origin: purple membrane[8], artemia cysts[9], and maize-seed components[10]. In all these cases a critical hydration for the onset of a biological function was already known, and our data showed the percolative threshold for proton motion to be correlated with the emergence of biological function[11].

2. LYSOZYME

Lysozyme was chosen because of the rather complete picture available of the hydration dependence of various physical properties[2-4]. The hydration of this molecule is step-wise and consists of at least three stages: i) from 0 to 60 water molecules per lysozyme molecule; at this stage water molecules interact only with charged groups of the protein. ii) from 60 to 220 waters; clusters of mobile waters tend to form around polar groups on the surface of the protein, leaving unfilled portions. iii) from 220 to about 300 waters ; the unfilled portions are gradually hydrated until monolayer coverage is reached. At a coverage near 160 waters per lysozyme molecule the enzymatic function emerges and increases with hydration.

Biologically Inspired Physics, Edited by L. Peliti
Plenum Press, New York, 1991

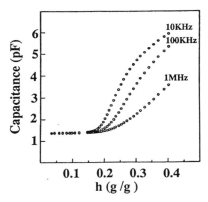

Fig. 1. Capacitance of the composite capacitor: lysozyme powder, pH=5.3, from [14].

In a first investigation[5], dielectric losses were measured for samples of lysozyme that had isopiestically hydrated with H_2O and D_2O. We detected a clear isotope effect. This showed that the d.c. conductivity associated to the observed hydration dependent relaxation had a protonic character, and the pH dependence of the conductivity suggested the involvement of ionizable side-chains on the protein surface in the protonic conduction. On these grounds we were able to picture the conduction process as a mean free path limited random walk of a proton on the surface of a single macromolecule.

Next, in a second paper[6], we investigated the hydration dependence of the capacitance in the low hydration region. In lysozyme powders there is a sharp increase of the capacitance at a threshold water content $h_c = 0.150 \pm 0.016$ g of water per g of protein, followed by saturation at increased hydrations (see fig. 1). We interpreted this threshold as a percolation threshold. In fact, the hydration level for monolayer coverage of the protein surface is $h_m = 0.38 \pm 0.04$ and if we estimate the percolation threshold as $p_c = h_c/h_m$ we obtain $p_c = 0.40 \pm 0.04$. This experimental value is in agreement with the value 0.45 ± 0.03 predicted by percolation theory for a two-dimensional network[12,13]. The threshold at $h_c = 0.15$ was observed to be the same for both D_2O- and H_2O- hydrated samples, to be constant from pH 3 to pH 8 and frequency independent. All these observations confirmed the percolative character of the transition, due only to the number of water molecules acting as interconnected conducting sites on the surface of the protein. The percolative transition occurs in the middle of hydration stage ii), where from heat capacity data one can rule out the presence of phase transitions.

When the lysozyme molecule was complexated with substrate the threshold was detected at a higher hydration; near $h_c = 0.20$ g of water per g of protein. This hydration level, is close to the one critical for the onset of the enzymatic activity[4], and strongly supports the idea that protonic percolation is involved in the catalytic process, because the presence of substrate perturbs the percolative paths through the active site of the protein.

All the preceding observations were based on the scrutiny of capacitance data at different frequencies, ranging from 10 kHz to 10 MHz. From the frequency dependence of the dielectric response of the composite capacitor one can derive the d.c.

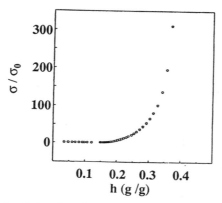

Fig. 2. Normalized conductivity for the same sample as in fig. 1; σ_0 is the limiting low hydration value.

conductivity σ of the powders, which displays a similar sharp increase at the same critical hydration. We have then interpolated, near and above h_c, the hydration-dependent d.c. conductivity by a power law, as usually done in critical phenomena:

$$\sigma(h) - \sigma(h_c) = K(h - h_c)^t.$$

We obtained[7] $t = 1.29 \pm 0.01$ for both H_2O- and D_2O- hydrated samples, where the ratio K_{H_2O}/K_{D_2O} was found to be $\sqrt{2}$, within experimental errors, confirming two facts: first, that the critical exponents t are in close agreement with the theoretical predictions for a 2D percolative process and, second, that the conduction is protonic, being affected by the square root of the carrier mass in the kinetic K factors. In figures 1,2 and 3 we show, for a particular sample at pH 5.3, the appearance of the hydration threshold, the parabolic character of the conductivity near this threshold, and the limiting value of the critical exponent[14]. Recently the hydration dependent percolation of protons in lysozyme powders has been investigated by TSDC (Thermally Stimulated Depolarization Currents)[15] spectroscopy. By this technique it was possible to detect the appearance of a peculiar relaxation, exactly above the critical hydration threshold $h_c = 0.15$. This relaxation was attributed to the protonic percolation previously detected by our technique.

In closing this section we want to point out that the results of percolation theory used above have been theoretically derived for infinite planar lattices. Finite size effects in 2D planar lattices with a number of sites comparable to the number of hydration sites existing on the surface of a single protein molecule have been studied[3], with the conclusion that the standard percolation theory still holds for such a small flat area. However one would like to see the theory extended to a nearly spherical surface, and possibly to a non-uniform distribution of sites, to cope with the real surface of the protein.

3. OTHER BIOLOGICAL MATERIALS

The experimental technique developed for lysozyme has been extended to other biomaterials of increasing complexity, where a critical hydration threshold was already known to control the onset of a biological function.

In lyophilized fragments of purple membrane from *Halobacterium halobium* [8] a similar transition in the protonic conductivity was found at a critical hydration of $h_c = 0.046$ g of water/ g of membrane. This system is known to display a typical photoresponse above a hydration level of 0.05. The experimentally found critical exponent t=1.23, for the hydration dependence of the protonic conductivity near the transition threshold, is characteristic of a 2D percolative process, as mentioned above for lysozyme. The threshold is frequency independent, neither h_c nor t are affected by deuteration and the ratio of the K constants is 1.38, not far from the square root of the mass ratio H/D, as expected for a classical kinetic process involving protons.

Artemia cysts (containing about 4000 cells per cyst) have also been studied by a similar technique[9]. This system is capable of undergoing repeated reversible cycles of dehydration and rehydration. Its physical and physiological behaviour has already been assessed. The dielectric response of the cysts over the frequency range 10 kHz - 10 MHz was investigated in our laboratory at hydration levels ranging from near zero up to 0.5 g of water/ g of dry material. The conductivity displayed an abrupt increase above h=0.35, an hydration at which metabolism is first initiated in this system, and the critical exponent of the conductivity was 1.65, in agreement with theory for a 3D process.

Similar studies in maize-seed components carried out in our laboratory have been recently reported[10]. Also in this case the conductivity was found to be due to proton displacements, as revealed by the ratio of the K constants, and displayed a percolative character. The critical exponent was found to be 1.23, a typical value for a 2D percolation. It is noteworthy that, in this system, near the hydration level $h = 0.10$, the physiological process of gas exchange, which precedes seed germination, is known to set in.

4. CONCLUDING REMARKS

The emphasis of the percolation model is on randomness of the arrangements of the conducting elements. This statistical viewpoint is particularly important for membranes, for which conduction and transport are the principal biological activities.

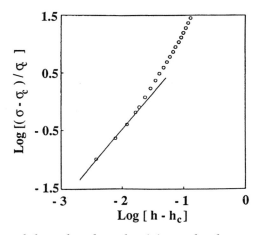

Fig. 3. Log - Log plot of the reduced conductivity vs $h - h_c$, same sample. The slope is 1.25 ± 0.02, from ref.[14]

It may not be necessary to have a relatively rigid geometry and a definite pathway for conduction. Conduction can follow from statistical assemblies of elements only partially filling a surface or pore and might be controlled by subtraction or addition of a few elements, moving the fractional occupancy backward or forward through the transition region [8].

We are convinced that the percolation model can have several applications to describe the emergence of function in biological systems, where disorder is present at a microscopic scale. A viewpoint from physics is that long range connectivity must be a necessary condition in order to let several disordered subunits stimulate and control each other in the living process. Long-range connectivity between subunits must be established via stochastic mechanisms, and the presence of a threshold should be expected according to this model.

As a matter of fact the percolation threshold h_c for proton conductivity, detected in four disordered biological systems in our laboratory, was found to coincide with the emergence of biological function. In these nearly dry biosystems long-range water-assisted proton transfer does control the emergence of function.

Table 1. Critical exponents of percolation conductivity, adapted from ref[11]

2D Systems

glass spheres and silver-coated spheres	1.25 ± 0.10
lysozyme powders, low hydration	1.29 ± 0.05
purple membrane fragments	1.23 ± 0.05
corn embryo pellets	1.23 ± 0.05
theory, by finite size-scaling	1.26 ± 0.05
theory, by transfer matrix	1.28 ± 0.03

3D Systems

amorphous cermet film	1.9 ± 0.2
amorphous carbon and teflon powder	1.8 ± 0.2
microcrystalline acetanilide	1.72 ± 0.05
artemia cysts	1.65 ± 0.05
theory, by finite size-scaling	1.87 ± 0.04
theory, by series expansion	1.95 ± 0.03

In table 1 some data in the literature are displayed together with our results. It is very satisfactory to see that the scaling law for conductivity is followed with such accuracy in quite different materials, both for electrons and protons. The importance of table 1 is that it offers complete evidence for the validity of the percolation model in different disordered materials of a living and nonliving origin.

We also note that, because of the experimentally proved[16] equivalence between percolation theory and fractal geometry, at the critical threshold the onset of function is achieved thanks to the presence of an "infinite" fractal. Thus the old binomium, widely used in biology, $function - form$, loses its validity here, because an extended fractal is "formless". Moreover the presence of fractals on both sides of the critical threshold prevents this functional transition from being described in terms of symmetry breaking.

All previous considerations are grounded on the behaviour of nearly dry biological systems, where hydration essentially controls biological function. The application of percolation theory to other disordered biosystems, which exhibit a discontinuity in a functional property without displaying a phase change, seems quite reasonable. The main problem is to detect the physical variables which control the onset of biological function (as water level does for nearly dry biological systems). We are convinced that once a drastic reduction of the complex pattern of interactions has been made, the percolation model will successfully describe the emergence of function, if the connectivity among the disordered subsystems is also controlled by the same variables.

ACKNOWLEDGMENTS

This work has been supported by EEC, under contract n. SCI 0229-C and by INFM (Istituto Nazionale di Fisica della Materia).

REFERENCES

1. G. Careri and A. Giansanti, Dielectric properties of nearly dry biological systems at megahertz frequencies, in: *Membranes, Metabolism, and Dry Organisms* A. C. Leopold, ed., Cornell University Press, Ithaca (1986).
2. J. A. Rupley, E. Gratton and G. Careri, Water and globular proteins, *Trends Biochem. Sci.* 8:18 (1983).
3. J. A. Rupley and G. Careri, Protein hydration and function, *Adv. Protein Chem.* in press (1990).
4. J. A. Rupley, P. H. Yang and G. Tollin, Thermodynamic and related studies of water interacting with proteins, in: *Water in polimers, ACS Symposium Series vol. 127* S. P. Rowlands, ed., American Chemical Society, Washington (1980).
5. G. Careri, M. Geraci, A. Giansanti and J. A. Rupley, Protonic conductivity of hydrated lysozyme powders at megahertz frequencies, *Proc. Natl. Acad. Sci. U. S. A.* 82:5342 (1985).
6. G. Careri, A. Giansanti and J. A. Rupley, Proton percolation on hydrated lysozyme powders, *Proc. Natl. Acad. Sci. U. S. A.* 83:6810 (1986).
7. G. Careri, A. Giansanti and J. A. Rupley, Critical exponents of protonic percolation in hydrated lysozyme powders, *Phys. Rev.* A37:2703 (1988).
8. J. A. Rupley, L. Siemankowski, G. Careri and F. Bruni, Two dimensional protonic percolation on lightly hydrated purple membrane, *Proc. Natl. Acad. Sci. U. S. A.* 85:9022 (1988).
9. F. Bruni, G. Careri and J. S. Clegg, Dielectric Properties of artemia cysts at low water contents, *Biophys. J.* 55:331 (1989).
10. F. Bruni, G. Careri and A. C. Leopold, Critical exponents of protonic percolation in maize seeds, *Phys. Rev.* A40:2803 (1989).
11. G. Careri, Emergence of function in disordered biomaterials, in: *Symmetry in nature, a volume in honour of L. A. Radicati di Brozolo*, Scuola Normale Superiore, Pisa (1989).
12. R. Zallen, *The Physics of Amorphous Solids*, Wiley, New York (1983).
13. D. Stauffer, *Introduction to Percolation Theory*, Taylor and Francis, London (1985).

14. G. Careri, A. Giansanti and J. A. Rupley, Detection of protonic percolation on hydrated lysozyme powders, in: *Disordered Solids: structures and processes*, B. Di Bartolo ed., Plenum Press, New York (1989).
15. P. Pissis and A. Anagnostopoulou Konsta, Protonic percolation on hydrated lysozyme powders studied by the method of thermally stimulated depolarisation currents method, *J. Phys. D: Appl. Phys.* 23:932 (1990).
16. R. F. Voss, R. B. Laibowitz and E. I. Allessandrini, Fractal (scaling) clusters in thin gold films near the percolation threshold, *Phys. Rev. Lett.* 49:1441 (1982).

PROBLEMS IN THEORETICAL IMMUNOLOGY

Gérard Weisbuch

Laboratoire de Physique Statistique
Ecole Normale Supérieure
24 rue Lhomond, F 75231 Paris Cedex 5, France

1. INTRODUCTION

Immunology [1] is not a new field of research: the first vaccinations against small-pox were done by the end of the seventeenth century by Jenner. But the practice of vaccination, although much more intricate than described in elementary textbooks, remained largely empirical. The important successes of experimental methods derived from immunology in molecular biology never had much counterpart from the theoretical point of view. Only a few models of infection, based on population dynamics, have been proposed. Until quite recently most immunologists considered that clinical conditions could be explained by the presence or absence of some specific macromolecules or cell types. Such simple approaches have in fact been sufficient to handle efficiently a number of clinical and experimental problems. It is only quite recently that the self/non-self recognition problem led N. Jerne[2] to formulate his network hypothesis, and that people realized that the consequences of the hypothesis could only be checked through some effort in theoretical modeling[3,4]. Before presenting our contribution let us review the important characteristics of the immune response that are indispensable to the understanding of our model.

2. ELEMENTARY PROCESSES (immunology for pedestrians)

2.1. *Recognition*

Foreign species invading our body are called antigens. Each different antigen triggers a specific response from the immune system. The immune system maintains a large variety of macromolecules, the immunoglobulins or antibodies, and cells, called lymphocytes, among which only a few are actually able to react with a given antigen. There are of the order of 10^8 different antigen specificities (or idiotypes) for lymphocytes and antibodies. A lymphocyte is able to produce only one type of antibody. The specificity of interaction process between an antigen and antibodies

Biologically Inspired Physics, Edited by L. Peliti
Plenum Press, New York, 1991

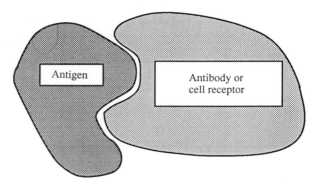

Fig. 1. Recognition of an antigen by an antibody or a cell receptor. The two macromolecules recognize each other and are able to interact if their shapes are
complementary.

or cells is based on recognition: the antigen is recognized (i.e. interacts with) by
macromolecules which have an external shape complementary to the shape of the
antigen (see figure 1). This indispensable recognition process is the first step of a
succession of events that finally results in the elimination of the antigen.

2.2. *Clonal proliferation*

The cells of the immune system, the lymphocytes, have on their membranes
macromolecular receptors similar to the antibodies, which are able to recognize the
antigens. If a recognition process occurs the cell changes its state, i.e., its biological function. A cell previously at rest differentiates into an active cell which can
for instance synthetize antibodies and secrete them in the blood or it can secrete
small proteins that act as signals for other lymphocytes (interleukins). Furthermore,
activated cells can proliferate. Since they are the only ones to do so, their relative
number increases with respect to cells at rest which do not proliferate. This selective recognition mechanism which is necessary for proliferation is the basis for the
specificity of the immune response. The term clonal selection means that the cell
population specific of the antigen derives from a single cell (or a few individuals)
that first recognized the antigen.

2.3. *Elimination of the antigen*

Once proliferation is large enough, which might take a few days, the antigen
is eliminated by various cellular or molecular mechanisms: phagocytosis, cell lysis,
complex formation... These three successive processes are now well understood by
immunologists.

3. THE COLLECTIVE BEHAVIOR

The purpose of our model is to explain certain collective behaviors of the immune system at the clinical level, and among them, memory, self/non-self recognition
and some failures of the immune system such as auto-immune diseases.

3.1. *Vaccination and memory*

Introducing a given antigen in the body is called antigen presentation. Under certain circumstances the first presentation of an antigen can be memorized: the secondary response, obtained when the antigen is presented for the second time is greater than the primary response. More antibodies specific for the antigen are produced more rapidly than during the primary response. This phenomenon is used in vaccination.

3.2. *Self-nonself discrimination*

The 10^8 different idiotypes that we mentioned earlier are such a large variety that any antigen of any shape can be recognized by at least one idiotype. One says that the immune repertoire is complete, which means that the immune system is able to deal with any foreign antigen. Since any shape is recognized, how does the immune system makes the difference with the shapes of the macromolecules used by our own body? How come that the immune system does not attack our own cells? The absence of the immune response is called tolerance and one has then to explain the tolerance of the self.

3.3. *Auto-immune diseases*

Tolerance of the self is not always achieved and it sometimes happens that the immune system does attack the cells of the body. We do know of terrible auto-immune diseases that are very difficult to treat: juvenile diabetes, multiple sclerosis, rheumatoid arthritis, lupus erythematosis... In these diseases, antigens of the self are the target of the attacks of the immune system[1].

Theories different from the network theory we are going to discuss further have first been proposed to explain the above behaviors, but none of them is fully satisfactory. For instance the existence of memory cells that would be created after the first antigen presentation could explain vaccination; but these long-life time memory cells were never observed. Tolerance of the self has been tentatively explained by the clonal elimination theory: clones that would recognize cells of our body would be eliminated soon after birth, and only lymphocyte populations that could not recognize the self would remain present in the immune system. Although some clonal elimination processes might indeed exist, the perfect discrimination, necessary if this were the only mechanism for self tolerance, is impossible because of thermodynamics. Furthermore, idiotypes dangerous for the self exist even in healthy individuals. The reader should be aware that the model we shall now discuss is far from being accepted by the whole communauty of immunologists, and that there still exists a number of defenders of the opposite theories that we mentioned above.

4. THE IMMUNE NETWORK HYPOTHESIS

The following hypothesis has been proposed by N. Jerne in 1974 [2]:

The same recognition mechanisms that work for foreign antigen, also work for cells of our body and lymphocytes and antibodies of the immune system.

The new thing is the idea that, since the components of the immune system are

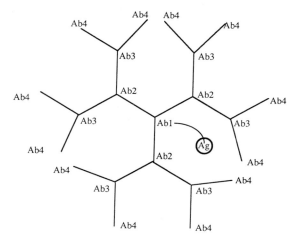

Fig. 2. An idiotypic network. Antigen Ag is recognized by clone Ab_1 which is itself recognized by clones Ab_2, which are recognized by clones Ab_1 and Ab_3 and so on.

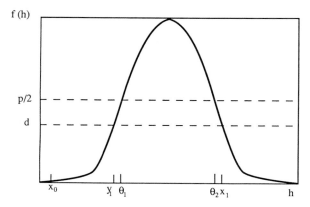

Fig. 3. f(h), the proliferation function of the field h acting on a clone x_0, x_1 and x_2 are steady state populations defined in the text.

recognized by some other components, we are in the presence of a network of interacting immune species (see figure 2). The elements of the immune network are the population of clones that express a given cell receptor on their membrane and that are able to secrete the corresponding antibody. Interactions among the elements of the net are anti-idiotypic interactions that can exist among macromolecules of complementary shapes. The connectivity of the network is then rather small. The number of network elements is 10^8 in the human body. But each element can represent a large population of identical cells. When this static representation of the network was first proposed by Jerne it was criticized by a kind of percolation argument: if antigen is recognized by Ab_1 then Ab_1 is going to proliferate and it will be recognized by Ab_2. Then if Ab_1 is recognized by Ab_2, Ab_2 is going to proliferate and it will be recognized by Ab_3 and so on. The whole network will start to proliferate and specificity of the immune response will be lost. Of course this argument only supposes positive interactions and the actual behavior of the network can only be understood if a more precise description of the interactions among the elements is given.

5. A MATHEMATICAL MODEL[5]

This model is one of the most simple model of Jerne network that can be conceived, while still exhibiting interesting immune properties. The dynamical behavior of each element i obeys a differential equation that describes the time evolution of the population x_i of clone i:

$$\frac{dx_i}{dt} = s + x_1(f(h_1) - d) , \tag{1}$$

where s is the source term corresponding to cells incoming from the bone marrow, the function $f(h_i)$ defines the rate of proliferation, and d specifies the rate of cell population decay. For each clone i we consider the total amount of anti-idiotypic stimulation as a linear combination of the populations of other interacting clones j. We call h_i the field acting on clone x_i:

$$h_i = \sum_j J_{ij} x_j , \tag{2}$$

where J_{ij} specifies the affinity between clones x_i and x_j. The choice of a J matrix defines the topology of the network. The proliferation function $f(h_i)$ has a stimulatory and a suppressive part with a threshold for activation (i.e. θ_1) and one for suppression (i.e. θ_2). Within the stimulatory part, increasing the idiotypic field increases proliferation; within the suppressive part increasing the field decreases proliferation. Similar dose response curves are found in receptor crosslinking [6]. Fig. 3 displays the bell-shaped $f(h)$ based [7] upon two Michaelean saturation functions with the two thresholds θ_1 and θ_2 ($\theta_2 >> \theta_1$), and a maximum proliferation, p:

$$f(h) = \left(\frac{h}{\theta_1 + h} \right) \left(1 - \frac{h}{\theta_2 + h} \right) = \frac{h\theta_2}{(\theta_1 + h)(\theta_2 + h)} \tag{3}$$

5.1. *The two population problem*

Before handling the case of a network of 10^8 clones, let us discuss the simple case of two interacting populations, one specific of the antigen that we call idiotype x, and the other one specific for the idiotype, the anti-idiotype, y. The interaction constant J_{ij} is set to 1. The system of equations (1) now becomes:

$$\frac{dx}{dt} = s + x(f(y) - d) , \tag{4a}$$

$$\frac{dy}{dt} = s + y(f(x) - d) , \tag{4b}$$

After presentation of the antigen, first x, and then y increase and by the time the antigen is eliminated they reach some value that we take as the initial conditions for the solution of system (4). We are then interested in the attractors of system (4). The two isoclines $\dot{x} = \dot{y} = 0$ intersect in five points of which only three are stable.

1. The virgin configuration is obtained when the proliferation function is very small. It has coordinates:

$$x_0 = y_0 = s/d$$

2. In the limit of small s, the other solutions correspond to intersections of the proliferation function with the decay term (see figure 3). Data from experimental immunology allow to consider s/d as small as compared to θ_1. Typically s/d can be between 1 and 10, and θ_1 is of the order to 1000. In the immune configuration, the idiotype has a large population and exercises a large suppressive field on the anti-idiotype which has a small population and exercises a small excitatory field on the idiotype. Small or large field solutions are computed by approximating $f(h)$ by either $ph/\theta_1 + h$ or $p\theta_2/\theta_2 + h$ and by considering s/d as small with respect to θ_1. One obtains:

$$x_1 = \frac{p'\theta_2}{d} \quad , \quad y_1 = \frac{d\theta_1}{p'} \quad ,$$

where $p' = p - d$.

Attractivity towards the equilibrium is obtained by solving system (4) linearized in the vicinity of x_1, y_1. From the solution of the characteristic equation attenuated oscillations towards the attractor are then expected and indeed observed (figure 4 and 5).

3. The tolerant attractor is obtained by exchanging x_1 and y_1 (figure and 5).

4,5. The two symmetrical solutions with either suppressive or excitatory fields are unstable. Computer simulations indeed verify these predictions (figure 4).

5.2. *The attractors of the large network*

The immune network is defined by the J_{ij}'s. Unfortunately our knowledge of the actual J_{ij}'s is very restricted although some experiments try to measure affinity matrices for panels of antigens [8,9]. Several attempts were also made to predict them from theoretical reasoning (shape space [10,11], or fitting strings of bits [12]). Of course, the spin glass specialists used random J_{ij}'s [13,14]. For reasons of simplicity, we have chosen a Cayley tree architecture [5] with m connections per site (see figure2), and

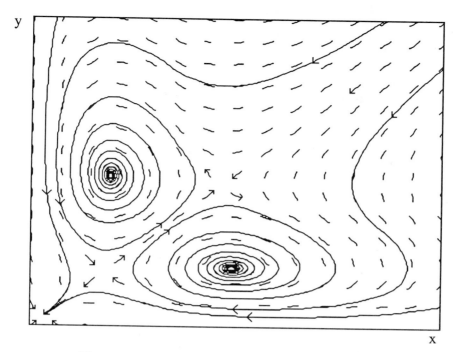

Fig. 4. Trajectories of system (4) in the $x - y$ plane.

J_{ij}'s which can only be 0 (no interaction) or 1 (maximum interaction). The root of the tree is selected by antigen presentation. We have checked that this very simple scheme is generic for a much larger class of networks. The different fields for clones 1,2,3 ... i are given by:

$$h_1 = mx_2 \quad ,$$

$$h_2 = x_1 + (m - 1)x_3 \quad ,$$

$$h_3 = x_2 + (m - 1)x_4 \quad ,$$

$$\cdots\cdots\cdots$$

$$h_i = x_{i-1} + (m - 1)x_{i+1}.$$

Our problem is to classify the different attractors of the network and to interpret the transitions from one attractor to another one under the influence of antigen presentation. Observations from immunology restrict our choice to localized attractors. Let us start with the most simple virgin configuration, corresponding to the hypothetical case where no antigen has yet been presented and all populations are s/d (all proliferation functions are 0). After presentation of the first antigen, memorization is obtained if some populations of the network reach stable populations different from s/d. We also want that when successive presentations of different antigens are made, the network remembers all these antigens: memorization of one antigen should be robust enough not to be destroyed by the presentation of another antigen. Such a scheme works only if the response to any given antigen remains localized in the vicinity of the idiotypic clone(s). In such a case, to each antigen correspond in the network a patch of clones that are modified by the presentation. As long as patches corresponding to different clones do not overlap, all these clones can be memorized. Once the idea of localized non interacting attractors is accepted, everything is simplified: instead of solving 10^8 equations, we only have to solve a small set of equations for those interacting clones with large populations, supposing that those neighboring clones which do not belong to the set have populations s/d.

Vaccination

Let us take for instance the case of antigen presentation on clone Ab_1 which results in excitation of clones Ab_2, clones Ab_3 remaining at their virgin level (see figure 5, left). In analogy with the two clones problem, we expect a low field solution h_L for Ab_1, and a large suppressor field solution h_S for Ab_2. The field equations allow to compute the populations:

$$h_1 = mx_2 = h_L = \frac{d\theta_1}{p'} \quad ,$$

$$h_2 = x_1 + (m - 1)\frac{s}{d} = h_S = \frac{p'\theta_2}{d}.$$

Of course, the solution remains localized only if the field h_3 on x_3 is much less than h_L, otherwise x_3 would also proliferate:

$$h_3 = \frac{h_L}{m} + \frac{(m-1)s}{d} < h_L$$

This is only possible if m is larger than 1 (a one dimensional network would then give rise to percolation; multiple connectivity is essential to localization) and if the following inequality is fulfilled:

$$\frac{p'ms}{d^2} < \theta_1.$$

It is in fact equivalent to saying that the field due to m virgin clones should be less than the proliferation threshold.

Tolerance

Another localized attractor corresponds to tolerance (see figure 5, right). A strong suppressive field acts on Ab_1 due to Ab_2's, Ab_2's proliferate due to a low field provided by Ab_3's, but Ab_4's remain nearly virgin. The field equations allow to compute the populations:

$$h_2 = x_1 + (m - 1)x_3 = h_L = \frac{d\theta_1}{p'} \quad ,$$

which gives x_3 by neglecting x_1 which is small.

$$h_3 = x_2 + \frac{(m-1)s}{d} = h_S = \frac{p'\theta_2}{d}.$$

and then:

$$h_1 = mh_S$$

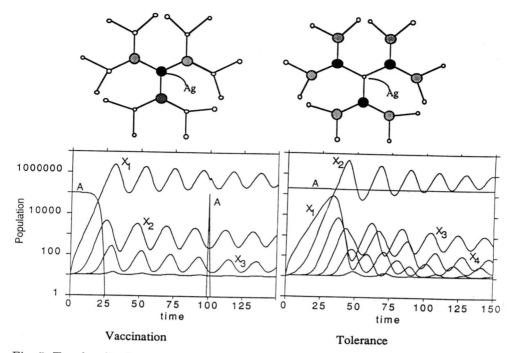

Fig. 5. Two localized attractors of the immune network corresponding to vaccination (on the left) and tolerance (on the right). On the above nets, black circles correspond to large suppressive populations (of the order of θ_2), gray circles to medium excitatory populations (of the order of θ and small white circles to small virgin or oversuppressed populations (of the order of s/d). The two lower graphs are time plots of the logarithm of the populations (numerical simulations results). A is antigen concentration. Vaccination as a memory effect is illustrated on the left plot: a second antigen presentation with the same dose at time $t = 100$ results in a faster elimination than during the first presentation. Slow decay of large antigen concentration results in tolerance.

which, injected in equation (1), shows that x_1 is of the order of s/d. h_4 small and x_4 nearly virgin.

Percolation

Some parameter set-up also allow excitation to propagate from the initial nodes accross all the lattice. In such a case lation is observed, which corresponds to a non-localized dynamics (figure 6). We of course do not expect this to be a healthy condition, but in some auto-immune diseases such as lupus erythematosis a large number of different auto-antibodies is detected as if there were percolation of the immune response.

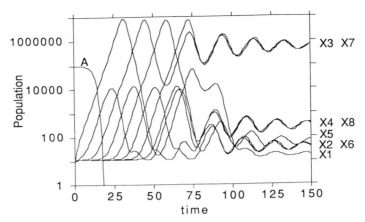

Fig. 6. Time diagram for percolation. After antigen presentation, clones 1 and 2
return to virgin state, but clone 3 and all clones with number multiple of
four apart such as 7 are excited at the suppressive level. Clone 4,8,12... are
sustaining the excitation of 3,7,13...

5.3. *First conclusions*

The model predicts localized attractors that can be interpreted in terms of
vaccination or of tolerance. Since we are dealing with localized attractors the generalization of the model to networks with loops is pretty simple. Differential systems
equivalent to system (1) can be analysed by the methods that we have developped
here. Non-identical J_{ij}, networks with loops... give similar results. p, the maximum
number of meaningful attractors can be evaluated by computing the probability that
p non related antigens (and then arbitrarily positioned on the network) have non
overlapping patches. The following scaling law for p is derived [11]: $p = a\sqrt{N}$.

Let us now shortly review two recent applications of more elaborate versions of
the simple model.

6. ATTAINING THE ATTRACTOR: DYNAMICS WITH ANTIGEN

Now that we have characterized the attractors of the dynamics, we would like to
be able to predict which attractor is obtained according to the conditions of antigen
presentation. We know from immunology that, according to the age of the individual
(newborn or adult), and the conditions of antigen presentation (high or low dose,
presence of adjuvant, intravenous or subcutaneous injections), vaccination, tolerance
or no response is obtained. In the model the influence of the antigen on clone Ab_1 is
taken into account by the addition of a new term in h_1:

$$h_1 = mx_2 + Ja.$$

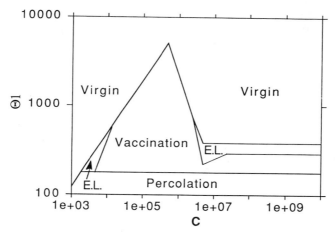

Fig. 7. Diagram of the attractor attained by the dynamics as a function of the first proliferation threshold θ_1 and C, the parameter of antigenic stimulation.

Antigen concentration a is decreased because of interaction with Ab_1:

$$\frac{da}{dt} = -kax_1$$

The nature of the attractors reached according to the parameters of the computer simulations are summarized on figure 7. The kinetic constant k and the initial antigen concentration are the parameters that define antigen presentation: theoretical analysis show that they can be combined into a unique relevant parameter C defined as Ab_1 concentration at the time of antigen elimination.

By further simplifying the present model one is able to derive the scaling laws that separate the different regimes in the parameter space. The idea is to replace the proliferation function f(h) by a window automaton which is 0 when the field is below θ_1 and above θ_2, and p in between.

$$f(h) = \begin{cases} p & \text{if } \theta_1 < h < \theta_2 \\ 0 & \text{otherwise} \end{cases}$$

Populations excited from the virgin level have then either exponential decay or increase (with respective exponents d and $p - d$) according to the values of the field. The times between successive excitations can then be estimated and from them the transitions between regimes computed.

259

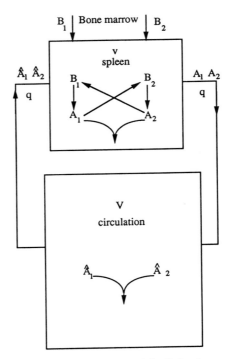

Fig. 8. A two-compartment model of the immune system.

7. SECONDARY LYMPHOID ORGANS

The simple model can be made closer to immunological reality by taking into account the fact that antibodies dynamics do not follow instantaneously the variation in cell populations, due to different lifetimes of cells (days) and antibodies (weeks)[17]. Unfortunately, simple extensions of sytem (1) to include antibody dynamics tend to display chaotic behavior for a "physiological" range of parameters[18].

Fortunately enough, when one takes also into account the fact that the immune system is not performing in a well stirred vessel, but rather in different compartments (blood and lymph circulation, lymphoid organs such as bone marrow, thymus, spleen, lymph nodes...) stable attractors are restored [19]. Figure 8 displays the extension to six species of the two populations model. B_1 and B_2 are the idiotype and anti-idiotype cells populations only present in the spleen. They experience fields due to antibodies A_2 and A_1 that are present in the spleen. Antibodies leave the spleen with flux constant q for blood general circulation where they have concentrations $\hat{A}_1$ and $\hat{A}_2$. All six species have decay times. Antibodies are secreted by B cells; they also decay because of recombination with anti-idiotypic antibodies. The relevant parameters of dynamics are ν, the volume ratio between spleen and general circulation, and δ and λ, the ratios of antibodies decay and inverse residence time of antibodies to cell decay. For small values of λ, corresponding to large residence time, stability is observed for δ larger than .21 as in the one-compartment model[18]: antibodies and cells decay times are not too different. The transition corresponds to the condition $\delta + \lambda = .21$, which

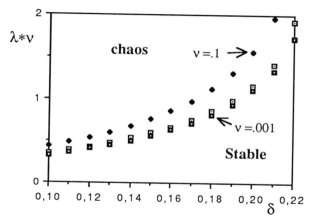

Fig. 9.   Regime transitions from stable attractors to chaos for fast mixing: $\lambda*\nu$ corresponds to a rate of mixing which varies as a function of δ, the relative decay rate of antibodies.

implies that both decay mechanisms are equivalent. At large values of λ, antibodies re-enter the spleen before any significant decay inside general circulation: the relevant parameter for the transition to chaos at large λ is $\lambda*\nu$ as seen on figure 9.

8. CONCLUSIONS

We now have a simple model of the immune network which still exhibits a very rich dynamical behavior. A comparison with neural nets, in their Hopfield version[20] for instance, reveals a similar approach by the way of the analysis of attractors. Among differences we might note the fact that in the immune nets the attractors are localized rather than distributed, learning occurs by antigen presentation which results into a local change of configuration of the attractor rather than a change of the J_{ij}'s, and the capacity of the networr varies as $\sqrt{N}$ rather than N. In accordance with biological facts, those attractors reached under antigenic stimulation (vaccination, tolerance, percolation) depend upon the conditions of the stimulation (dose, adjuvant, ...) and upon the parameters of the network (to be related to the age of the animal). Modeling the immune system is not only describing a new dynamical system. It is important to better understand this dynamics in order to devise protocols of intervention in auto-immune diseases[21].

ACKNOWLEDGMENTS

We thank H. Atlan, R. de Boer, B. Derrida, A. U. Neumann and A. S. Perelson for helpful discussions. GRIND[22] software was used for numerical simulations. The Laboratoire de Physique Statistique is associated with CNRS (URA 1306) and we acknowledge financial support from INSERM grant 879002.

REFERENCES

1. Hood, Weissman, Wood and Wilson, *Immunology*, Benjamin, Menlo Park (1984).
2. Jerne, N.K. Towards a network theory of the immune system *Ann. Immunol. (Inst. Pasteur)* 125C, 373-389 (1974).
3. A.S. Perelson, Ed., *Theoretical Immunology. SFI Studies in the Science of Complexity, Vol.III*, Addison-Wesley, Redwood City, CA (1988).
4. Immunol. Rev., *110, issue on Immune network theory (1989).*
5. Weisbuch G., R. de Boer and Perelson, A.S., J. Theo. Biol.,to appear (1990).
6. A.S. Perelson, *in: Cell Surface Dynamics: Concepts and Models.* (Perelson, A.S., DeLisi, C. and Wiegel, F.W., Eds.) pp. 223-275. Marcel Dekker, New York (1984).
7. R.J. De Boer and Hogeweg, P. , *Bull. Math. Biol.* 51, 223-246 (1989).
8. J.F Kearny and Vakil, M., *Immunol. Rev. 94*, 39-50 (1986).
9. D.S. Holmberg, Forgren, S., Ivars, F. and Coutinho, A., *Eur. J. Imunol. 14*, 435-441 (1984).
10. L.A. Segel and Perelson, A.S. pp.321-344, in: *Theoretical Immunology, Part Two*, ed. by A.S. Perelson, Addison Wesley (1988).
11. Weisbuch G., *J. Theor. Biol.* , 143(4),507-522 (1990).
12. J. D. Farmer, Packard, N.H. and Perelson, A.S. *Physica* 22D,187- 204 (1986).
13. G. Parisi, pp.394-406, in: *Chaos and Complexity*, ed. R. Livi, S. Ruffo, S. Ciliberto and M. Buiatti, World Scientific (1988) .
14. G. Weisbuch, pp.53-62, in: *Theories of immune networks*, (Atlan, H. Ed.). Springer, Berlin (1989).
15. R.J. De Boer and Hogeweg, P., *Bull. Math. Biol.*, 51, 381-408 (1989b).
16. A. Neumann and G. Weisbuch, submitted to Bull. Math. Biol.(1990).
17. J. Stewart and F.Varela, *Immunol. Rev.*, 110, 37-61 (1989).
18. R. De Boer, I.G. Kevrekidis and A.S. Perelson (1990).
19. A.S.Perelson and G. Weisbuch (1990).
20. J.J.Hopfield, *Proc. Natl. Acad. Sci. USA.* 79, 2554-2558 (1982).
21. I.R. Cohen pp. 6-12, in: *Theories of immune networks*, (Atlan, H. Ed.). Springer, Berlin (1989).
22. R.J. De Boer, *GRIND: Great Integrator Differential Equations*, Bioinformatics Group, University of Utrecht, The Netherlands (1983).

SPECIALIZATION, ADAPTATION AND OPTIMIZATION IN DILUTE RANDOM ATTRACTOR NEURAL NETWORKS

D. Sherrington and K.Y.M. Wong

Department of Physics, University of Oxford, 1 Keble Road
Oxford, OX1 3NP, United Kingdom

1. INTRODUCTION

In a biological context one is used to examples in which the properties of organisms are adapted to provide efficient operation in their particular environments, with otherwise similar systems differing in appropriate details when their environments are different. In this talk we demonstrate a similar phenomenon in some simple neural networks. In particular, we consider the performance of dilute attractor neural networks for associative memory with respect to optimal performances as characterized by two different measures, retrieval overlap and size of the basin from which retrieval is possible. We believe however that our conclusions are more broadly applicable [1].

2. NETWORK CHARACTER AND GENERAL OPERATION

The networks we consider consist of a large number N of simple formal binary neurons ($\sigma_i = \pm 1$; firing/nonfiring), interconnected randomly with an average of C neurons driving any one (and equivalently any neuron providing direct input to an average of C others). The dynamics of the networks are determined by a set of local rules, the state of neuron i at time t being determined by those of the neurons driving it, via local rules coded in the network parameters. The principal interest is however in the global behaviour. The networks are used to store uncorrelated patterns

$$\{\xi_i^\mu\} = \{\pm 1\} \ ; \ \mu = 1, \ldots, p \ ; \ p = \alpha C \tag{1}$$

and the global state of the network is characterized by its overlap with the patterns

$$m^\mu = N^{-1} \sum_i \sigma_i \xi_i^\mu \tag{2}$$

We concentrate on the dilute, but thermodynamic, limit, so that correlation effects are unimportant in the dynamics [2]. Taking the example of synchronous dynamics (extension to asynchronous dynamics is straightforward), a system started with a finite

Biologically Inspired Physics, Edited by L. Peliti
Plenum Press, New York, 1991

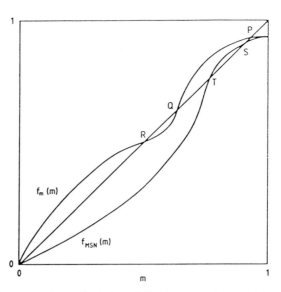

Fig.1. Retrieval curve $f_{MSN}(m)$ for finite-temperature retrieval of a synaptic network which is maximally stable at zero temperature, together with a schematic plot of the retrieval envelope $f_m(m)$ of a set of noise-trained synaptic networks in a parameter region where there are two non-zero stable fixed points.

overlap with only a single pattern (infinitesimal with the rest) evolves according to an iterative map

$$m(t+1) = f(m(t)) \tag{3}$$

where $f(m)$ is a function of the network parameters, known as the retrieval function. The stable fixed points

$$m^* = f(m^*) \; ; \; \mathrm{d}f/\mathrm{d}m|_{m=m^*} < 1 \tag{4}$$

give the retrieval overlaps and the unstable fixed points

$$m_B = f(m_B) \; ; \; \mathrm{d}f/\mathrm{d}m|_{m=m_B} > 1 \tag{5}$$

give the boundaries of the basins of attraction. This is illustrated in Fig. 1 where $f_{MSN}(m)$ is such a retrieval function (more particularly f_{MSN} refers to finite temperature retrieval of a synaptic network which is maximally stable at zero temperature; see for example [3]), O and S indicate its stable fixed points and T is an unstable fixed point; the details of the origin of the actual curve f_{MSN} will be explained later. For m greater than the value associated with T the system iterates asymptotically towards the higher stable fixed point value, whilst for smaller initial m the iteration is towards zero overlap.

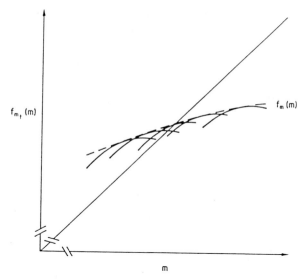

Fig.2. Schematic plot of a family of retrieval curves $f_{m_t}(m)$ and their envelope $f_m(m)$.

3. OPTIMIZATION: GENERAL PRINCIPLES

Now let us consider the possibility of global modification of the network parameters. Symbolically we characterise the modification by a parameter x. This x may correspond to some environmental parameter which we use to tune the learning stage of the network, e.g. training noise, but for the moment we keep x unspecified so that the discussion is entirely general. The tuning of the network to a particular value of x results in a unique system, whose retrieval function is characterized by $f_x(m)$. The issue we address is the optimal choice of x.

Let us first consider optimization of the retrieval overlap. This is obtainable from a consideration of the envelope $F(m)$ which touches the set of $f_x(m)$ from above. The maximum retrieval overlap over the set of $\{x\}$ is obtained from the highest stable fixed point of the retrieval envelope

$$m^* = F(m^*) \; ; \; \mathrm{d}F/\mathrm{d}m|_{m=m^*} < 1 \; ; \; \max m^* \; . \tag{6}$$

The corresponding optimal x is given by the retrieval function $f_x(m)$ for which

$$f_x(m^*) = m^* \; . \tag{7}$$

We shall denote the corresponding x by x^*. Assuming that there is no degeneracy among the $f_x(m^*)$, any other $x \neq x^*$ would yield a stable fixed point

$$m_x^* = f_x(m_x^*) \; ; \; \mathrm{d}f/\mathrm{d}m|_{m=m_x^*} < 1 \; ; \; \max m_x^* \tag{8}$$

which is always less than the fixed point m^* of the envelope $F(m)$, i.e.

$$m_x^* < m^* \; . \tag{9}$$

Fig. 2 illustrates this property.

Let us now turn to the issue of the optimal basin of attraction, that is maximizing the minimum m from which retrieval is possible. Since the basin boundary overlap is given by Eqn. (5) one immediately notes from Fig. 2 that the network with maximum retrieval overlap does not in general have the optimal basin size. On the other hand, the envelope $F(m)$ does determine the basin-optimized network, but from the unstable fixed point

$$m_B = F(m_B) \; ; \; \mathrm{d}F/\mathrm{d}m|_{m=m_B} > 1 \; ; \; \max m_B \; , \tag{10}$$

with the corresponding x_B determined by

$$f_{x_B}(m_B) = m_B \; . \tag{11}$$

Particular networks are therefore 'specialized', in the sense that a network having the best retrieval overlap does not have the widest attractor, and vice versa. Above we have discussed two extreme examples. Clearly one can follow an appropriate middle path by choosing a network with an intermediate value of x.

Furthermore, the envelope $F(m)$ indicates the possibility to consider 'adaptive' networks which change their tuning parameters during retrieval, starting with an x-value ensuring a large enough basin, but changing during retrieval to a value giving the greatest retrieval overlap.

Another interesting possibility arises when the retrieval envelope $F(m)$ has more than one stable fixed point, as illustrated by the curve labelled $f_m(m)$ in Fig. 2. In this example, the overlap associated with fixed point P can only be achieved by a continuously adaptive network starting with an overlap greater than that associated with point Q. If, however, the initial overlap is low, there exists also a second class of adaptive networks which can achieve the stable fixed point R from a starting situation with lower overlap. This is another example of potential specialization such as, allegorically speaking, might be utilized by a "predatory bird" which hunts at night and received poor input data, as compared with one which hunts during the day and received better images (say, an "owl" as compared with an "eagle").

4. OPTIMIZATION: A SPECIFIC EXAMPLE

Thus far we have been very general. To be more specific let us consider networks operating synaptically with a stochastic local updating rule

$$\sigma_i(t+1) = \mathrm{sign}\left(\sum_j J_{ij}\sigma_j(t) + Tz\right) \tag{12}$$

where the $\{J_{ij}\}$ are the synaptic efficacies, and z is a Gaussian random variable of variance 1 and T is the 'temperature'. We further choose the $\{J_{ij}\}$ so as to maximise $f(m_t)$ for this system, where m_t will then be called the training overlap. Considering m_t as the tuning parameter, the retrieval function $f_{m_t}(m)$ is then given by [4]

$$f_{m_t}(m) = \int \mathrm{d}\Lambda \rho_{m_t}(\Lambda) g_m(\Lambda) \tag{13}$$

where

$$g_{m_t}(\Lambda) = \mathrm{erf}\left(\frac{m_t\Lambda}{\sqrt{2(1 - m_t^2 + T^2)}}\right) \; , \tag{14}$$

$$\rho(\Lambda) = \int \frac{dt}{\sqrt{2\pi}} \exp(-t^2/2)\delta(\Lambda - \lambda(t)) \;, \tag{15}$$

and $\lambda(t)$ is the inverse function of $t(\lambda)$ defined by

$$t(\lambda) = \lambda - \gamma g'_{m_t}(\lambda) \tag{16}$$

with γ a constant determined by the condition

$$\int \frac{dt}{\sqrt{2\pi}} \exp(-t^2/2)\Big(\lambda(t) - t\Big)^2 = \alpha^{-1} \tag{17}$$

choosing the value giving the largest value of $g_{m_t}(\lambda) - (\lambda - t)^2/2\gamma$ in cases where $\lambda(t)$ is multi-valued. $f_{m_t}(m)$ is of course maximized as a function of m_t at $m_t = m$. This is an example of adaptation, since a network optimized in the training environment parametrized by m_t has a better performance than any other network when operating in the same retrieval environment. The envelope function $F(m)$ is therefore $f_m(m)$. The upper curve in Fig. 1 and the curves of Fig. 2 are for this case.

5. PHASE DIAGRAM OF ADAPTATION

As is evident from Eqn. (17), retrieval curves and hence the retrieval envelope vary with the number of stored patterns, as well as with the retrieval 'temperature' T. Beyond critical values of T, no network capable of retrieval exists (no solution of Eqn. (4) with $m^* > 0$.) Lower values may separate regions of different retrieval/adaptation characters.

The fixed points of the retrieval envelope determine the behaviour of a network, if it is allowed to gradually adapt its performance to optimum during its retrieving operation, a process we term self-adaptation. In Fig. 3 we exhibit the phase diagram for self-adaptation for the networks of Eqn. (12). In region I the retrieval envelope $F(m)$ has a single stable fixed point and a complete basin of adaptation ($m_B = 0$). The stable fixed point goes continuously to zero on the line DF. In region II the retrieval envelope has an additional pair of stable and unstable fixed points, so that it permits the co-existence of strong and weak adaptive networks (i.e. the "eagle" and the "owl"). On lines AE and ED there are discontinuous transitions to the single stable fixed point behaviour of Region I. Line DB marks the continuous disappearance of the weaker adaptive retriever. In region III one has adaptation with a high asymptotic overlap but with only a narrowed basin of adaptation and a discontinuous transition on DC to the non-retriever region.

6. CONCLUSION

Within a simple general iterative map framework appropriate to the dynamic evolution of a dilute random attractor neural network, we have demonstrated specialization and adaptation for different optimization requirements. We have also presented specific results for a synaptic network retrieving with Gaussian synaptic noise.

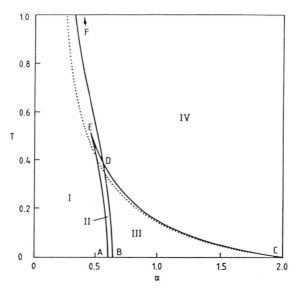

Fig.3. Retriever phase diagram for adaptation in temperature-storage capacity space. The retrieval phase boundary of the maximally stable network is also shown for comparison (dotted line).

ACKNOWLEDGMENTS

We thank H. Horner, M. Evans and D. Amit for useful discussions. Financial assistance from the U.K. Science and Engineering Research Council is gratefully acknowledged.

REFERENCES

1. K.Y.M. Wong and D. Sherrington, Optimally adapted attractor neural network in the presence of noise, *J. Phys. A* in press (1990).
2. B. Derrida, E. Gardner and A. Zippelius, An exactly solvable asymmetric neural network model, *Europhys. Lett.* 4:167 (1987).
3. D. Amit, M. Evans, H. Horner and K.Y.M. Wong, Retrieval phase diagrams for attractor neural networks with optimal interactions, *J. Phys. A* 23:3361 (1990).
4. K.Y.M. Wong and D. Sherrington, Training noise adaptation in attractor neural networks, *J. Phys. A* 23:L175 (1990).

COMPUTATIONAL STRATEGY IN THE PREMOTOR CORTEX OF THE MONKEY: A NEURAL NETWORK MODEL

Paolo Del Giudice[1], Emilio Merlo Pich[2], Giacomo Rizzolatti[3]

1. Istituto Superiore di Sanità - Laboratorio di Fisica INFN - Sezione Sanità Roma
2. Istituto di Fisiologia Umana - Università di Modena
3. Istituto di Fisiologia Umana - Università di Parma

1. INTRODUCTION

Experimental data obtained through single cell recording of neurons in the premotor cortex of monkeys during goal-directed motor acts show a remarkable correlation between the firing of given neurons and specific characteristics of the movement.

In particular, neurons located in area F4 fire in response to target objects located in certain space sectors, and become active during reaching movement towards these objects[1]; on the other hand neurons belonging to area F5 fire during different types of grasping movements towards target objects of various sizes, independently of their space location[2](see fig.1).

The existence of such selective neurons may suggest that an at least partially local information processing takes place in these areas.

In view of the interpretation of these data, neural network models may turn out to be useful in identifying the possible computational strategy underlying the observed firing patterns.

In our work, three-layer, feed-forward neural network models have been trained with back propagation[3] to simulate various aspects of visuomotor information processing taking place in goal-directed motor acts (see also ref. 4).

In the following two sections we list our results, deferring the discussion to the conclusions.

2. MANIPULATION AND ROTATION MODEL

In the first model (see fig.2 for details) we simulate the choice of appropriate grasping modes relative to objects of various sizes and orientations.

Size and orientation are defined with respect to an "object space" in which the objects are embedded and in which manipulation takes place, thus providing a scale relation between the object and the hand.

Biologically Inspired Physics, Edited by L. Peliti
Plenum Press, New York, 1991

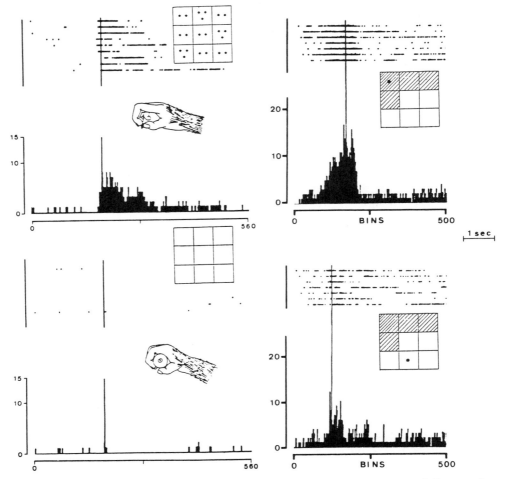

Fig. 1. Studies of the responses of two neurons in premotor cortex of the monkey during goal - directed motor acts. Target objects of small or large size are presented to the monkey in 9 different positions, schematically represented by the 3x3 square in the picture. In the upper half of each panel discharges of the same neuron is shown, each row representing a trial; the cumulative frequency of discharge is shown in the histogram just below. The vertical bar indicates in both cases the instant when the hand touches the object. In the left two panels, responses of the same F5 area neuron are shown during precision grip (small objects, upper panel) and whole - hand prehension (large objects, lower panel). Activity of F5 neuron appears to be selective for small objects, irrespective of their space locations. Stars in the 3x3 squares represent firing of the neuron during prehension for the different space locations of the object. In the right two panels responses of the same F4 area neuron are shown during reaching movements towards the region of external space indicated by the dashed squares (upper panel) or in any other region (lower panel). The dots indicate the actual end point of the reaching movement. This neuron is selective for specific space location, independently of the object size.

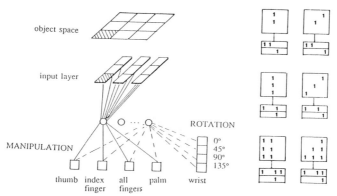

Fig. 2. Description of the manipulation and rotation model. On the left the network architecture is shown: it consists of 3 layers and each unit projects to all the units in the subsequent layer. The object space is mapped into the input vector as described in the text . The two groups of output units code grasp modes and wrist rotation. On the right we show a subset of the examples used for training; in each picture the upper 3x3 square is the input and the lower two small rectangles illustrate the corresponding correct output, the upper one representing the hand effectors and the lower one representing the wrist rotations. From the top to the bottom examples of small, medium and large size vertcal (90°) and oblique (135°) objects are shown. The output configurations for small object correspond to precision grip, for medium-size objects correspond to finger prehension and for large size objects correspond to whole-hand prehension.

Input units code the bidimensional object space, taking on the value 1 if the corresponding element of the object space is occupied by the object, and 0 otherwise.

Output units are divided into two groups: the first four stand for the various hand effectors (fingers, palm, etc.) collectively involved in the various type of grasping; the remaining four code the wrist rotations, each of them indicating a different orientation.

In fig.3 we show some of the results obtained for this model with 6 hidden units. The activation patterns, the receptive and the projective fields (see fig. 3 for definitions) for three of the hidden units are shown, the results from the other three being qualitatively similar .

Two units turn out to be selective for the size of the object (units 1 and 3), while unit 2 responds selectively to vertical and horizontal orientations. The observed selectivity reflects itself in the organization of the receptive and projective fields.

Interestingly, the receptive fields of hidden units selective for size, show a center-sorround organization; the weights connecting the input unit corresponding to the central element of object space with the hidden unit selective for small and medium size is excitatory while all the sorrounding weights are inhibitory.

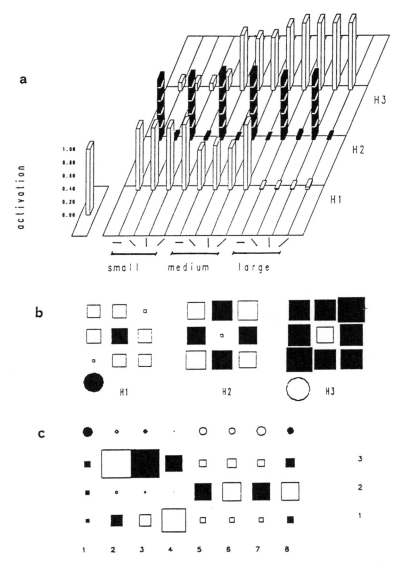

Fig. 3. Results for the manipulation and rotation model. a. Histogram of activation of three hidden units. Each histogram bar represents the mean activation corresponding to a class of objects of a given size and orientation (oriented bars stand for different orientations). b. Receptive fields of the hidden units; squares represent the connection strenght of the weights from input units to a given hidden unit, the side of the square being proportional to the absolute value of the weight. Excitatory weights are black and inhibitory weights are white; circles stand for the biases. c. Projective fields of the hidden units, i.e. the connection strenght of the weights from hidden units (represented in rows) to the output units (1 to 4 : hand effectors, 5 to 8: wrist rotation).

The converse is true for the hidden unit selective for large objects; in both cases, the bias is approximatively equal to the central weight. For the orientation selective units, the receptive fields show the excitatory weights arranged according to the preferred directions.

These organizations of receptive fields are well known in physiology as main features of cortical neurons located in the primary visual and other sensory areas (this point is discussed in more details in ref. 5).

While receptive fields are connected with the kind of internal coding of the input developed by the network, projective fields are related to the relevance of a given hidden unit for the correct performance. In this case, the organization of the projective fields appears to be consistent with the selectivity observed in the activation patterns.

A more complete analysis of the projective fields implies observing the performance of the network after lesions of single hidden units; an account of this study as well as data concerning generalization properties of these networks will be published elsewhere.

3. MANIPULATION AND TRANSPORT MODELS

In these models we consider embedding the object space in external space.

In a first model, the external space is represented as a square of 5x5 elements, each associated with a unit in the input layer (see fig. 4a).

The 3x3 object space appears in nine different, partially overlapping sectors of the external space.

The output units are divided into two groups; the first four units code grasp modes in the same way as in the previous model, while the remaining nine code the position of the end points of the reaching movement, each representing one of the nine possible sectors.

The results shown in fig.5 refer to a model with nine hidden units. Units 1 and 7 show an activation pattern essentially invariant with respect to the position of the object in external space. In particular unit 7 appears to be selective for small objects, irrespective of their space locations. Other units show preferential activation for certain sectors of external space; among them, units 4 and 9 which appear in the figure. For units insensitive to space locations, the receptive fields' organization is essentially homogeneous, with inhibitory weights and positive biases, lacking the kind of structure observed in the previous model; this is due to the overlapping of the space sectors and to the fact that these units are mainly activated for small objects.

For units responding preferentially to certain space sectors, irrespective of the size of the object, the receptive fields show excitatory weights located congruently with the preferred sectors.

Projective fields are basically consistent with the selectivity observed.

A second model has been developed, differing from the previous one for the coding of input stimuli in the input layer.

Here we attempt to take a first step towards an improvement of the biological plausibility.

A preprocessing of the visual information is inserted consisting in coarse coding the input stimuli by input units with overlapping receptive fields; this is roughly reminiscent of one of the mechanisms involved in early information processing in the retina and visual cortex related to spatial resolution and accuity[6].

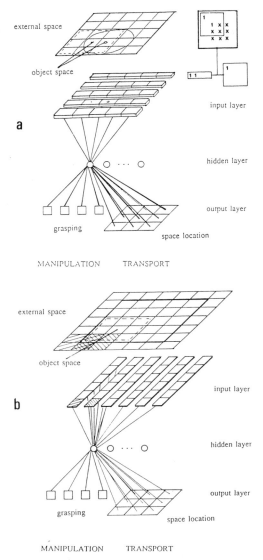

Fig. 4. Description of the two transport and manipulation models. a. The first model described in the text. In the upper right, one of the examples used for training is shown: it is a small object which occupies the first space sector, whose object space is delimited by the smaller square. The crosses indicate all the other possible positions of the central element of the object space. The small rectangle and square just below represent the corresponding correct output (manipulation and transport respectively). b. Schematic description of the second model (see text).

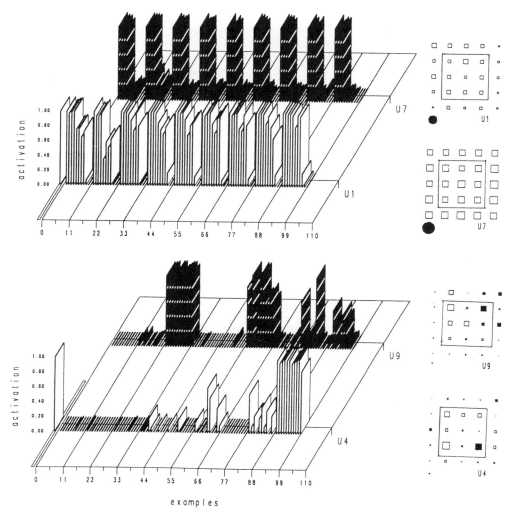

Fig. 5. Activation patterns and receptive fields for the first manipulation and transport model. Eleven objects of different sizes (4 small, 4 medium size, 3 large) were presented in the 9 different space sectors summing up to a total of 99 examples. Receptive fields are shown on the right; the square drawn inside each receptive field includes the 9 different positions of the central element of the space sectors. The conventions are the same as in Fig. 3

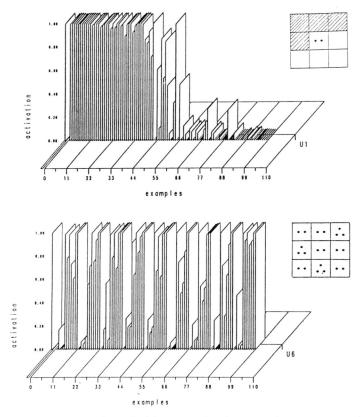

Fig. 6. Activation patterns of two hidden units in the second transport and manipulation model. The 99 examples are the same as in the first model, coarse coded
as explained in the text. In the 3x3 squares on the right are represented
the space sectors for which the unit is coherently activated (dashed areas)
and those for which the unit is incoherently activated (stars). By coherent
activation we mean activation present for all the objects in a given sector,
irrespective of their size.

In this model the external space consists of 7x7 elements. The input layer
consists of 6x6 elements; each input unit has a receptive field of 2x2 space elements,
with a partial overlap (2 elements) with the receptive fields of adjacent input units.
An input unit is activated to 1 when at least one element of its receptive field is
activated by the presence of an object.

As in the previous model, the object space appears in 9 different overlapping
sectors of the external space (see fig.4b).

Despite the different input representation, the results concerning the pattern
of activation of the hidden units are qualitatively similar to those observed in the
previous model.In fig.6 we show the activation pattern of unit 1 which is mostly

selective for a set of space sectors, and that of unit 6, which appears insensitive to space location, but moderately selective for large objects.

For all the models described, several different random initial conditions for the weights have been adopted, which produced qualitatively consistent results, thus showing that the features observed in the final configurations developed by the networks are a 'statistically typical' consequence of the constraints imposed by the task.

4. DISCUSSION

Despite the oversimplification and the limits of neural networks, (see ref. 7 for a general discussion) the models we described show a remarkable similarity with the biological system they refer to, as far as the computational strategy is concerned.

We would like to stress that no biological plausibility is implied for the specific learning algorithm we used; to the extent to which learning can be thought of as a trial and error procedure which may be described in abstract terms as the minimization of some cost function, what interest us are the features characterizing the most part of the minimum error configurations, for a given architecture.

In the various models described in this paper, the computational process required to implement the specified input-output maps led to the emergence of selective (or 'locally tuned') hidden units. On the other hand, selective neurons have been observed in inferior area F4 and F5 of the premotor cortex of the monkey, as mentioned in the introduction. Furthermore, neurons have been observed in inferior parietal lobe, a region strictly connected with premotor cortex, which appear to be selective for different object orientations (Sakata, personal communication).

This analogy may suggest that a computational strategy similar to the one observed in our models could be relevant in the emergence of selective neurons involved in goal directed motor acts. The present paper confirms and extends previous finding obtained using different models[4].

Our work is in the same line with several previous ones which showed the capability of neural networks to simulate some known computational properties of the nervous system[7,8,9]; however, the usefulness of model building is greatly enhanced when it can be used as a heuristic tool for further experimental investigation.

In view of this, in our case it is tempting to consider the possibility that the structures observed in the receptive fields in our models could motivate further experiments aimed to a better understanding of the relevant connectivity patterns in the premotor cortex, at the physiological level, and of internal representation of external space necessary to visuo-motor integration developed in this area, at the cognitive level.

REFERENCES

1. M. Gentilucci, L. Fogassi, G. Luppino, M. Matelli, R. Camarda, G. Rizzolatti, *Exp. Brain Res.* 71:475 (1988)
2. G. Rizzolatti, R. Camarda, L. Fogassi, M. Gentilucci, M. Matelli, *Exp. Brain Res.* 71:491 (1988)
3. D.E. Rumelhart, J.L. Mcclelland, *"Parallel Distributed Processing vol.1"*, MIT Press, Cambridge Ma (1986)

4. E. Merlo Pich, S. Favi, M. Gentilucci, G. Rizzolatti, *"Second Italian Workshop on Parallel Architectures and Neural Networks"*, E.R. Caianiello ed., World Scientific Pub., Singapore (1990), pp.27-41
5. P. Del Giudice, E. Merlo Pich, G. Rizzolatti, in preparation
6. G.E. Hinton, J.L. Mcclelland, D.E. Rumelhart, in *"Parallel Distributed Processing vol.1"*, D.E. Rumelhart J.L. Mcclelland eds., MIT Press, Cambridge Ma (1986), pp. 11-109
7. T.J. Sejnowski, C. Koch, P.S. Churchland, *Science* 241: 1299 (1988)
8. D. Zipser, R.A. Andersen, *Nature* 331: 679 (1988)
9. S.R, Lehky, T.J. Sejnowski, *Nature* 333: 452 (1988)

MECHANISMS OF BIOLOGICAL PATTERN FORMATION

Hans Meinhardt

Max-Planck-Institut für Entwicklungsbiologie
Spemannstr. 35, D-74 Tübingen, FRG

1. INTRODUCTION

The development of a higher organism with all its differentiated cells in complex but precise spatial arrangement is one of the most spectacular events in living systems. This process must be encoded in the genes. The similarity of identical twins provides some intuition about how precisely the final pattern is determined. The reference to the genes, however, does not provide an explanation of how spatial pattern of an organism is generated, since during cell division, as the rule, both daughter cells obtain the same genetic information. The question is then how different genetic information can be activated in different regions of the (originally more or less homogeneous) mass of cells that differently programmed cells arise in a controlled spatial neighborhood.

Most information about how pattern formation is achieved has been derived from perturbations of normal development: for instance, by removal of some parts or by their transplantation into an unusual environment. The regulation observed after such a perturbation provides some insights of how normal development is controlled. One experiment of this type is the early fragmentation of an embryo. In sea urchins, this can lead to the formation of two complete organisms. Such a regulation demonstrates clearly that the pattern of the adult organism is not necessarily present in a mosaic-like, latent form in the fertilized egg but that the pattern is generated during development in a self-regulating manner.

The regulatory events observed after an experimental perturbation do not provide a direct insight into the molecular mechanisms on which development is based, but they allow the construction of dynamic and molecularly feasible models. The mathematical formulation requires coupled partial differential equations for the production rates of the molecules involved as functions of the other molecules present. The total time course can be calculated by an integration of these equations. Since the as yet hypothetical reactions are necessarily non-linear, the demonstration that the regulatory features of the models agree with the experimental observations requires numerical integrations with a computer. Making hypotheses, checking them by simulations and, if necessary, modifying the hypotheses allows us to attempt to reconstruct the basic principles on which development must be based. We have devel-

Biologically Inspired Physics, Edited by L. Peliti
Plenum Press, New York, 1991

oped several models for different developmental situations which can be summarized as follows:

(i) Primary gradients as well as periodic structure can be generated by local self-enhancement and long-range inhibition (Gierer and Meinhardt, 1972).

(ii) Stable cell states result from genes which feed back directly or indirectly on their own activation (e.g. transcription). Each gene competes with the alternative genes for activation. Positive feedback and competition has the consequence that only one of the alternative genes remains active within one cell. By appropriate coupling with gradients that are generated by mechanism (i) or (iii), a position-dependent gene activation results (Meinhardt, 1978).

(iii) Borders between regions in which different genes are active act as organizing regions for the determination of substructures. Pairwise substructures, such as legs and wings, are initiated at the intersections of two borders, one resulting from a patterning in the anteroposterior, the other from a patterning in the dorsoventral dimension (Meinhardt, 1980, 1983a,b).

(iv) Ordered sequences of cell states result from a mutual long-range activation of cell states which locally exclude each other. This allows a controlled neighborhood of cell states. Missing elements can be intercalated (Meinhardt and Gierer, 1980).

This article contains a brief description of these mechanisms. A more detailed account including computer programs can be found elsewhere (Meinhardt 1982).

2. GENERATION OF A PRIMARY PATTERN BY AUTOCATALYSIS AND LATERAL INHIBITION

Pattern formation from almost homogeneous initial conditions is by no means unique to living systems. The formation of high sand dunes (despite the permanent redistribution of the sand by the blowing wind) or of sharply contoured rivers (despite the fact that the rain is homogeneously distributed) provides a few examples. Common in all these inorganic pattern formation is that small deviations from a homogeneous distribution have a strong feedback such that the deviation grows further. We have proposed that primary embryonic pattern formation is accomplished similarly by the coupling of a short range self-enhancing (autocatalytic) process with a long range inhibitory effect that acts antagonistically to the self-enhancement (Gierer and Meinhardt, 1972; Gierer, 1981; Meinhardt, 1982). A simple molecular realization would consist of an "activator" whose autocatalysis is antagonized by a long-range inhibitor. The production rate of the activator (a) and inhibitor (b) can be formulated in the following set of couplet differential equations:

$$\frac{\partial a}{\partial t} = \frac{\rho a^2}{h(1 + \kappa a^2)} - \mu a + D_a \frac{\partial^2 a}{\partial x^2} + \rho_0 \tag{1a}$$

$$\frac{\partial h}{\partial t} = \rho a^2 - \nu h + D_h \frac{\partial^2 h}{\partial x^2} + \rho_1 \tag{1b}$$

Fig. 1 shows numerical solutions of this set of equations (the saturation term $1 + \kappa a^2$ will be discussed later). The simulations demonstrate that the interaction Eq. 1 has properties which are basic for the explanation of biological pattern for-

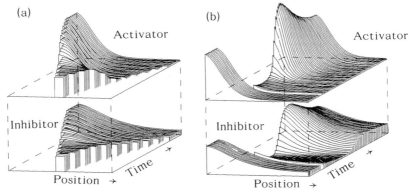

(a)   (b)

Activator   Activator

Inhibitor   Inhibitor

Position →   Time   Position →   Time

Fig. 1. Formation (a) and regeneration (b) of a graded concentration profile. Assumed are an autocatalytic substance, the activator (top) and its highly diffusible antagonist, the inhibitor (Eq.1a,b). Both concentrations are plotted as functions of position and time. Assumed further is a linear array of cells (to enable a space-time plot) that grows at both margins. If a critical size (the range of the activator) is exceeded, random fluctuations are sufficient to initiate pattern formation. A high concentration appears at a marginal position of the field since a central maximum would require space for two activator slopes for which no space is available at the critical size. A polar pattern results that can be maintained upon further growth. (b) After separation into an activated and a non-activated half, the activator maximum and thus the gradient regenerates in the non-activated part after the decay of the remnant inhibitor.

mation: a pattern emerges whenever the size of the field exceeds the range of the activator. A high activator concentration is formed at one end of the field. Thus, the system described by Eq. 1 is able to generate "positional information" (Wolpert, 1969). The graded activator or the graded inhibitor distribution can activate different genetic information in different parts of the tissue in an ordered sequence (see Fig. 4).

An important feature of many developing systems is their ability for regeneration. The removal of some parts of an embryo are compensated by pattern regulation. This is a property of the activator-inhibitor system since, for instance, after removal of the activated region, the remnant inhibitor decays until a new maximum is formed via autocatalysis (Fig. 1).

The self-enhacement does not require a molecule with direct autocatalytic regulation. Autocatalysis can be a property of the system as a whole. For instance, if two substances, a and b, exist and a inhibits b and *vice versa*, a small increase of a above the equilibrium leads to a stronger repression of the b production and thus to a further increase of a, and so on, in the same way as if a would be autocatalytic. The same holds for b. Together, a and b form a switching system in which either a

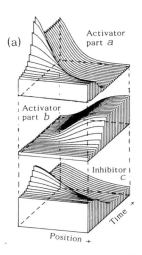

(a)

Activator part a

Activator part b

Inhibitor c

Position →

Time

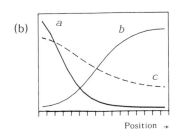

(b)

a

b

c

Position →

Fig. 2. Realization of autocatalysis by an inhibition of an inhibition. If two sub-stances, a and b, inhibit each other's production in a non-linear way (Eq. 2), a small increase of one of the substances above the steady state leads to its further increase. Both substances together have the property of self-enhancement, such as required for pattern formation (part a and part b of the activator). The long-range inhibitor (c in Eq.1) is assumed to antago-nize the repression of b-production by the a-molecules. Thus, only inhibitory interactions are involved in this scheme. (a) Generation of graded distribu-tions in a linear array of cells as function of time. The pattern is initiated by random fluctuations. (b) Concentrations at the final stable steady state. a and b are distributed in counter-gradients; the inhibitor distribution c has the same polarity as the a distribution but is shallower due to the higher diffusion rate.

or b is high. The switch of the lambda phage between the lytic and the lysogenic phases is based on such an inhibition of an inhibitor (Ptashne et al., 1980). To allow pattern formation, a long-range signal is required which interferes with the mutual competition of the two substances. For instance, if a has won the a-b competition in a particular region, b must win in the surroundings. A possible realization would be that the a molecules control the production of a substance c which, in turn, either inhibits a or promotes b production. These modes are equivalent since in competing systems, a self-limitation is equivalent with a support of the competitor. A more symmetrical pattern arises if the long-range help is reciprocal. In Eq. 2a-c an inter-action is described in which the diffusible antagonistic substance c is produced under control of the a molecules and undermines the repression of b production by the a molecules.

$$\frac{\partial a}{\partial t} = \frac{\rho}{\kappa + b^2} - \mu_a a + D_a \frac{\partial^2 a}{\partial x^2} + \rho_0 \qquad (3a)$$

$$\frac{\partial b}{\partial t} = \frac{\rho}{\kappa + a^2/c^2} - \mu_b b + D_b \frac{\partial^2 b}{\partial x^2} + \rho_0 \qquad (3b)$$

$$\frac{\partial c}{\partial t} = \gamma a - \mu_c c + D_c \frac{\partial^2 c}{\partial x^2} \tag{3c}$$

No direct autocatalytic interaction is assumed. As shown in Fig. 2, this leads to countergradients of the a and b molecules as well as to a shallower graded c distribution.

3. FORMATION OF STRIPE-LINE DISTRIBUTIONS REQUIRE SATURATION OF SELF-ENHANCEMENT AND ACTIVATOR DIFFUSION

Stripes, i.e. structures that have a long extension in one dimension and a short one in the other, is a very frequent type of pattern, generated in many situations during embryogenesis.

Formation of stripes will occur if activator autocatalysis saturates at relatively low activator concentrations (Meinhardt, 1988). Due to this saturation, more cells remain activated although at a lower level. Stripe formation requires, in addition, a modest diffusion of the activator. Due to this diffusion, activated regions tend to occur in large coherent patches since, if a cell is activated, the probability is high that the neighboring cell becomes activated too. On the other hand, it is necessary that activated cells are close to non-activated cells into which the inhibitor can diffuse, otherwise no activation above average would be possible. The two seemingly contradictory requirements, coherent patches and proximity of non-activated cells are satisfied if a stripe-like pattern is formed (Fig. 3). Each activated cell is bordered by other activated cells, but non-activated cells are not too far away. The mutual activation of cell states discussed below also has the capability of stripe formation.

4. SEGREGATION OF CELL POPULATIONS

A very common patterning process is the segregation of an originally homogeneous cell population into two different cell types. The segregation of neuroectodermal cells into neuroblasts and ventral ectodermal cells in *Drosophila* (see Campos-Ortega, 1988; Artavanis-Tsakonas, 1988) or the formation of prestalk and prespore cells in Dictyostelium (Williams et al., 1989) may serve as examples. Common to both systems is a very good regulation of the proportion of the two cell types. Removal of either prespore or prestalk cells leads to the restoration of the correct ratio due to reprogramming. In the insect system, ablation of neuroblast cells causes ectodermal cells to take over the fate of the deleted cell (Doe and Goodman, 1985). Transplantation of marked ectodermal cells from the neuroectodermal region into a younger host can cause a switch into the neuronal pathway (Technau and Campos-Ortega, 1985).

Molecules have been identified that play a decisive role in the formation of these patterns. In the slime mold, a molecule called DIF is required for prestalk formation (Kay et al., 1989). In the insect system, about six genes are required for the formation of ectodermal cells (see Campos-Ortega, 1988; Artavanis-Tsakonas, 1988); among them are the genes *Notch* and *Delta*. It has been regarded as a surprise and counterintuitive for a morphogenetically active molecule that both molecules,

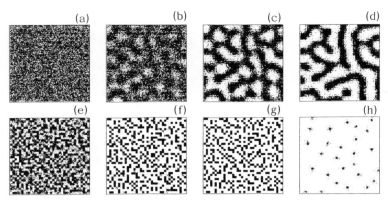

Fig. 3. Stages in the formation of a stripe-like pattern and segregation of cell types. Assumed is an activator-inhibitor system (a-d). A saturation of autocatalysis at low activator concentration which limits the competition among neighboring cells. More cells remain activated, although at a lower level. Slight diffusion of the activator leads to the tendency to form coherent activated regions. Stripes are the most stable pattern since activated cells have activated cells in the neighborhood but, nevertheless, non-activated cells are nearby into which the inhibitor can escape (e-g). Without activator diffusion, a segregation into activated and non-activated cells take place. The number of activated cells is a certain proportion of the total number of cell. The ratio depends on the saturation of the activator production (κ in Eq.1). (h) For comparison, without saturation but otherwise the same parameters and initial conditions as in Fig. 3 (a-d), a bristle-like pattern results. The maximum concentration is about ten times higher than in Fig. 3 (d).

DIF and the *Notch* product appear to be almost homogeneously distributed. However, according to the model discussed below substances with such distributions are necessary component.

A model for segregation must have the following features: (i) In a certain proportion of the cells a particular gene becomes activated. The other cells remain either in a ground state or activate an alternative gene. (ii) The proportion of the two cell types is regulated. Removal of one cell type leads to a reprogramming of some of the remaining cells, such that the correct ratio becomes restored. (iii) The two cell types appear more or less at intermingled positions.

With appropriate parameters, the activator-inhibitor mechanism can reproduce these features. The known experimental and genetic data are compatible with such an interpretation. As discussed above, if the autocatalysis of the activator saturates at low activator concentrations, the number of activated cells reaches a certain proportion of the non-activated cells. Since the local activator increase is limited due to the saturation, the degree of competition between neighboring cells is limited too. Thus, neighboring cells can remain activated independent of the range of inhibitor.

If the activator is non-diffusible, the decision whether a cell will become activated is to a large extent independent of the decision made by a neighboring cell. Only on average must the correct ratio be maintained. The chance to become activated is not increased by an activated neighbor. As long as no other constraints are superimposed, random fluctuations are decisive. Activated and non-activated cells emerge at intermingled positions (Fig. 3e-g). An obvious realization of cell-restricted self-enhancement would be the involvement of genes that have positive feedback on their own transcription.

Such a system is able to regulate. If, for instance, too few cells are activated, the level of the inhibitor is lower than that required for an equilibrium. More and more cells switch from the non-activated into the activated state. This leads to an increase of the inhibitor concentration until no further switching is possible. At this stage the correct ratio is obtained. The reverse argument is valid if too many cells are activated.

5. CELL DETERMINATION AND REGION-SPECIFIC GENE ACTIVATION

In higher organisms, the pattern generated by a reaction-diffusion mechanism is necessarily transient since, due to growth, the polar pattern cannot be maintained over the whole expanse of the growing organism. This requires that, at an appropriate stage, the cells make use of position-specific signals, i.e. that they become determined for a particular pathway by activating particular genes. Afterwards the cells maintain this determination whether or not the evoking signal is still present.

The simplest system with a long term memory would consist of a substance with a threshold behavior. A possible mechanism would consist of a substance that feeds back on its own production rate in a nonlinear way. The feedback saturates at high concentrations. With appropriate parameters, only two stable states are possible. Eq. 3 provides an example.

$$\frac{\partial g}{\partial t} = \frac{cg^2}{(1 + \kappa g^2)} - \mu g + m \tag{3}$$

If m is above a certain threshold, only the high stable state is stable. The system remains at this state even if later on m becomes small (Meinhardt, 1976).

More important is the region-specific activation of several genes under the control of a morphogen gradient. To see which types of molecular interactions are required for selective activation of few alternative genes, it is helpful to realize that the activation of a particular gene has many formal similarities to the formation of a pattern. In pattern formation, a particular substance is produced at a particular location but this production is suppressed at other locations. Correspondingly, determination requires the activation of a particular gene and the suppression of the other alternative genes of a set. Gene activation may thus be regarded as a pattern formation in the gene space.

In analogy to the activator-inhibitor system for pattern formation, one can formulate the following set of equations for activation of particular genes via their gene products g_i ($i = 1...n$, n is the number of alternative pathways) (Meinhardt, 1978, 1982).

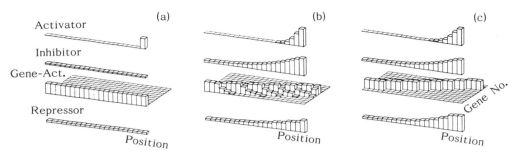

Fig. 4. Activation of genes by an activator-inhibitor system. The inhibitor is assumed
to act as morphogen. The genes are assumed to be a set of feedback-loops
which compete via a common repressor r (Eq. 4). This has the consequence
that only one gene of the set can be active in a particular cell. (a) Initially,
gene 1 is turned on in every cell. (b,c) The morphogen (m) is assumed to
provoke the transition from one gene to the next, the repressor (r) slows down
this transition. The repressor concentration increases with the activation of
a subsequent gene (since in Eq. 4 it has been assumed that $c_{i+1} > c_i$). Thus,
the activation of higher and higher genes comes to rest at a particular gene (c).
Which one it is depends on the local morphogen concentration. The result
is an ordered sequence of gene activity in space. Although the positional
information is graded, the activation of the genes is an all-or-nothing event
(after Meinhardt, 1978).

$$\frac{\partial g_i}{\partial t} = \frac{c_i g_i{}^2}{r} - \mu g_i + \frac{\delta m g_{i-1}}{r} \tag{4a}$$

$$\frac{\partial r}{\partial t} = \sum_i c_i g_i{}^2 - \nu r \tag{4b}$$

Each gene product g_i is autocatalytic, but also produces and reacts upon the
repressor r. The last term in Eq. 4a describes a possible influence of the graded
morphogen concentration m which provides the positional information. This could
be, for instance, an activator or inhibitor gradient (Eq. 1; Fig. 1). Under the driving
force of the morphogen, the cells switch from one activated gene to the next. The
number of steps depends on the morphogen concentration. Since each gene feeds
back on its own activation, a gene remains active, independent of whether the signal
is present or not. Due to the competition via a common repressor (or via a direct
negative influence of the alternative genes), only one gene of the set can be active in a
particular cell. The result of this "interpretation" of positional information is that in
groups of neighboring cells a particular gene is active. An abrupt transition from one
activated gene to the subsequent one takes place between neighboring cells despite
the smooth distribution of the morphogen. The determinations form an ordered
sequence in space.

Evidence for such a two-step mechanism (positional information plus interpretation) has been obtained for the body segments of insects (Meinhardt, 1977), for the digits of vertebrates (Tickle et al, 1975), and for the sequence of segments in insect legs (Meinhardt, 1980, 1983). It is remarkable that these are always segmented structures. Meanwhile, many genes with autocatalytic properties (autoregulation) have been found (for review see Serfling, 1989), so that the predicted principle, the maintenance of the determined state by feedback of a gene on its own activity, appears to be a more general process.

Of course, the structure of a higher organism is much more complex than can be achieved by the interpretation of a single gradient or two orthogonal gradients. A reliable finer and finer subdivision can be achieved in hierarchical way if the borders generated by one process provide a scaffolding for a subsequent process. A hierarchical model for pattern formation in *Drosophila* is discussed in detail elsewhere (Meinhardt, 1986,1988). The initiation of substructures such as legs and wings at intersections of borders will be discussed further below.

6. MUTUAL ACTIVATION OF CELL STATES THAT LOCALLY EXCLUDE EACH OTHER

In the mechanism described above, the cells do not communicate directly with each other to obtain the correct determination. They measure the local concentration and behave accordingly in a slavelike manner. In other developmental systems, for instance in the body or leg segments of insects (Fig. 5), a direct communication between the cells appears to be involved in obtaining a particular determination. This requires some modifications of the gene-activation mechanism discussed above.

We have seen that self-activation of genes coupled with a mutual repression of the alternative genes leads to stable states of determination. If two (or more) such states not only exclude each other locally but activate each other mutually over long ranges, these cell states stabilize each other in a symbiotic manner. Both cell states need each other in a close neighorhood while the local exclusiveness assures that the two states do not merge or overlap (Meinhardt and Gierer, 1980). Eq. 5 describes a possible interaction.

$$\frac{\partial g_i}{\partial t} = \frac{c_i g_i'^2}{r} - \mu g_i + D_g \frac{\partial^2 g_i}{\partial x^2} \tag{5a}$$

with

$$g_i' = g_i + \delta^- s_{i-1} + \delta^+ s_{i+1}$$

$$\frac{\partial s_i}{\partial t} = \gamma(g_i - s_i)D_s \frac{\partial^2 s_i}{\partial x^2} \tag{5b}$$

$$\frac{\partial r}{\partial t} = \sum_i c_i g_i'^2 - \nu r \tag{5c}$$

It is a feature of these interactions that they can also generate stripes. Stripes have necessarily a very long border between cells of different cell types. A particular cell type is close to cells of the other type. This facilitates the mutual support of the

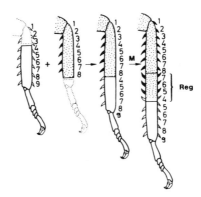

Fig. 5. An experimentally induced pattern discontinuity and its repair: Graft experiment with parts of the femur segment of cockroach legs (Bohn, 1970). A large distal fragment, denoted arbitrarily 4-9, is grafted on a large proximal stump (1-8). After one or two molts, the pattern discontinuity is repaired by intercalary regeneration (7-5 Reg.). The spines of the intercalate have a reversed orientation which indicates a polarity reversal. It is a property of the regulatory system to restore a normal neighborhood of structures. The system is unable to restore the natural pattern. A simulation of this process is shown in Fig.6.

two cell states by the diffusible substances s_i in Eq. 5). A stripe-like arrangement is therefore especially stable.

The mutual activation scheme can comprise several cell states. For instance, state 1 activates state 2, state 2 activates state 3 and so on until state n. The system can be cyclic in that state n activates state 1. This system is able to generate self-regulating sequences. For instance, if cell types which are not usually neighbors are juxtaposed, the intervening sequences become intercalated. In this way a normal neighborhood of cell states is restored. This need not necessarily be the natural sequence of structures. The intrasegmental pattern of insect legs is a biological system in which, after manipulation, a correct neighborhood but a non-natural pattern can result (Fig. 5).

These equations are very similar to the Hypercycle-equation proposed by Eigen and Schuster (1977) for the early evolution of genetic complexity. This similarity is not accidental. In both cases, autocatalytic feedback loops are involved. Genes provoke either their own replication or their own activation respectively. The mutual help is required in both cases to assure that the different autocatalytic loops remain in equilibrium with each other and that the best-replicating (or activating) loop does not overgrow and suppress all the others. The difference of the proposed equations for the generation of a sequence of structures in space consists on the one hand in the local action of autocatalysis and mutual repression and in the long range of the mutual help on the other. This leads to the predictable ordering of the "elements of the hypercycle" in space.

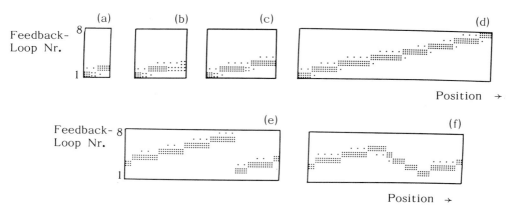

Feedback-Loop Nr.

Position →

Feedback-Loop Nr.

Position →

Fig. 6. Generation of sequences of structures and their repair. (a-d) Computer simulation with Eq. 4 in a growing array of cells. Subsequent structures are added whenever sufficient space is available. (c) If, by a manipulation, groups of cells become juxtaposed which are usually not neighbors, the pattern discontinuity is smoothed out. This intercalation can lead to supernumerary structures and polarity reversal (compare with the experiment Fig. 5; after Meinhardt and Gierer, 1980).

7. THE THREEFOLD SUBDIVISION OF SEGMENTS

Segments clearly resemble a periodic structure. The formation of a segment border does not require the transition from one segmental quality to the next. After deletion of the genes of the *Bithorax* gene complex, all abdominal segments obtain mesothoracic character, but the number of segments remains unchanged (Lewis, 1978). The simplest periodic structure would consist of the alternation of two specifications, for instance A and P (such as suggested by the early subdivision of the thoracic segments into anterior and posterior compartments). However, in a sequence ...APAPAP... neither the polarity of the segments nor the segment boundary is determined (AP/AP/AP or A/PA/PA/P; the slash indicates the segment border). For that reason I have proposed that segments consist of the alternation of at least three states ...P/SAP/SAP/S..(Meinhardt, 1982, 1984). A segment border is formed whenever S and P cells are contiguous. A sequence of three states has a defined polarity. This model predicts that, if one of the three states is absent due to a mutation, the segment polarity will be lost and a symmetrical pattern will emerge. For instance, a loss of A would lead to a pattern ...S/P/S/P/S/... i.e. to a symmetrical pattern with the same number of denticle bands (S) but twice the normal number of segment borders. This corresponds to the mutation *patch* (Nüsslein-Volhard and Wieschaus, 1980). More recent experiments indicate that segmentation results from the repetition of four cell states, to be called there ..D/ABCD/ABCD/A.. The formation of the symmetrical pattern caused by the mutation mentioned above could be based on the insertion of an additional cell state at an ectopic position in order to produce a normal neighborhood. For instance, after a mutation of a gene that is required for the structure B, instead of a sequence ..ACD/ACD/.. a sequence ..A/DCD/A/DCD/.. will be formed, which is obviously symmetrical (see Martinez-Arias et al., 1988).

8. HOW TO DETERMINE LEGS AND WINGS?

Primary pattern formation such as described above must also control the position at which substructures such as legs, wings or antennae are to be formed. On the other hand, many experiments indicate that, after initiation, the formation of these structures is a more or less autonomous, self-regulating process. The question is then how a particular group of cells can be determined to form, for example a limb. Moreover, how is it achieved that a limb has a particular handedness (left or right) and a particular orientation with respect to the main axes of the developing embryo?

The determination of appendages clearly requires not only a preceding determination along the antero-posterior axis but also a pattern formation perpendicular to the first, along the dorsoventral axis. One could imagine that in this way a particular group of cells—specified like a particular field of a checkerboard—obtains the signal to produce, for instance, a leg, and that the further pattern formation proceeds as described above for the primary pattern formation: by autocatalysis and lateral inhibition. However, an analysis of the experimental data indicate another mechanism: not the homogeneous region that has obtained a particular specification forms the future substructure but the region that surrounds the border between different determinations (Meinhardt, 1980, 1983 a,b). A border can become the source region for a new morphogen if cells of the two types cooperate to produce a new morphogen m. For instance, if the cells of the type A produce a precursor and the P-cells the final product m, the m-production is restricted to the A-P border, since at a distance from the border, one of the necessary components is missing—either no precursor molecules or no cells which can produce the final morphogen are available.

The A-P border has a long extension and surrounds the embryo in a belt-like manner. To determine the position of a limb, not only a single border but the intersection of two borders is necessary. The intersection of two borders defines unique points which becomes, due to the required cooperation of the different cell types, a local source region of the morphogen.

9. HOW DOES A LIMB OBTAIN ITS HANDEDNESS?

The hypothesis that the intersection of two borders is the necessary condition for the formation of legs or wings solves the problem of handedness. The requirement of two intersecting borders is equivalent with the requirement that four quadrants or three sectors are close to each other (depending on whether the intersection of the two borders is of the X- or of the T-type). Such an arrangement necessarily has a handedness since it is different from its mirror image. Assuming a symmetrical dorsoventral pattern formation a particular intersection is always present twice, one is located on the left, the other on the right side of the body. Both intersections have the opposite handedness. For instance, anterior-dorsal, anterior-ventral and posterior are arranged in counterclockwise fashion in a left leg (see Fig. 7).

The boundary-model for limb development is supported by many experimental

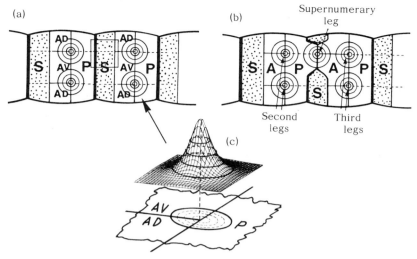

Fig. 7. Experimental evidence for the three-fold subdivision of segments and the A/P confrontation as precondition for leg formation: the experimental induction of a supernumerary leg in cockroaches (experiments of Bohn, 1974). (a) Ventral view of the thoracic segments of a cockroach according to the model: legs (concentric rings) are formed pairwise at the intersection of the A-P and the D-V compartment border. (b) Removal of an anterior portion of a segment (framed in Fig. a) leads, according to the model, to the removal of the S region that separates the P and the A region of two neighboring segments. After healing, a new PA/DV juxtaposition and thus an additional leg results. It has the opposite handedness and orientation compared with the other legs at the side of operation since the new A-P juxtaposition is inverted (P is anterior), and AV, P, AD are arranged clockwise instead of counterclockwise.

observations with insect (Meinhardt, 1980,1983b) and vertebrate limbs (Meinhardt, 1983a). Only one example should be provided here. According to the model of segmentation discussed above, segmentation results from the repetition of at least three cell states, called the S, A and the P regions. Bohn (1974) has removed small stripes of ectodermal tissue from the anterior region of a thoracic segment of a cockroach (Fig. 7). The result was the formation of an additional leg. This is not some sort of regeneration since, under normal circumstances, a leg would never have been formed at this position. In addition, this supernumerary leg has the reversed handedness (for instance, a left limb would be formed on the right side) and the opposite orientation in respect to the body axis (posterior leg side points anteriorly). According to the model, by such an operation, a portion of the S-region is removed that isolates one AP pair from the next. After wound healing, the P-region of one segment becomes juxtaposed to the A region of the following segment. The result is an additional P-A confrontation with a reversed anteroposterior polarity (P becomes anterior). Since the dorsoventral axis remains normal, this polarity change is accompanied by a change in the handedness. This result provides a strong support for the hypothesis that, on the one hand, the juxtaposition of differently determined tissues (A and P) is required for leg induction and, on the other, that these two tissues are separated by at least a third tissue (S).

10. CONCLUSION

Relatively simple molecular interactions are able to account for pattern formation during the development of higher organisms. The postulated main steps include the generation of positional information by a system of short range autocatalysis and long range inhibition and, under its control, the regional activation of different genes at particular locations. Borders between regions of different gene activities act as a scaffold to generate periodic sequences of cell states. Self-regulating sequences of cell states emerge if two or more cell states activate each other on long range but exclude each other locally. Segmentation requires the alternation of at least three cell states. The border between two of them is a precondition to form appendages such as legs or wings. Since the intersection of two borders give rise to a new positional information system, consecutively finer subdivisions are possible in a predictable way. The new structures have necessarily the correct location and orientation in relation to the already determined cells. Thus, a cascade of basically simple molecular interactions may be responsible for the pattern formation during embryogenesis.

REFERENCES

1. Artavanis-Tsakonas, S. (1988). The molecular biology of the Notch locus and the fine tuning of differentiation in *Drosophila*. *Trends Genetics* 4: 95-100.
2. Bohn, H. (1970). Interkalare Regeneration und segmentale Gradienten bei den Extremitäten von Leucophaea-Larven (*Blattari*). I. Femur und Tibia. *Roux Arch. Dev. Biol.* 165: 303-341.
3. Bohn, H. (1974). Extent and properties of the regeneration field in the larval legs of cockroaches *(Leucophaea maderae)*. I. Extirpation experiments. *J. Embryol. exp. Morph.* Vol. 31: 3, 557-572.
4. Campos-Ortega, J. (1988). Cellular interactions during early neurogenesis of *Drosophila melanogaster*. *TINS* 11: 400-405.
5. Doe, C.Q. and Goodman, C.S. (1985). Early Events in Insect Neurogenesis. II. The Role of Cell Interactions and Cell Lineage in the Determination of Neuronal Precursor Cells. *Devl. Biol.* 111: 206-219.
6. Eigen, M. and Schuster, P. (1977). The hypercycle. A principle of natural self-organization. Part A: Emergence of the hypercycle. *Naturwissenschaften* 64: 541-565.
7. Gierer, A. (1981). Generation of biological patterns and form: Some physical, mathematical, and logical aspects. *Prog. Biophys. molec. Biol.* 37: 1-47.
8. Gierer, A. and Meinhardt, H. (1972). A theory of biological pattern formation. *Kybernetik* 12: 30-39.
9. Kay, R.R., Berks, M. and Traynor, D. (1989). Morphogen hunting in *Dictyostelium*. *Development Suppl.* 107: 81-90.
10. Lewis, E.B. (1978). A gene complex controlling segmentation in *Drosophila*. *Nature* 276: 565-570.
11. Martinez-Arias, A., Baker, N.E. and Ingham, P.W. (1988). Role of segment polarity genes in the definition and maintenance of cell states in the *Drosophila* embryo. *Development* 103: 151-170.

12. Meinhardt, H. (1976). Morphogenesis of lines and nets. *Differentiation* 6: 117-123.
13. Meinhardt, H. (1977). A model of pattern formation in insect embryogenesis. *J. Cell Sci.* 23: 117-139.
14. Meinhardt, H. (1978). Space-dependent Cell Determination under the control of a morphogen gradient. *J. theor. Biol.* 74: 307-321.
15. Meinhardt, H. (1980). Cooperation of Compartments for the Generation of Positional Information. *Z. Naturforsch.* 35c: 1086-1091.
16. Meinhardt, H. (1982). *Models of biological pattern formation.* Academic Press, London.
17. Meinhardt, H. (1983a). A boundary model for pattern formation in vertebrate limbs. *J. Embryol. exp. Morph.* 76: 115-137.
18. Meinhardt, H. (1983b). Cell determination boundaries as organizing regions for secondary embryonic fields. *Devl. Biol* 96: 375-385.
19. Meinhardt, H. (1983c). A model for the prestalk/prespore patterning in the slug of the slime mold *Dictyostelium discoideum. Differentiation* 24: 191-202.
20. Meinhardt, H. (1984). Models for positional signalling, the threefold subdivision of segments and the pigmentation pattern of molluscs. *J. Embryol. exp. Morph.* 83:(Supplement) 289-311.
21. Meinhardt, H. (1986). Hierarchical inductions of cell states: a model for segmentation in *Drosophila. J. Cell Sci. Suppl.* 4: 357-381.
22. Meinhardt, H. (1986). The threefold subdivision of segments and the initiation of legs and wings in insects. *Trends Genetics* 2: 36-41.
23. Meinhardt, H. (1988). Models for maternally supplied positional information and the activation of segmentation genes in *Drosophila* embryogenesis. *Development* 104: (Supplement), 95-110.
24. Meinhardt, H. and Gierer, A. (1980). Generation and regeneration of sequences of structures during morphogenesis. *J. theor. Biol.* 85: 429-450.
25. Nüsslein-Volhard, C. and Wieschaus, E. (1980). Mutations affecting segment number and polarity in *Drosophila. Nature* 287: 795-801.
26. Ptashne, M., Jeffrey, A., Johnson, A.D., Maurer, R., Meyer, B.J., Pabo, C.O., Roberts, T.M. and Sauer, R.T. (1980). How the lambda repressor and Cro work. *Cell* 19: 1-11.
27. Serfling, E. (1989). Autoregulation, a common property of eucaryotic transcription factors? *Trend Genetics* 5: 131-133.
28. Technau, G.M. and Campos-Ortega, J.A. (1985). Fate-mapping in wild-type *Drosophila melanogaster.* II. Injections of horseradish peroxidase in cells of the early gastrula stage. *Roux's Arch. Dev. Biol.* 194: 196-212.
29. Tickle, C., Summerbell, D. and Wolpert, L. (1975). Positional signalling and specification of digits in chick limb morphogenesis. *Nature* 254: 199f.
30. Williams, J.G., Jermyn, K.A. and Duffy, K.T. (1989). Formation and anatomy of prestalk zone of *Dictyostelium. Development Suppl.* 107: 91-97.
31. Wolpert, L. (1969). Positional information and the spatial pattern of cellular differentiation. *J. theor. Biol.* 25: 1-47.

LIQUID CRYSTALLINE ORDER IN COLLAGEN NETWORKS

Françoise Gaill

CNRS Biologie cellulaire.
Ivry (France)

INTRODUCTION

Many tissues which are covering the surface of animals show fibrillar organizations which have been called twisted plywoods[1] or helicoids[2]. This arrangement is similar to that of molecules in cholesteric liquid crystalline phases[1], but the liquid character is often abolished by the presence of molecular crosslinks. Cholesteric and other liquid crystalline-like organizations are encountered in extracellular matrices composed of various components[3] but we will only consider in this paper the case of the collagen networks.

ORGANIZATION OF COLLAGEN NETWORKS

Three main types of fibrillar organization have been observed in the collagenous extracellular matrix (Fig.1): 1- unidirectional plywood where all the fibrils are parallel; 2- orthogonal plywoods, where the structure is made of superimposed layers of fibrils presenting two orthogonal directions. In this case the plywood is a discontinuous twisted plywood; 3-continuously twisted plywoods which corresponds to the cholesteric organization of the fibrils. In twisted plywood organizations (Fig. 1), all molecules lie parallel in a given plane and their direction rotates continuously from one plane to the next by a constant angle[1].

ORTHOGONAL PLYWOOD, BLUE PHASE AND DISCRETE HELICOID

The cuticle of annelid has already been described as an orthogonal plywood[4], but cuticular fibrils are composed of subunits, the microfibrils, which present a double twist[5, 6] (Fig. 2). This situation occurs between the molecules of the blue phase which appears during the transition between the isotropic and the cholesteric phase. The double twist situation is a simple way to regulate the fibril diameter[5].

Biologically Inspired Physics, Edited by L. Peliti
Plenum Press, New York, 1991

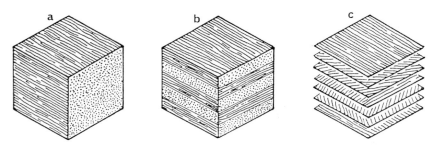

Fig. 1. Three common arrangements in fibrillar network of extracellular polymers: a, unidirectionnal alignement; b, orthogonal plywood; c, twisted distribution of fibrils.

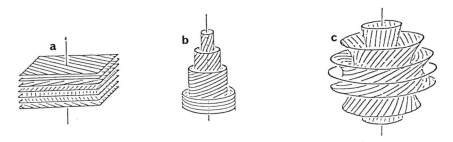

Fig. 2. Planar (a) cylindrical (b) and toroidal twist (c).

Fig. 3   A. Orthogonal plywood in the cuticle of annelid: tranverse section in *Paralvinella grasslei*. Fibrils are seen either in cross or in tranverse sections. Microvilli are orthogonal to the cuticle thickness(TEM) (G:25000).
B. Tangential sections of annelid cuticle (*Paralvinella grasslei*). Fibrils are helical cylinder as they present in all plane section a sinusoidal course. Fibrils are composed of subunits, the microfibrils. The fibrillar pattern is changing from the right part to the left: it is a consequence of the curvature of the worm surface (TEM)(G:25000).
C: Nested arc-like pattern in oblique section of the cuticle of vestimentifera (*Riftia pachyptila*)(TEM)(G:30000).
D: Collagen type I: cholesteric aggregates obtained in vitro. The twist between the fibrils is toroidal (G:25000)(with courtesy of Denèfle and Lechaire).

296

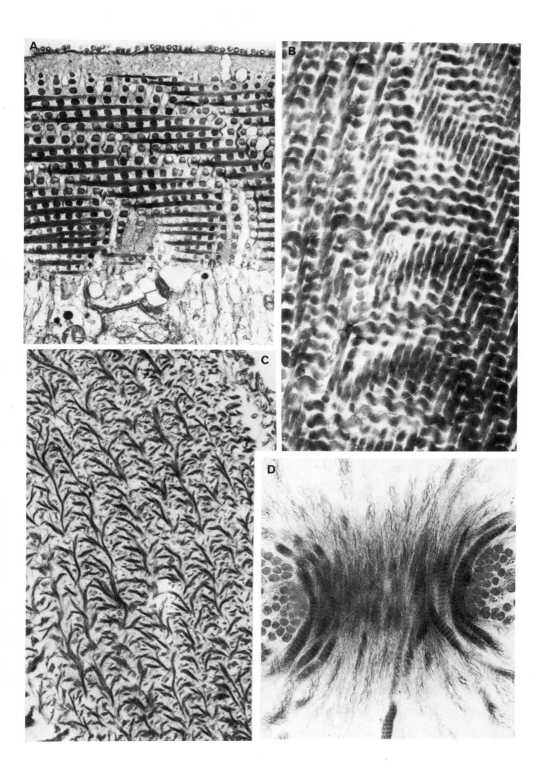

Cuticular fibrils of hydrothermal vent annelids (Fig. 3 A) are helical cylinders (Fig. 3 B) and the supercoil of the fibril allows a relaxation of the molecular twist[5]. This helicity is observed from the molecular level to the macroscopic one[7]. The helical course of the fibril around the worm play a part in the variation of shape of the worm which can contract or extend itself[7].

Nested arcs-like patterns have been described in the cuticle of other worms, the vestimentiferan[8] (Fig. 3 C). Even in tranverse section the fibrillar network seems to be an orthogonal one, it has been demonstrated that this network is a discontinuous twisted plywood made of three directions of fibrils[9]. In this case, the twist is $\pi/3$. Such an organization stabilizes the cuticle and reduces its deformation. This situation is observed in the upper part of the worm which is non elastic, whereas the most posterior parts, which are contractile, present an orthogonal pattern[9].

FIBRILLAR NETWORKS OF TYPE I COLLAGEN

The same fibrillar organizations occur with both cuticular and type I collagen but there is a difference in the intrafibrillar organization. All the molecules of type I collagen are parallel to the fibril axis and the lateral packing of the molecules seems to be an hexagonal one[10] even this model is still debated[11]. If we accept the model of the hexagonal packing of the molecules in the type I collagen fibril, the cholesteric organization of the compact bone[12] is a discrete cholesteric organization of hexagonal domains. It means that crystalline domains may form cholesteric organizations.

Cholesteric aggregates (Fig. 3 D) presenting a new type of twist, a toroidal one have been observed in collagen gels obtained in vitro[13]. True cholesteric phases have only been obtained with sonicated molecules of type I collagen[14]. Such results demonstrate that a self assemby process occurs in vitro between the molecules of collagen and one can assumes that the first step of the plywood assembly corresponds to a cholesteric organization of the molecules. The second step is the compaction of molecules which lead to the formation of fibrils. The fibril formation corresponds to an increase of the polymer concentration leading to the hexagonal packing of the molecules. It is known that when the concentration of polymer increases in liquid crystalline phases, the hexagonal order extends and prevents a continuous cholesteric twist[15]. The twist then relaxes in a discrete way between domains of hexagonal order creating twist boundaries. In the case of type I collagen, the domains correspond to the cylindrical fibrils and the discontinuous twist between the hexagonal domains is a consequence of the compaction process.

Whatever the origin of such discrete twist, one can consider that collagen fibril is always the result of a phase seggregation. In this point of view, the fibrillar network is an ordered array of phase seggregation. How this order can be explained and what is the relative part of physical and biological factors in this process is an opened question.

REFERENCES

1. Y.Bouligand, Twisted fibrous arrangement in biological materials and cholesteric mesophases, *Tiss. Cell*. 4: 189 (1972).
2. A.C. Neville, A pipe cleaner molecular model for morphogenesis of plant cell walls based on hemicellulose complexity, *J.theor. Biol.* 243: 131 (1988).
3. Y.Bouligand, M.M. Giraud-Guille, Spatial organization of collagen fibrils in skeletal tissues: analogies with liquid crystals, *in* Biology of invertebrates and lower vertebrate collagens, A.Bairati and R.Garronne eds, Plenum Publishing Corporation, N.Y.(1985).
4. L.W.Murray, M.L.Tanzer., The collagen of annelida, *in* Biology of invertebrates and lower vertebrate collagens, A.Bairati and R.Garronne eds, Plenum Publishing Corporation, N.Y. (1985).
5. F.Gaill, Y. Bouligand, Long pitches helices, *in* Biology of invertebrates and lower vertebrate collagens, A.Bairati and R.Garronne eds, Plenum Publishing Corporation, N.Y. (1985)
6. L.Lepescheux, Spatial organization of collagen cuticle in annelids: order and defects, *Biol. Cell*, 62: 17 (1988).
7. F. Gaill, Y. Bouligand, Supercoil of collagen fibrils in the integument of Alvinella, an abyssal annelid, *Tiss. Cell* 19: 625 (1987).
8. F.Gaill, D.Herbage, L.Lepescheux, Cuticle structure and composition of two hydrothermal vents invertebrates, *Oceanol. Act.*, 8: 155 (1988).
9. F.Gaill, D. Herbage, L.Lepescheux, A three directional plywood of collagenous fibrils: the cuticle of *Riftia Pachyptila*, Matrix, in press.
10. D.J.S. Hulmes, A.Miller, Quasi hexagonal packing in collagen fibrils, *Nature*, 282: 878 (1979).
11. J.A. Chapman J.A., Molecular organization in the collagen fibril, *in* Connective tissue matrix, D. W.L. hukins ed., Verlag chemie, 1985.
12. M.M. Giraud-Guille, Twisted Plywood Architecture of Collagen Fibrils in Human Compact Bone Osteons, *Calcif. Tissue Int*, 42: 167 (1988).
13. Y. Bouligand, J.P. Denèfle, J.P. Lechaire, M. Maillard, Twisted architecture in cell-free assembled collagen gels: study of collagen substrates used for cultures, *Biol. Cell*, 54: 143 (1985).
14. M.M. Giraud-Guille, Liquid crystalline phases of sonicated type I collagen, *Biol. Cell*, 67, 97 (1989).
15. F. Livolant, Y. Bouligand, Columnar textures presented by helical polymers, *J. Phys.* 47: 1813 (1986).

PHYSICAL APPROACHES TO BIOLOGICAL EVOLUTION

M.V. Volkenstein

Inst. Molecular Biology
Acad. Scienc. USSR
Moscow, URSS

1. PHYSICS AND BIOLOGY

There are physicists and biologists who consider physics and biology as incompatible fields of knowledge. Some years ago B.-O. Küppers edited a collection of papers[1]. This interesting book contains an old paper of Niels Bohr, who wrote about the complementarity of studies of an organism based on atomic-molecular treatment and of the studies of an organism as a whole entity. However at the end of his life Bohr considered this complementarity not as a fundamental principle but only as a practical one[2]. Polanyi emphasized the important role of information in biological phenomena but considered this notion as something alien to physics[3]. Heitler wrote about complementarity of living and non-living matter[4], Elsässer[5] and Wigner[6] tried to introduce the so-called biotonic laws, which have nothing to do with physics, but are characteristic of life.

Küppers treated this problem in the introductory paper in ref.1 and in his recent book devoted to information and the origin of life[7]. I agree with the point of view concerning the unity of physics and biology which he calls reductionist. However this word is not very good because it means that biology can be reduced to physics, hence physics is simpler than biology. The last statement can be successefully criticized.

On the other side there are biologists, who oppose biology to physics. In his important book[8] the eminent specialist in the theory of evolution Ernst Mayr considers physics as something totally alien to biology. He thinks that there are no laws in biology in the same sense as they exist in physics. "When they say that proteins do not translate information back to nucleic acids, molecular biologist consider this a fact rather than a law". I do not understand why. This non-translation is a real law and the base of another biological law, namely the law of non-inheritance of acquired characters.

"The last twenty-five years have also seen the final emancipation of biology from physical sciences". Here Mayr is mistaken again. Just the second half of our century witnessed a strong development of molecular biology, tightly connected with physics.

In his further presentation Mayr opposes Darwinism and physics: "It was Darwin more than anyone else who showed how greatly theory formation in biology differs in many

Biologically Inspired Physics, Edited by L. Peliti
Plenum Press, New York, 1991

respects from that of classical physics ... It is only since 1859 that the biological sciences begun to emancipate themselves from the dominance of the physical sciences". There were "many instances where physicalism has had a deleterious effect on development in biology".

In reality the situation is just the opposite. Darwin for the first time in science has shown the possibility of the formation of order from chaotic variability. Darwin can be considered as the founder of the new field of physics. This field developed mainly by Prigogine and later by Haken is called the physics of dissipative systems (i.e. of open systems which are far from equilibrium) or synergetics. Boltzmann has called his XIX century the century of Darwin. These words are not accidental. They mean that physics has to study the non-equilibrium, historical events in nature, and Boltzmann devoted his works to this most important goal.

It is rather strange that no one of the papers and books already quoted contains the definitions of physics and biology. Without such definitions the discussion of the interrelations of natural sciences loses its basis.

Evidently physics is the most general science which studies concrete forms of matter and field and the forms of their existence - space and time. Biology studies living nature, and chemistry studies the changes of electronic shells of atoms and molecules as the result of their interactions. We see that these definitions use different logical characters and therefore are uncompatible.

The definition of physics does not include the difference between living and non-living nature. But it does not mean that all natural sciences could be "reduced" to physics, which has the Aristotelian sense. It only means that every natural science has many levels of investigation and the deepest level is always physical. This level is achieved in chemistry, contemporary theoretical chemistry is completely based on physics, namely on thermodynamics and statistical mechanics, kinetics and quantum mechanics. But we are far from this level in biology because living nature is much more complicated than the non-living one.

Hence the search for the physical approaches to the most important problems of theoretical biology, such as the theory of evolution, is quite legitimate.

The struggle against "physicalism" or "reductionism" has always been non-constructive and non-scientific. The arguments of Heitler, Wigner, Mayr and many other scientists are based mainly on the insufficient knowledge of biology or physics. The so called philosophers in Soviet Union during many years considered "physicalism" in biology as anti-Marxist heresy. At the base of this ideology we see the Stalinist tendency to abolish every kind or creative work in science and art. It is much easier to destroy some part of natural science if it is separated from other parts. Lysenko was a strong "anti-reductionist", and I think that such an excellent biologist as Mayr should be very disappointed by this fact - it is rather unpleasant to formulate ideas which would be supported by Lysenko.

The main feature of biology which makes it different from usual physics is the historicity of every organism. Living systems change during their observation and keep the memory of their evolutionary origin. However there are also other fields of physics which are also historical, such as cosmology, astrophysics, geophysics. Historicity means the formation of new information, the evolutionary (or revolutionary) development due to the instabilities of the previous states. We have to cope with non-equilibrium phase transitions, with the appearance of order from chaos.

Contemporary physics contains new chapters which did not exist not so long ago. Among these are the physics of dissipative systems, tightly connected with the theory of information, cybernetics. The development of these new fields does not mean that the old physics has reached the limits of its validity as it was in the case of the theory of relativity or quantum mechanics. Synergetics does not contradict thermodynamics but means its broadening.

In the thirties the unification of Darwinism with population genetics was achieved. Darwinism is not a very nice word, it reminds us of philosophy or politics, but nothing can be done - the term is already established. This unification resulted in the formation of the so-called synthetic theory of evolution. Now we have to connect evolutionary theory with molecular biology, synergetics and the theory of information. Physics approaches evolutionary biology from two sides - from the atomic-molecular side, and from the phenomenological side of synergetics.

2. MOLECULAR BASIS OF EVOLUTION

In the course of the development of molecular biology it has been shown that the amino-acid contents of proteins (and, of course their primary structures) reflects molecular evolution. The nearer the species, the nearer are their homologous proteins. Hence, evolutionary trees can be built using the data about their amino-acid structure[9,10]. We know that only the primary structures of proteins are coded by the genomes, but natural selection acts on biological functions determined by the spatial structures of proteins. Of course, there must exist a correlation between the primary sequence and spatial structure - without such a correlation the whole molecular biology and molecular genetics would lose its sense. Really it has been shown that proteins denatured thermally can be renatured by cautious annealing. But there remains the question - whether this correlation is really unique.

During the last decades the Japanese geneticist Motoo Kimura has developed the "neutral theory of evolution", which is presented in his monograph, published in 1983[11]. It has been shown that a great proportion of point mutations in proteins, and correspondingly in nucleic acids, is neutral in the sense that these substitutions do not produce changes in the biological behavior of biopolymers, and are not under the pressure of natural selection. The main arguments in favor of the neutral theory are the existence of the "molecular clock" of evolution, and the presence of isoenzymes.

Molecular clock means that the rate of evolution of homologous proteins is approximately constant if it is calculated per year and not per generation. This rate can be estimated using paleontological data for the appearance of corresponding taxa. The evolutionary rates for different proteins differ very much as it is shown by the table.

TABLE
The rate of evolution - the number of substitutions per one amino-acid residue per 10^9 years

Fibrinopeptides	9.0
Pancreatic ribonuclease	3.3
Hemoglobin	1.4
Myoglobin	1.3
Animal lysozyme	1.0
Insulin	0.4
Cytochrome c	0.3
Histone H IV	0.006

Which is the meaning of those numbers, why is the rate for histone by three order of magnitude smaller than for fibrinopeptides?

Roughly speaking the protein consists of the active and passive parts. In an enzyme the active part contains the active site and its nearest surroundings. The chemical - electronic-changes together with the conformational events occur in this part of the protein. In the passive part - in the remaining moiety of the protein - only conformational changes occur. It has been shown that the dynamics of proteins is due to the interactions of the electronic and conformational degrees of freedom[12,13]. Evidently the substitutions at the active part are more dangerous than those at the passive part and, hence, less neutral. Kimura himself has shown that the rate of substitutions at the active part of hemoglobin is ten times smaller than that at the surrounding moiety. The numbers in the table are connected with the ratio of passive and active parts. The smallest rate of histone shows that practically the whole molecule is active - histone HIV interacts with DNA, with other histones and non-histone proteins at the chromosome.

The physical sense of neutralism is the degenerate correlation of the primary and spatial structure of the proteins. In many cases pronounced changes of the primary structure do not produce any important change of the spatial one. This has been shown for the first time in the paper of Lesk and Chothia[14], who compared the X-ray data for nine globins. Their primary sequences differ rather strongly but their spatial structures are very similar.

In the calculations made by Ptitsyn, who investigated the secondary structures of copolymers built by two kinds of residues - hydrophobic and hydrophilic - it has been shown that the amount of α-helices and β-strands in such a model is similar to that of real proteins. According to Ptitsyn the protein can be considered as an "edited" random copolymer[15]. Similar results have been obtained in the more elaborate calculations made by Ken Dill[16] and by Shakhnovich and Gutin[17]. The "editing" work is done by natural selection at molecular level, an important role in these events is played by metals, which are important cofactors of a multitude of enzymes. Contemporary proteins are the results of memorizing the accidental, random choice[18].

Together with neutral mutations exist the pseudoneutral ones. Such are the harmful substitutions compensated at the metabolic or genetic level[19]. Recently Kimura investigated the possibility of such a compensation by a second mutation on the same protein[20]. The violation of the compensation of pseudoneutral mutations can be a cause of extinction of species[21].

These considerations, based on an atomic-molecular treatment of biopolymers, show that the main part of point mutations cannot be responsible for speciation and macroevolution. This is also shown by comparison of homologous proteins of neighbor species. For instance the amino-acid contents and sequences of the most important proteins of humans and chimpanzees do not differ more than by 1%[22]. This means that it is not sufficient to consider only the structure of the proteins for the studies of genotype-phenotype relations. We have to know the answers to three other most important questions: how much, where and when this protein has to be produced in the ontogeny of a multicellular organism. The differences of the species are due to the regulation of the activity of genes, and to the pronounced changes in the structure of genome as a whole entity.

Two most important discoveries have been made during last decades concerning the structure and behavior of genomes of eukaryotes - the existence of an enormous quantity of DNA which does not code the biosynthesis of proteins and the non-constancy of the genome. The non-coding DNA has been considered as garbage, as egoistical material etc. However it seems now that this DNA is used in molecular evolution in many ways (cf, e.g. 23). The non-constancy of genome means that genes can change their positions at the chromosomes, can be transferred from one chromosome to another one, and to be introduced from outside in the forms of plasmids, transposons etc. One of the results of such events can be the non-symmetrical crossing-over. Hence, the Mendelian heredity based on the constant positions of genes at the chromosomes can be violated.

All these phenomena have been united in the concept of molecular drive proposed by Gabriel Dover[24,25]. The schematic presentation of these events is shown at Figure 1. We can think that at the molecular level speciation and macroevolution are due mainly to molecular drive. These processes are non-Mendelian.

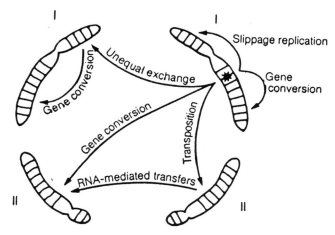

Fig.1   Schematic presentation of molecular drive-non-reciprocal exchange occuring within and between homologous and non-homologous chromosomes (pairs I and II).

One of the greatest problems of theoretical biology is not yet solved. This is the problem of relations between genotype and phenotype. We are used to think that the genome is a kind of the blueprint of the organism. In reality it is not. Taking to pieces a machine made be human hands it is possible to reconstruct the original blueprint. Such a procedure is totally impossible in the case of an organism - the genome is to be considered not as a blueprint but as a recipe. In the ontogeny occur processes of self-organization. They are especially complicated and realizable only in a dissipative system.

3. THE THEORY OF MANFRED EIGEN

The molecular treatment of evolution presented in the previous section considers the multicellular organisms with very big genomes and limited populations. The situation of viruses and prokaryotes - bacteria and some algae - is quite different. Their genomes are comparatively small but the populations can be enormous.

In 1971 Manfred Eigen has published his first theoretical paper devoted to self-organization of macromolecules, to prebiotic evolution[26]. This outstanding work is directly connected with the problem of the origin of life. It has both the physical and philosophical sense. The first problem which has been solved by Eigen was the construction of the corresponding theoretical model based on reasonable physical assumptions. Such a model consists of an open system where the processes of copolymerization and decay of polymers occur. The first process proceeds autocatalytically - polymer chains serve as templates for further synthesis of themselves. Polymerisation occurs with mistakes - the system possesses mutability. The system is open and far from equilibrium - there is a flow of monomers through the semipermeable walls of the vessel. Hence the metabolism is also taken into account. The energy required for the synthesis is provided by the energetically enriched monomers.

Plenty of polymers with various sequences of monomeric units are formed in such a system. They possess different selective values determined by the rates of polymerization and by the "quality factor" which depends on the probability of mutations. Therefore there occurs a competition of different polymers and Darwinian natural selection - the polymers with the highest selective value grow at the expenses of others. This well known theory uses non-linear differential equations but similar results can be obtained also by stochastic methods.

Classical Darwinism considered a single species as the winner in the struggle for existance. Eigen's theory introduces the very important notion of quasi-species[27-33]. Not a single "master copy" of the polymer (e.g. RNA) is selected but a collection of its mutants. In real living nature we have to do with competing hypercycles - the cycles are formed by two complementary chains of nucleotides, in the hypercycle enzymes are produced, which catalyze the replicative synthesis of RNA or DNA (polymerases)[27].

For the treatment of the quasi-species model the multi-dimensional sequence space can be introduced and investigated. Every sequence of the length v of the copolymer built by k different units corresponds to a point in a space of dimension v, formed by k equivalent points. This space is a hypercube with k^v points. The quasi-species is clustered in a definite region of this space determined by the fitness landscape due to interactions of macromolecules with surrounding medium. There are ridges of mountains at this landscape corresponding to the higher fitness. Mutants produce mutants, those which do not differ much from those of quasi-species are positioned nearer to it than the sequences containing many mutations. The distribution of macromolecules at the space of sequences is asymmetrical and chains with new advantageous mutations are positioned rather far from the initial quasi-species. If the frequency of errors (the mutation rate) is high enough, there occurs a phase transition similar to a phase transition of the second kind and described by a theory similar to the theory of the bidimensional Ising model for the ferromagnetic - paramagnetic transition. The threshold of this transition characterizes the loss of the previous quasi-species which becomes non-stable. Hence a new quasi-species can be formed[33].

Only the mutants forming the quasi-species can be considered as neutral in this model. The population is very great and fills some part of the sequence space. As the mutants produce new mutants there is a bias of substitutions, the evolution is directed or guided, and proceeds along the ridges to the higher peaks at the landscape.

Eigen introduced two new notions in the theory of evolution - quasi-species and sequence space.

It is very important that this model did not remain a purely theoretical one. It could be proved experimentally. There has been studied the evolution of the nucleic chain of the Q_β-phage in the test-tube. The theory allows to make quantitative predictions in such causes which have been totally corraborated by Biebricher[34,35]. It can be shown that some human viruses are near the threshold of transition, and therefore can change their structure rather rapidly. This concerns such important viruses as that of influenza and AIDS.

It has to be emphasized that the fitness landscape itself must depend on the populations climbing the mountains. These processes require higher approximations of the theory, which have not yet been studied.

We see that molecular physics (and chemistry) of single macromolecules (and of such systems as viruses and bacteria) differs from the physics of eukaryotes, of multicellular organisms reproducing in the sexual way. The neutral theory of Kimura is applicable only in the second case, if the genomes are very great but the populations are comparatively small. The greater the population, in relation with the number of the units in RNA or DNA, the more restricted is its random behavior. The evolution becomes channeled or guided. Small changes of the medium can produce phase transitions shifting the center of gravity in sequence space.

The difference of these two cases is connected with different genotype-phenotype relations. In the case of macromolecules or viruses the natural selection acts directly on genotypes. In multicellular organisms molecular genetics exists at the level of genotype but the phenotypes are selected. However it is no doubt that many ideas of Eigen's theory can be used also in this case.

The importance of the theory of Eigen cannot be exaggerated. After a century of Darwinian theory a real physical approach to the problems of biological evolution has been worked out, and it has been shown that the normal classical physics is sufficient for the understanding of the most complicated biological phenomena such as natural selection. The theory of Eigen plays a central role in the contemporary stage of the theory of evolution. As we shall see it is directly connected with the theory of information.

4. PHENOMENOLOGICAL THEORY OF GRADUAL AND PUNCTUAL EVOLUTION

After this approach to evolution at the basis of molecular physics and chemistry let us try to understand it using the ideas of the physics of dissipative systems or synergetics.

Every cell, every organism, every species, and the biosphere as a whole are dissipative systems which are open and far from equilibrium. Just because of these features evolution is possible. Darwin, who studied evolution of course not at the molecular level, considered it as a gradual process. Darwin explained the lack of many intermediate forms in the paleontological record by its incompleteness. Already in 1921 in his "Nomogenesis" the outstanding Russian biologist L.S.Berg wrote: "The birth and death of individuals, species, ideas is a catastrophic process. The appearance of all these categories is preceded by a long hidden period of development, proceeding on the base of some laws. Then at once a jump, *saltus*, sets in, expressed by their appearance, spreading over the earth's surface and conquest of the place under sun. The process of transition of a gas into liquid is a saltatorial change"[36].

The comparison of speciation with a phase transition is remarkable. A synergetic system can exist in a chaotic state. The physical definition of the chaotic system is the possibility to rise small fluctuations to macroscopic dimensions. Instabilities of state are followed by bifurcations, non-equilibrium phase transitions with pronounced changes of structure and behavior. Already in the simplest example of a Mendelian biallelic system the stationary dynamics possesses in many cases properties of phase transitions, as it has been shown in ref.37 (cf. also [40]).

Many facts testify in favour of saltationist speciation and appearance of higher taxa (macroevolution). Some rather complete paleontological lineages have been discovered which show that speciation and macroevolution occur during a time which is much shorter than the time of stasis, during which the species have no pronounced changes. Species can be stable during hundreds of thousands and millions of years, speciation occurs instantaneously from the geological point of view - during tens of thousands of years. This argument (which was already known to Darwin) has been stressed by Gould and Eldredge who proposed the theory of "punctuated equilibrium"[38,39].

Punctualism does not reduce only to the saltatorial evolution. According to this conception the total adaptation of a new species cannot occur immediately. And really we know that plenty of characters are not adaptive at all. Let us consider three examples. The red colour of blood is not itself an adaptive character -it has no sense to ask "what for the blood is red". In general in biology as in physics we have to ask "why", and not "what for", and of course we know why blood is red. The second example is the structure of external genitalia of the female of the spotted hyaena *Crocuta crocuta*. They are totally similar to the genitalia of male. What for? Surely it is not an adaptive character which is rather unconvenient for the poor animal. This situation has been studied intensively and it is known now that the female of the spotted hyaena is much bigger than the male, and this character has some adaptive meaning. This is due to the shift of the balance of sexual hormones, the increase of the content of androsterone produces the described effect as a secondary character. The third example is the most important for humans. The Cro-Magnon men of the Paleolithic belonged to the same species and sub-species as we - they were Homo Sapiens Sapiens. To what purpose did these men need a brain able for abstract thinking, for building the theory of evolutlon or the theory of relativity? It was not a directly adaptive character, a secondary result of human evolution. Punctualism is directly connected with the non-adaptationism of many characters.

At the other side it is connected with the neutralism of many point mutations in biopolymers. We have seen that the neutral mutations do not play any role in speciation and macroevolutlon. The polnt mutations are gradual.

There is a triad presenting the connections between punctualism, non adaptationism and neutralism (Fig.2)[40].

We have already seen that according to Eigen's theory evolution in the case of macromolecules or viruses is punctual and occurs as phase transitions.

The statement of phase transitions at speciation and macroevolution is purely thermodynamical - time is not taken into account - speciation and macroevolution can be more or less rapid. There are lineages which are more or less punctual.

Speciation is a kind of bifurcation. Let us present the gradual and punctual divergence of species by corresponding physicomathematical models. At Fig.3a gradual speciation is shown schematically, at Fig.4a the punctual scheme. At Fig.3b simple non-linear equation can be compared with Fig.3a.

$$\frac{dx}{dt} = x^3 - \lambda x \tag{1}$$

Here x means some quantitative morphological character, λ the evolutionary parameter changing with time. The stationary solutions are

$$x_{st} = 0, \pm \lambda \tag{2}$$

The first solution is stable at $\lambda \leq 0$, the second and the third are stable at $\lambda > 0$. At Fig.3b and 4b the stable solutions are shown by continuous lines, and the unstable solutions by the broken lines.

Punctual speciation can be described by the more complicated equation of the fifth order

$$\frac{dx}{dt} = x[\lambda - x^2 (x^2 - a^2)] \tag{3}$$

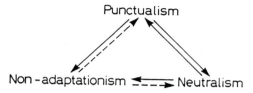

Fig.2   The triad of the most important features of evolution.

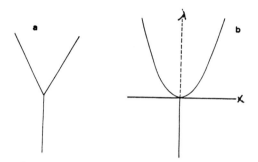

Fig.3   a. The scheme of gradual speciation.
          b. Mathematical presentation of gradual speciation.

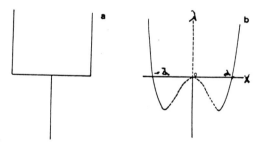

Fig.4   a. The scheme of punctual speciation.
          b. Mathematical presentation of punctual speciation.

where a is an additional parameter. The stationary solutions of (3) are

$$x_{st} = 0, \pm \left[\frac{a^2}{2} \pm \left(\frac{a^4}{4} + \lambda\right)^{1/2}\right]^{1/2} \tag{4}$$

The corresponding picture is shown at Fig.4b. The situation is like that of the Van der Waals gas. When the vertical line $x_{st}=0$ reaches the value $\lambda=0$ the system chooses the stable solutions $\pm a$. Further increase of λ means the increase of x_{st} along the parabola of the fourth order, i.e. much steeper than in the case of 3b[41].

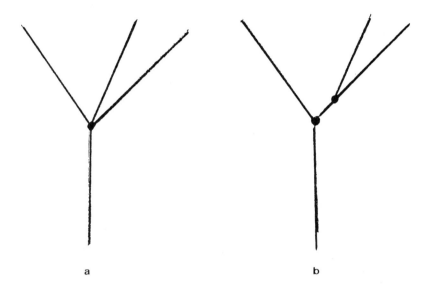

a   b

Fig.5   a. The apparent three branches of triple speciation.
              b. Real sense of triple speciation.

Further analysis shows that if the speciation occurs really as bifurcational branching, the probability of more than double branching must be very low. If the biologists draw a triple speciation (Fig.5a) it means in reality that there are two points of double branching positioned rather near to one another (Fig.5b)[42].

Of course this presentation is not a real theory. Equations (1) and (3) are taken "from the sky". However they represent the reasonable approach to the problems of evolution, and it is possible to hope that non-linear equations describing evolution will be obtained on the basis of the theory of molecular drive.

5. THE DIRECTIONALITY OF EVOLUTION

Some physicists, who are interested in biology platonically, but are rather far from it, think that the theory of Darwin is obsolete and wrong because it cannot explain how the material and time could be sufficient for the evolutionary building of the complicated biosphere. These ideas are erroneous. The material for selection is quite sufficient because it is presented by the multitude of recombinants of the parent genomes expressed in tens of percentage in every population. Concerning the time of evolution, this difficulty does not exist because evolution is directioned, channeled.

Not every kind of changes are possible in an evolving species, but only those, which are compatible with the structure of the previous organisms and limited possibilities of their change. Not everything can be done by natural selection, and it is no less important to know these limitations than the possibilities. Evolution is constrained.

Darwin understood it very well. At the very beginning of his immortal book Darwin wrote: "we clearly see that the nature of the conditions is of subordinate importance in comparison with the nature of the organism in determining each particular form of speciation; - perhaps of not more importance than the nature of the spark, by which a mass of combustible matter is ignited, has in determining the nature of the flames"[43].

These words of Darwin are usually forgotten because the further exposition in the "Origin of Species" is mainly devoted to the spark and not to the flame.

The same idea has been developed by Gould, who wrote[44]: "Organism are not billiard balls, struck in deterministic fashion by the cue of natural selection, and rolling to optimal position on life's table. They influence their own destiny...". In another paper of the same author[45] we read: "... the constraints of inherited form and developmental pathways may so channel any change that even though selection induces motion down permitted paths, the channel itself represents the primary determinant of evolutionary direction". But already in 1921 Berg expressed similar views[36].

As we have seen, the theory of Eigen and the corresponding experiments with viruses show that the evolution of macromolecules proceeds in a guided way being channeled by sets of sequential mutations. In the case of multicellular organisms the channelizing occurs in the ontogeny, and it is not an accident that the embryos of various classes of vertebrates possess similar characters, such as gills which are inherent to fishes. The ontogenic development provides serious constraints of the action of moving selection[46]. All earth's vertebrates possess four and only four limbs. Perhaps it would be more convenient to have six of them as the insects have. The number four cannot be considered as an adaptive character - adaptation occurs by the transformation of these limbs into different forms. Only legendary creatures, such as Assyrian winged bulls, centaurs or angels have six limbs. Why four limbs? Because the ancestors of the earth's vertebrates were the *Crosspterygii* fishes, which possessed four fins being the predecessors of the limbs.

The great artist Daumier has drawn the cartoons showing the evolution of the king of France Louis Philippe into a pear (Fig.6). The constraints are expressed by the transformation into a pear but not to an apple.

Ebeling and co-workers have studied evolution mathematically using the Fisher-Eigen equation for the distribution function in the space of phenotypic characters[47,48]. The same equation was used in our work[19], where we studied the evolution of the distribution function near the maximum of fitness. The theory allows to take into account the constraints which limit the possibilities of changes of the single characters. There is a "corridor" of the changes at the evolutionary landscape, evolution is directed, and it can be shown that this directionality accelerates the evolution very much.

The directionality of evolution can be treated with the help of the theory of Markov's chains. Constraints are expressed by small values of many non-diagonal terms in stochastic matrices.

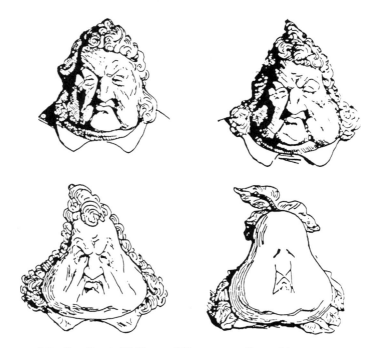

Fig.6   Louis Philippe of France transformed into a pear.

6. EVOLUTION AND THE THEORY OF INFORMATION

The theory of information is an important part of synergetics, of the physics of dissipative systems. The canonical theory of information treats only the amount of information in a message independent on the sense of this message. Information is contained in the message which can be perceived and memorized.

The well known connection between information and entropy is not trivial. The decrease of entropy means increase of order, increase of information. If we drop a coin one bit of information is obtained. The minimal entropical pay for one bit is k ln 2, where k is the Boltzmann constant. But the real pay is many orders of magnitude greater - the coin can be very big. Information is equivalent to the small difference of two great amounts of entropy[50]. The loss of entropy determines the value of energy of the system which can be expressed by the dimensionless value.

$$W(E,X,t)=1 - \exp \{[S(E,X,t) - S_{eq}(E,X)]/k\} \qquad (5)$$

where S is the entropy, E - energy, X- other extensive parameters, t- time[51]. Loss of entropy means gain of information, and of the value of energy. We see that the amount of information is really connected with entropy and the notion of information belongs to physics. Polanyi erroneously opposed information to physics[31]. Of course the exact equivalence of one bit of information and k ln 2 of entropy is valid only for microinformation which cannot be memorized. Hence it is not the true information. However the amount of information, its reception and memorizing, and its creation have definite physical meaning.

We meet with information in all branches of biology and human activity. According to von Weizsäcker information is that what produces information - the measure of the amount of structure[52]. History does not exist besides information, historical sciences study the temporal changes of the systems which are far from equilibrium and open, the dissipative systems. This concerns first of all biology, cosmology, all human sciences[53].

The creation of information means the memorizing of accidental choice[54]. Complicated systems, such as living organisms, can be formed and maintained only as a result of creation of new information and of its storage which occurs at two levels - at the level of genetics and in the memory in the direct sense of the word[53]. New information is created at the birth of every organism, whose genome is the randomly chosen recombinant of the parents genomes. New information is created at the formation of every new species or higher taxon. Natural selection is a creative process in the same sense as art or literature.

The first attempt to apply the theory of information to biological evolution belongs to the outstanding Russian evolutionist Schmalhausen[55]. At the end of his work Schmalhausen wrote: "the contemporary theory of information does not possess any methods of estimation the quality of information, but it has a decisive significance in biology". Küppers writes about the semantics of information[53]. This word can be used in the same sense as quality or value of information. The corresponding problems have been discussed in the papers[56,57].

The selective value in the theory of Eigen has the sense of the value of information contained in the "text", the primary sequence of the macromolecule.

The value of information cannot be defined independently of its reception. Reception and memorizing of information are irreversible processes due to the non-stability of initial state of receptor which turns to the new more stable state obtaining information. The value depends on the preliminary supply of information possessed by the receptor.

Real applications of the theory of information in biology require the estimations of the value. The tentative measure of it is the irreplaceability. The less redundant is the information, the more valuable it is[50,56,57]. It could be shown that the irreplaceability increases in the course of evolution[12,50,58,59]. There exist some attempts to apply the theory of information in biology without consideration of semantic information[60-62]. These books did not produce any effect in science. The books[61,62] contain many errors.

Only information remains when we die - our genes in children, our creations in art and science, memory of us. The science of life and death at the end is the physics of information - biologically inspired physics.

7. CONCLUSIONS

The relations between physics and biology are discussed. The "anti-reductionist" ideas are criticized. There are two possible physical approaches to biological evolution - the atomic-molecular approach and the phenomenological one based on synergetics and the theory of information. The neutral theory of molecular evolution developed by Kimura is presented, its physical meaning is discussed. Additional arguments in favour of the neutral theory are presented. At the level of macromolecules and such comparatively simple systems as viruses and bacteria the neutral theory cannot be applied. Here the theory of molecular self-organisation and evolution developed by Eigen shows quantitatively the guided character of evolution, the phase transitions at speciation. For the first time a real physical theory of these phenomena is established and proved experimentally. The concept of molecular drive is described. The treatment of living systems as dissipative systems is given, speciation and macroevolution can be considered as bifurcations, as non-equilibrium phase transitions. The punctual evolution corresponds to this treatment, punctualism, neutralism of point mutations, and the non-adaptationism of many characters are interconnected. The informational aspects of evolution are discussed. Historical sciences have to do with information.

REFERENCES

1. B.-O.Küppers, ed. Leben=Physik+Chemie? Piper; Munchen, Zurich 1987.
2. N.Bohr, Naturwissenschaften, 50, 21, 1963.
3. M.Polanyi, Science, 160, 1308-1313, 1968.
4. W.Heitler, Abhandlungen der Mat.-Naturwiss. Klasse der Akademie der Wissenschaften und Literatur in Mainz, 1, 3-21, 1976.
5. W.Elsässer, The Phisical Foundation of Biology. Pergamon, Oxford, 1958.
6. E.Wigner, Symmetries and Reflexions. Indiana Univ. Press. Bloomington, Indiana, 1987.
7. B.-O.Küppers, Information and the Origin of Life. The MIT Press; Cambridge Mass., London 1990.
8. E.Mayr, The Growth of Biological Thought. Harvard Univ. Press, Cambridge Mass. 1982.
9. C.Anfinsen, The Molecular Basis of Evolution. John Wiley and Sons, N.Y. 1959.
10. R.Dickerson, L.Geis, The Structure and Action of Proteins. Harper and Row, N.Y., Evanston, London, 1969.
11. M.Kimura, The Neutral Theory of Molecular Evolution. Cambridge Univ. Press, London, N.Y., Melbourne, Sidney, 1983.
12. M.Volkenstein, Biophysics, Mir, Moscow, 1983.
13. M.Volkenstein in: The Fluctuating Enzyme. G.Welsh ed. J.Wiley and Sons. N.Y. etc., 403-419, 1986.
14. A.Lesk, S.Chothla. J.Mol.Biol. 136, 225-270, 1980.
15. O.Ptitsyn in Conformations in Biology, Eds. R.Srinivasan, R.Sarma, Adenine Press, N.Y., 49-60, 1983.
16. Kit Fan Lan, Ken Dill. Theory for Protein Mutability and Biogenesis. Proc. Nat. Acad. Sci. US 87, 638-642, 1990.
17. E.Shakhnovich, A.Gutin. Proteins, 1990, in press.
18. O.Ptitsyn, M.Volkenstein. J.Biomol.Structure and Dynamics, 4, 137-156, 1986.
19. M.Volkenstein, Mol.Biol.(Russian), 19, 55-66, 1985.
20. M.Kimura, J.Genet. (India), 64, 7-19, 1985.
21. V.Volkenstein, T.Rass. Doklady Acad.Sci. USSR, 295, 1513-1517, 1987.
22. M.King, T.Jukes, Science, 164, 788-793, 1969.
23. E.Trifonov, Bull.Math.Biol., 51, 417-431, 1989.
24. G.Dover, Nature, 299, 111-117, 1982.
25. G.Dover, Trends Genet., 2, 159-165, 1986.
26. M.Eigen, Naturwissenschaften, 58, Heft 10, 1971.
27. M.Eigen, P.Schuster, The Hypercycle. Springer-Verlag, Berlin, Heidelberg, N.Y., 1979.
28. B.-O. Küppers, Molecular Theory of Evolution. Springer-Verlag, Berlin, Heidelberg, N.Y., 1985.
29. M.Eigen, Chemica Scripta, 26B, 13-26, 1986.
30. M.Eigen, Stufen zum Leben. Piper. München, Zürich, 1987.
31. M.Eigen, Cold Spring Harbor Symposia on Quant. Biology, 52, 307-320, 1987.

32. M.Eigen, R.Winkler-Oswatitsch, A.Dress. Science, 85, 5913-5917, 1988.
33. M.Eigen, J.Mc Caskill, P.Schuster. Adv.Chem.Phys., 75, 149-263, 1989.
34. C.Biebricher, Cold Spring Harbor Symposia on Quant. Biology, 52, 299-306, 1987.
35. C.Biebricher - the paper at this conference.
36. L.Berg, Nomogenesis or Evolution Determined by Law. MIT Press, Cambridge Mass., 1969.
37. B.Belintsev, M.Volkenstein, Doklady Acad. Sci. USSR, 235, 205-210, 1977.
38. S.Gould, N.Eldredge, Paleobiology, 3, 115-151.
39. N.Eldredge, S.Gould, Nature, 332, 211-212, 1988.
40. M.Volkenstein, Bio Systems, 20, 289-304, 1987.
41. M.Volkenstein, M.Livshits, Bio Systems, 23, 1-5, 1989.
42. M.Livshits, M.Volkenstein, Bio Systems, in press.
43. Ch. Darwin, The Origin of Species. The final words of the first section of the first chapter.
44. S.Gould, Paleobiology, 6, 119-130, 1980.
45. S.Gould, Science, 216, 380-387, 1982.
46. P.Alberch, American Zoologist, 20, 653-667, 1980.
47. R.Feistel, W.Ebeling, Bio Systems, 15, 291-296, 1982.
48. W.Ebeling, A.Engel, B.Esser, R.Feistel, J.Stat.Phys., 37, 369-384, 1984.
49. M.Volkenstein, M.Livshits, Doklady Acad.Sci. USSR, 301, 731-734, 1988; Bio Systems - in preparation.
50. M.Volkenstein, Entropy and Information, Verlag Harry Deutsch, Frankfurt am Main, 1990.
51. W.Ebeling, M.Volkenstein, Physica A, 163, 398-402, 1990.
52. C.von Weizsäcker. Vorword to 53.
53. B.Küppers, Information and the Origin of Life, MIT Press, Cambridge Mass., London, 1990.
54. G.Quastler, The Emergence of Biological Organization. Yale Univ. Press, New Haven, 1964.
55. I.Schmalhausen, Cybernetical Problems of Biology (Russian), Nauka, Novosibirsk, 1967.
56. M.Volkenstein, D.Chernavsky, J.Soc.Biol.Struct., 1, 95-108, 1978.
57. M.Volkenstein, D.Chernavsky in Self-Organization, Autowaves and Structure far from Equilibrium, V.Krinsky ed., Springer-Verlag, B., Heidelberg, N.Y., Tokyo, pp. 252-261, 1984.
58. M.Volkenstein, J.Theor.Biology, 92, 293-299, 1981.
59. M.Conrad, M.Volkenstein, J.Theor.Biol., 92, 293-299, 1981.
60. L.Gatlin, Information Theory and the Living Systems. Columbia Univ. Press, N.Y., London, 1972.
61. J.Wicken, Evolution, Thermodynamics and Information. Oxford Univ. Press, N.Y., Oxford, 1987.
62. D.Brooks, E.Wiley, Evolution as Entropy, Univ. of Chicago Press, Chicago, London, 1988.

QUANTITATIVE ANALYSIS OF SELECTION AND MUTATION IN SELF-REPLICATING RNA

Christof K. Biebricher[1], M. Eigen[1] and William C. Gardiner, Jr.[2]

1. Max-Planck-Institut für Biophysikalische Chemie,
   D-3400 Göttingen, Germany
2. Department of Chemistry, University of Texas, Austin, TX 78712

1. INTRODUCTION

The neo-Darwinian theory, based on Mendelian genetics[1], brought the first quantitative analysis of evolution in the first half of this century. It was often claimed, however, that while Darwinian theory is excellent for explaining natural phenomena, it fails to make quantitative predictions. Given the complexity of biological expression of highly sophisticated living organisms, selective or fitness values can of course only be determined from the outcome of selection. Furthermore, many assumptions have to be made to calculate gene frequencies and their alteration, and in many cases the assumptions are not realistic in the ever-changing environments.

The basic laws of molecular evolution were set forth by one of the authors[2] in a mathematical form applicable to any vegetatively self-reproducing system. SPIEGELMAN[3,4] and coworkers introduced a system particularly suitable for the study of evolution *in vitro*: RNA replication with the help of the viral enzyme $Q\beta$ replicase. $Q\beta$ replicase is highly discriminating in accepting RNA as template, and thus the phenotypic expression of self-replicating RNA that determines its fitness is just its efficiency in replication by $Q\beta$ replicase[5]. In this paper, we show that quantitative analysis of evolution is possible for this system and that mathematical expressions describing the observed phenomena can be derived.

2. RNA-REPLICATION

2.1. *The chemical reaction mechanism*

The biochemical details of RNA replication have been described previously[6-9] Let us assume conditions which allow the reaction dynamics to be described in the simplest form, primarily the use of saturating, buffered monomer concentrations and

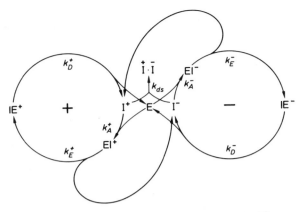

Fig. 1. Simplified three-step mechanism of RNA replication.The steps are described in the text. From [28].

high enzyme concentrations, and neglect mutations. The chemical reaction mechanism can then be adequately described by the irreversible cross-catalytic cycle depicted in Fig. 1 [10,11]. If we furthermore assume a palindromic RNA sequence, with indistinguishable plus and minus strands, the mechanism shown in Fig. 1 reduces to a single autocatalytic cycle. The latter assumption is not realistic, but the rate constants in optimized fast-growing RNA species have been found not to differ greatly between the complementary cycles, so that essentially equal concentrations of plus and minus strands are present.[12,13]

Let $[^iI]$ be the molar concentration of free palindromic RNA species i. Free RNA binds to replicase E (with second-order rate constant ^ik_A) to form the active replication complex $[^iEI]$, which synthesizes and liberates the replica RNA strand (with first-order rate constant ^ik_E). The remaining inactive complex $[^iIE]$ must dissociate into its components (with first-order rate constant ^ik_D) before the enzyme can again serve as replicase. The template-enzyme complexes $[^iE_c] = [^iEI] + [^iIE]$ as well as the free complementary strands constitute replication intermediates analogous to different development or age classes of organisms; the sum $[^iI] + [^iEI] + [^iIE]$ corresponds to the total population of RNA species i, $[^iI_o]$.

It is straightforward to set up the dynamic equations:

$$d[^iEI]/dt = {^ik_A}[E][^iI] - {^ik_E}[^iEI]$$
$$d[^iIE]/dt = {^ik_E}[^iEI] - {^ik_D}[^iIE]$$
$$d[^iE_c]/dt = {^ik_A}[E][^iI] - {^ik_D}[^iIE] = -d[E]/dt$$
$$d[^iI]/dt = {^ik_E}[^iEI] + {^ik_D}[^iIE] - {^ik_A}[E][^iI]$$
$$d[^iI_o]/dt = {^ik_E}[^iEI]$$

The reproduction term A_i of species i is the relative rate of increase in population due to replication, $(1/[^iI_o])d[^iI_o]/dt$, and we obtain from the last equation $A_i = {^ik_E}[^iEI]/[^iI_o]$. We can express this result in biological terms: The relative rate of population increase is equal to the proportion of the population in the reproducing age times its birth rate.

Reproduction rates alone are usually not rate-determining in biological systems. In animal populations, newborn individuals undergo long maturation periods before they reach reproductive age. Nascent free RNA strands, however, can bind rapidly to enzyme and are thus immediately able to reproduce. On the other hand, the parent strands have long regeneration periods before they can reproduce again, because the dissociation of the inactive complex ^iIE is the slowest and thus rate-limiting step. Fibonacci's famous two-step growth system where rabbits reproduce and mature over equal time intervals has an analog for RNA: it is what happens when the reproduction and regeneration steps have equal rate constants.

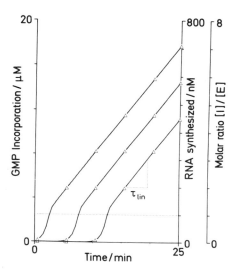

Fig. 2. Measurement of replication rates from incorporation profiles. A simulated profile is shown; experimental profiles are shown in [8,28]. The overall replication rate $^i\varrho$ for the linear growth phase is calculated from the slope of the profiles; the replication rate $^i\kappa$ for the exponential phase is calculated from the displacement Δt on the time axis after dilution of the template by a factor F_{dil} by $^i\kappa = \Delta t^{-1} \ln F_{dil}$.

2.2. *Growth phases*

Experimentally, it is convenient to measure overall replication rates, determined by radioactive nucleotide incorporation measurements. Profiles such as those shown in Fig. 2 are obtained. Two growth phases are clearly distinguished: an *exponential* one where enzyme is in excess and a *linear* one where enzyme is saturated with template. In the exponential growth phase, the intermediates and the total population show after an equilibration period coherent growth[14]:

$$d \ln [^iI]/dt = d \ln [^iEI]/dt = d \ln [^iIE]/dt = d \ln [^iI_o]/dt = {^i\kappa}$$

We therefore immediately get a constant $A_i = {}^i\kappa$. In the linear growth phase, where virtually all of the enzyme is bound to template, a steady state is established where the intermediate concentrations do not change and the flux through each step is equal to the total flux v [14]:

$$v = A_i[{}^iI_o] = {}^ik_A[{}^iI][E] = {}^ik_E[{}^iEI] = {}^ik_D[{}^iIE] = {}^i\varrho[{}^iE_c] \approx {}^i\varrho[E_o]$$

where $[E_o] = [E] + [{}^iE_c]$ is the total enzyme concentration, free and bound. In the linear growth phase, v and ${}^i\varrho$ are constants, while A_i is variable, because $[{}^iI_o]$ grows with time.

Both ${}^i\kappa$ and ${}^i\varrho$ are functions of ${}^ik_A$, ${}^ik_E$ and ${}^ik_D$. At replicase concentrations in excess of 100 nM, template binding is found to be much faster than the two other steps. Under these conditions, ${}^i\kappa$ and ${}^i\varrho$ are to good approximation functions of ${}^ik_D$ and ${}^ik_E$ alone. Since ${}^i\kappa$ and ${}^i\varrho$ can be determined independently, ${}^ik_E$ and ${}^ik_D$ values can be inferred[13].

2.3. Double strand formation

At high concentrations of free RNA strands a loss of template by formation of double strands occurs, because the single-stranded RNA strands are complementary to each other. Double strands, being unable to reproduce, must therefore not be counted as part of the total population of the species. We assume in the following that they neither inhibit nor enhance replication of free strand, an assumption which is not entirely realistic, because double-stranded RNA can bind to enzyme to inhibit the self-replication process[10]. Ignoring this effect we have for the relative template loss term D_i the loss by double strand formation we obtain

$$D_i = 2\,{}^{ii}k_{ds}[I^+][I^-]/[{}^iI_o] \qquad (= \frac{1}{2}\,{}^{ii}k_{ds}[{}^iI]^2/[{}^iI_o])$$

where the equation in parantheses part applies for the palindromic case.

A_i contains the fraction $[{}^iE_c]/[{}^iI_o]$, D_i the fraction $[{}^iI]/[{}^iI_o]$, both quantities being variables. We can define an apparent (steady state) enzyme binding constant

$$^iK_{IE} \equiv \frac{[{}^iE_c]}{[E][{}^iI]} = \frac{{}^ik_A}{{}^i\varrho}$$

that allows us to calculate these ratios; note that $[{}^iI_o] = [{}^iE_c] + [{}^iI]$. In the linear growth phase, $[{}^iE_c] \approx [E_o]$ and thus $[{}^iI][E] = {}^iK_{IE}[E_o] = \text{constant}$[14].

The net relative production of RNA is $E_i \equiv A_i - D_i$. For the exponential growth phase, the loss term is very small and $E_i \approx {}^i\kappa$. In the early linear growth phase, we obtain

$$E_i = {}^i\varrho\frac{[{}^iE_c]}{[{}^iI_o]} - \frac{1}{2}\,{}^{ii}k_{ds}\frac{[{}^iI]^2}{[{}^iI_o]}$$

The quadratic form of the loss term eventually leads to a steady state ($E_i = 0$) where synthesis and loss are balanced, ${}^i\varrho[{}^iE_c] = \frac{1}{2}\,{}^{ii}k_{ds}[{}^iI]^2$. The concentration of free strands, increasing in the linear growth phase as $d[{}^iI]/dt = d[{}^iI_o]/dt$, is in the steady state (denoted with a tilde) $[{}^i\tilde{I}] = \sqrt{2\,{}^i\varrho[{}^iE_c]/{}^{ii}k_{ds}}$. In this stationary growth phase, only the concentration of double strands increases, with $d[{}^{ii}\mathrm{II}]/dt = \frac{1}{2}\,{}^i\varrho[E_o]$.[12]

2.4. *Flux*

Evolution experiments usually involve a large number of generations, and populations would rise to astronomical sizes unless the excess production E_i is removed. For RNA evolution experiments this is done by diluting the growing mixture into fresh medium, either continuously in a flow reactor, or – experimentally more conveniently – by serially diluting aliquots, after incubation for time periods t_{tr}, by a dilution factor F_{dil} into fresh medium [3]. Since the growth conditions change with increasing population density, it is generally necessary to adjust F_{dil} in order to avoid large changes of the population size[10,13]. An apparatus for maintaining constant growth conditions automatically has been built[15].

3. COMPETITION AMONG RNA SPECIES

3.1. *Selective values*

Competition of two or more RNA species can be readily observed experimentally[13]. Since competing species usually differ in their physical-chemical properties, they can be separated without problems. Mutation can be neglected entirely in the dynamic equations, because the species barriers make interconversion of the species very slow. The different species do not interfere with each other except in sharing the environment and competing for common resources.

When the growth rates of the different species are of the same order of magnitude, equilibration to steady states as described above is nearly complete and the above equations remain valid except for the approximation $[\,^iE_c] \approx [E_o]$, which must be replaced by $\sum_i[\,^iE_c] \approx [E_o]$. Loss by double strand formation also occurs; when the species have unrelated sequences, they can not form stable mixed double strands.

In the *exponential growth* phase, the growth of the species is unaffected by the presence of other species. Their relative populations change if the growth rates of the different species differ. To observe this experimentally, the enzyme concentration must be kept in excess of the growing RNA concentration by continuous or serial dilutions. It is convenient to introduce relative populations $x_i \equiv [\,^iI_o]/\sum_j[\,^jI_o]$. Advantageous species get enriched and disadvantageous depleted in the population according to

$$\varsigma_i \equiv d\ln x_i/dt = E_i - \bar{E} \qquad \text{and} \qquad x_i(t) = x_i(0)e^{(\,^i\kappa - \bar{E})t}$$

where $\bar{E}$ is the average excess growth ($\bar{E} = \sum_j x_j E_j$). We call $\varsigma_i(t)$ the *selection rate value*.

In classical evolution theory, the concept of *fitness* is often used to describe success in selection. Quantitative definitions of fitness vary depending on the model chosen. In discretionary models, synchronized reproduction during generation times is assumed. The numbers of individuals of a certain genotypes are counted in each generation and the population change factor of type i, λ_i, is often defined as the absolute fitness, calculated as the probability of survival to reproduce times the number of offspring produced or—in sexual reproduction—the number of gametes supplied for the successful hatching of offspring[16]. After n generations, the number of individuals of type i is $^iN_n = {}^iN_o\lambda^n$. A relative fitness is defined by dividing the absolute fitness by the factor of total population increase per generation. For RNA reproduction, we obtain then, if τ is the average generation time, for the absolute fitness $e^{A_i \tau} e^{-D_i \tau} = e^{E_i \tau}$ and for the relative fitness $e^{A_i \tau} e^{-D_i \tau} / e^{\bar{E} \tau} = e^{\varsigma_i \tau}$. The average generation time is the time required to complete a replication cycle and is thus $\tau = 1/{}^i k_A[E] + 1/{}^i k_E + 1/{}^i k_D$. For overlapping generations, Fisher introduced the Malthusian parameter ${}^i r$ as fitness with the growth law ${}^i N = {}^i N_o e^{{}^i r t}$. Therefore, ${}^i r = \ln \lambda_i / \tau$. The latter value is fully compatible with the selection term used in the molecular evolution theory[2,17−19] and we define thus fitness as $\mathcal{F}_i = E_i \tau$ in an evolving population.

3.2. Selection in the exponential growth phase

In the *exponential growth* phase, $E_i = {}^i \kappa = $ constant and $\bar{E}(t) = \sum_j {}^j \kappa x_j(t)$ is the average excess production. E_i is of course measurable in the absence of other species and is equal to the Malthusian parameter of the classical Fisher's theorem. During the course of the selection $\bar{E}(t)$ increases and the selection rates of all species drop. Therefore, more and more species get depleted until finally only the fittest remains: $\tilde{E} \rightarrow E_{max}$ and $x_{max} \rightarrow 1$, while all other species disappear $x_{i \neq max} \rightarrow 0$. This case has been extensively treated by others[2,17−19] and will not discussed further here. Experimental observations and computer simulations have confirmed the above conclusions.

3.3. Selection in the linear growth phase

Conditions for exponential growth are rarely fulfilled in biology, however. Instead, limited resources set an upper bound to population growth; growth rates are accordingly not constants, but decrease with increasing population[20]. Indeed, in the *linear growth* phase of RNA replication, the selection behaviour of self-replicating RNA species is found to be fundamentally different from that of the exponential growth phase. Selection is more dramatic and need not be correlated at all to the Malthusian parameter. Experiments have shown that, in the linear growth phase, competition between the RNA species MNV-11 and MDV-1 results in rapid selection of MDV-1 despite the fact that MDV-1 has much lower overall replication rates in both phases[13]. Analysis of the underlying molecular process revealed that MDV-1 deprives MNV-11 of newly released free enzyme because its rate constant for enzyme binding is higher.

In the linear growth phase, the net excess RNA production rates $E_i(t)$ are not constants: they change with time depending on the concentrations of all species present. They vanish in the steady state of double strand formation and may even assume negative values when a species loses enzyme to a competitor. In contrast to the exponential phase, where $\bar{E}$ increases during selection, $\bar{E}$ might be positive, zero or negative in the linear growth phase, because the total RNA population may stagnate or even diminish. The definitions of the selection rate values ς_i remain valid; thus $E_i = A_i - D_i$, $\varsigma_i = E_i - \bar{E}$, $\mathcal{F}_i = E_i \tau$, $\tau = 1/\bar{\varrho}$, $A_i = {}^i \varrho [{}^i E_c] / [{}^i I_o]$ and $D_i = {}^{ii} k_{ds} [{}^i I]^2 / [{}^i I_o]$. A_i, D_i and $\bar{E}$ are not constant and a stable 'age distribution' as described in section 2 is not obtained when macroscopic selection takes place. The selection rate values can be calculated numerically, but the formulas for them are complicated and not particularly instructive. The experimental results are fully

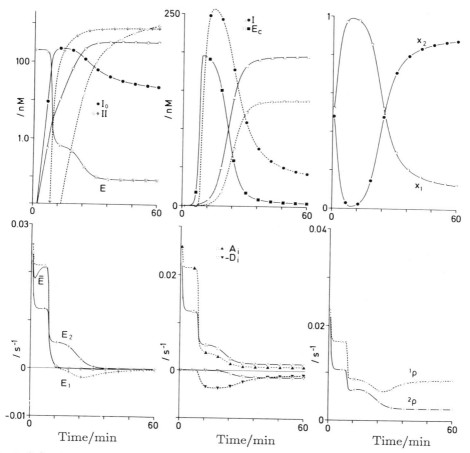

Fig. 3. Selection during growth of RNA (computer simulation, full mechanism). Standard rate constant values (see glossary) were assumed except $^2k_A = 4\times$ standard, $^2k_D = 1/4\times$ standard values. Both species are present at the beginning at 1 pM and grow to saturation. In the exponential phase (0–8 min), the smaller ^2k_D value is detrimental: species 1 (filled symbols) grows more rapidly to saturate the enzyme and enter a steady state of double strand formation (8–12 min) where its selection value vanishes. Species 2 (open symbols) continues to grow exponentially with a lower rate (8–12 min, see text). Macroscopic selection takes place where species 2 conquers the majority of the enzyme (12–40 min). Note the total population decrease ($E_1, \bar{E} < 0$). Eventually, an ecosystem is formed where E_1, E_2 and $\bar{E}$ disappear.

supported: there is no selection for a high growth rate; instead the $^iK_{IE}$ values are quite important for the calculation: higher $^iK_{IE}$ values mean lower proportions of free strands (diminished destruction) and higher "affinity" to replicase.

In some cases, however, simple equations do pertain. If one species d has saturated the enzyme and is in steady state of double strand formation, the selective value of a species i present in minute amounts can be determined. Its rate of double strand formation can be neglected, $\bar{E} \approx E_d$, the concentration of enzyme is determined by the steady state of the dominant species d and at least a partial equilibration of the intermediate composition has been formed. Then $\varsigma_i = {}^ik_A[\tilde{E}] = {}^ik_A / {}^dk_A \sqrt{{}^d\varrho[\mathrm{E_o}]\,{}^{dd}k_{ds}/2}$, a constant since the concentration of free enzyme is determined by the dominant species. The minor species is thus growing exponentially while the dominant species is still stationary. The equation predicts, and experiments confirm, that rapid selection can be observed when the minor variant has a higher binding rate to enzyme and when the dominant species has a high rate of double strand formation.

3.4. *Change of selection by RNA growth*

In typical test tube experiments, the RNA grows from small concentrations to saturation through the various growth phases. The selection criteria change during the growth process and selection rate values change. The rather complicated selection behaviour is illustrated by a computer simulation of a typical competition experiment with two species (Fig. 3). One (species 2, open symbols) – usually the one with the higher chain length – has a lower ${}^2k_D$ but a higher ${}^2k_A$. In the exponential phase, the selection values are constants. While the difference in the ${}^ik_A$-values does not contribute to the overall replication rate, the difference of the ${}^ik_D$-values causes species 1 to have a much higher selection value. During the amplification in the exponential phase, the difference in the selection values causes enrichment of species 1 by several orders of magnitude. It saturates the enzyme and reaches the steady state of double strand formation where its selection value vanishes. Species 2 continues to grow exponentially at the steady state of free enzyme concentration as described above. A period follows where macroscopic selection can be observed and can be quantitatively characterized experimentally[21]. In this period $\bar{E}$ changes; under the chosen conditions (typical for experimental systems) $\bar{E}$ is negative and the total population drops, because loss by double strand formation exceeds synthesis of new strands[13]. The loss rate is strongly dependent on the concentration of the species; during macroscopic selection, the loss rate of the disadvantageous mutant decreases and the loss rate of the advantageous mutant increases; the absolute values of the selection rates thus decrease with time.

3.5. *Ecosystem equilibrium*

Eventually, both species reach a steady state where their selection rate values disappear. If the two species do not form mixed double strands, i.e., ${}^{ij}k_{ds} = 0$, then a balanced coexistence between species is obtained where the population ratio of the two species remains constant, because fluctuations of the steady state concentration are immediately counteracted. The steady state concentrations can be readily determined. The ratios of free and bound strand concentrations are

$$\frac{[{}^1\mathrm{I}]}{[{}^2\mathrm{I}]} = \frac{{}^1k_A\,{}^{22}k_{ds}}{{}^2k_A\,{}^{11}k_{ds}} \qquad\qquad \frac{[{}^1\mathrm{E_c}]}{[{}^2\mathrm{E_c}]} = \frac{({}^1k_A)^2\,{}^{22}k_{ds}\,{}^2\varrho}{({}^2k_A)^2\,{}^{11}k_{ds}\,{}^1\varrho}$$

Since $[{}^1\mathrm{E_c}] + [{}^2\mathrm{E_c}] \approx [\mathrm{E_o}]$, the absolute concentrations can also be calculated.

4. RNA EVOLUTION AS INTERPLAY OF MUTATION AND SELECTION

4.1. *Experimental analysis*

The interplay of mutation and selection has been studied extensively in bacterial cultures, using phenotypic markers for the detection of mutants[22-26]. These genetic studies brought much insight; however, they were restricted to the few mutations producing measurable phenotypic alterations.

The studies of selection among different species, described above, can be extended to selection among different mutants or types within a species; the laws remain the same. The experimental realization of the study of selection and mutation on the RNA level, however, is much more difficult. First, the physical and chemical properties of different mutants differ very little or not at all. Second, no experimental setup can be devised to rule out selection forces unless the number of replication rounds is restricted to one. Third, methods must be developed to analyse extremely small amounts of RNA. Fourth, deterministic equations do not adequately describe mutations and particularly not rare mutations. Even though RNA molecules can be readily cloned on the RNA level[27,8], analysis of different clones does not give satisfactory results, because a mutant distribution is rapidly re-formed during amplification of the clones. The methods we used to solve these problems, to be published elsewhere, include retrotranscribing RNA into double-stranded DNA, cloning into plasmids, amplifying the clones in vivo with the unbiased and very accurate DNA replication apparatus, and determining the sequences of the clones[28]. By this method we obtain a mutant frequency (x_i) landscape in the multi-dimensional sequence space[29]. This mutant frequency landscape is generated by mutation and selection. Often, the mutant frequency landscape has been correlated directly to a *mutation frequency* (μ_i) landscape where a peak in the population of a certain mutant is caused by a so-called "hot spot", an especially high mutation rate. However, even though differences in mutation rates definitely exist, this view is clearly wrong. Populations build up over many generations and selection forces affect the formation of the mutant frequency landscape more than mere mutation rates[29].

Additional information is necessary to get a full picture. We do this by determining the selective values of the different RNA mutants. Experimentally, this is done by cloning the cDNA behind an appropriate promoter, amplifying the DNA clones *in vivo* and synthesizing the RNA *in vitro* again by transcription from the DNA clone. By this method we obtain a homogeneous RNA mutant population: even though the accuracy of transcription is not higher than the accuracy of RNA replication, transcription always uses the master strand DNA for synthesizing RNA. Kinetic analysis of the replication of the mutant population leads then to selective values and we obtain a fitness landscape in the sequence space[30].

Computer simulations of evolution by mutation and selection were done as in the selection case, with the following modifications:

i Contributions to the RNA sysnthesis by mutation are taken into account.
ii Probability arrays for production of correct and mutant copies during replication are introduced.
iii Double strand formation occurs not only between homologous strands, but also between strands of different mutants.

iv Replication of types present in concentrations below one RNA strand per test-
tube aliquot is suppressed.

4.2. Evolution parameters

The general dynamic equation describing the evolution in a population of the
mutant i is

$$\mathrm{d}x_i/\mathrm{d}t = (W_{ii} - \bar{E})x_i + \sum_{j\neq i} w_{ij}x_j$$

$W_{ii} = A_i Q_{ii} - D_i$ is termed the *intrinsic selective value*, the relative selection rate $\varsigma_i = W_{ii} - \bar{E}$. The w_{ik} terms are the contributions of mutations to the population growth of i [2,17]. Only the fraction Q_{ii} of the synthesis rate $A_i = {}^i k_E[{}^i EI]$ leads to progeny of the same type, in the rest $(Q_{ji}, j \neq i)$ different mutants are produced. In experiments with self-replicating RNA, Q_{ii} is near unity. The error rate of Qβ replicase has been determined to be 3×10^{-4} [31,32] while the chain length of small self-replicating RNAs is below 100. The error introduced by neglecting erroneous copies is thus only about 3 %. Viruses, including Qβ itself, usually operate with small Q_{ii} values. There is an upper bound for the amount of information which can be stably maintained; above this error threshold the information evaporates[18,19]. Viruses are found to thrive just below the error threshold[29].

At a given time only the part of the population of each mutant $[{}^i EI]/[{}^i I_o]$ is in the reproductive age. It might appear more convenient to use the overall reproduction rates applying to the population of type i, as discussed in section 2.1. However, the steady state described there usually does not apply when the rate of formation by mutation is substantial, since mutation always produces nascent free strands, thus causing a "youth bias" in the population.

For the loss terms, contributions by formation of mixed double strands (het-eroduplices) can not be neglected; in fact, when mutants differ only in one or two sequence positions, no discrimination at all between the strands of different mutants should be expected. We introduce a rate constant matrix whose elements ${}^{ij}k_{ds}$ are the rate constants of double strand formation between plus strand of mutant i and minus strand of mutant j. In the palindromic case, assumed in most cases for simplicity, ${}^{ij}k_{ds} = {}^{ji}k_{ds}$. The destruction term is then $D_i = ([{}^iI]/[{}^iI_o]) \sum_j {}^{ij}k_{ds}[{}^jI]$. For the limiting case of no discrimination in double strand formation, all ${}^{ij}k_{ds}$ have the same value and $D_i = k_{ds}([{}^iI]/[{}^iI_o]) \sum_j [{}^jI]$.

The $\sum_{j\neq i} w_{ij}x_j$ term comprises the contributions to the synthesis of mutant i from erroneous copying of the other mutants. In chemical kinetics language they have the rate $\sum_{j\neq i} Q_{ij} {}^j k_E[{}^j EI]$. The relative rates of production of mutant i in the replication of mutant j are $w_{ij} = Q_{ij} {}^j k_E[{}^j EI]/[{}^j I_o]$, from which we derive by summation the *mutation rate* $\mu_i = \sum_{j\neq i} w_{ij}x_j/x_i$ as the relative population increase of mutant i contributed by mutation.

Finally, we call the sum of the selection and mutation rates the *evolution rate*

$$\xi_i \equiv \mathrm{d}\ln x_i/\mathrm{d}t = \varsigma_i + \mu_i$$

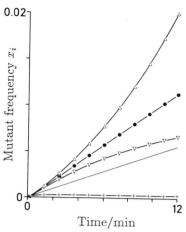

Fig. 4. Formation of mutants from a homogenous population (computer simulation
with the simplified mechanism). At the beginning, a homogenous population
of one type is present. Mutants develop with the error rate 3×10^{-4}. For
the advantageous mutant (Δ) $k_E = 1.1\times$ standard, for the disadvantageous
mutant (∇) $k_E = 0.9\times$ standard, for the neutral mutant ($\bullet$) standard values
and for the lethal mutant ($+$) $k_A = 0$. The calculated value for the neutral
mutant formation neglecting the youth bias is shown as thin line.

4.3. Mutation and selection in the exponential growth phase

4.3.1. *Measurement of mutation rate values.* The simplest case is a flow system
initially containing a homogenous clone of one type d (dominant) with $[{}^d I_o] < [E]$
and a flow rate $\phi = \bar{E}$ to maintain the condition of constant organization. This
situation is comparable to a bacterial clone growing in a chemostat[25,26]. In the
exponential growth phase of self-replicating RNA, destruction by double strand for-
mation can be neglected and the intrinsic selection rate is $W_{ii} = Q_{ii} \, {}^i k_E [{}^i EI]/[{}^i I_o]$.
While the quantities appearing in this equation can not be measured directly, In
coherent exponential growth a quasi-steady-state exists where the relative proportions of intermediates remains constant (Section 2.1) and ${}^i k_E [{}^i EI] = {}^i \kappa [{}^i I_o]$ and
$W_{ii} = Q_{ii} \, {}^i \kappa$. This quasi-steady-state is rapidly reached by the dominant mutant
d. At the very beginning, a new mutant i is produced exclusively by miscopying of
mutant d. Thus $d[{}^i I_o]/dt \approx \mu_i [{}^i I_o] \approx Q_{id} \, {}^d k_E [{}^d EI] = Q_{id} \, {}^d \kappa [{}^d I_o]$. During the initial phase, macroscopic growth is entirely dominated by the dominant mutant, such
that $\sum_j [{}^j I_o] \approx [{}^d I_o]$, $\bar{E} \approx {}^d \kappa$ and $x_d \approx 1$. Thus $dx_i/dt \approx Q_{id} \, {}^d \kappa$, i.e., the mutant
frequency x_i increases linearly with time[20]. The linear plot not only allows direct de-
termination of an individual mutation rate Q_{id}, but also supplies information about
the time period for which this measurement is valid.

The evolution rate drops with $\xi_i \approx \mu_i \approx Q_{id} \, {}^d \kappa (x_i)^{-1}$. As x_i increases, however,
the selection rate can no longer be neglected. If $\varsigma_i = 0$, i.e. if the mutant i is neutral
with respect to d, the linear increase of the mutant population continues until x_d

327

deviates considerably from 1, and the mutation rate can be taken from the slope of the linear increase of x_i. With larger positive or negative selection rates, deviations from linear growth appear when mutants have accumulated after a few replication rounds (Fig. 4). The deviations from linear increase of the mutant frequencies are easy to understand: the linear mutant production term gets modulated by autocatalytic growth of the mutant, i.e., $\mathrm{d}x_i/\mathrm{d}t = \varsigma_i x_i + Q_{id}\,^d\kappa$. We obtain $\mathrm{d}x_i/\mathrm{d}t = (A_i Q_{ii} - \bar{E})x_i + Q_{id}\,^d\kappa$ and, since $\bar{E} \approx\,^d\kappa$, $\mathrm{d}x_i/\mathrm{d}t = A_i Q_{ii}x_i + (Q_{id} - x_i)\,^d\kappa$. For a lethal mutant $A_i = 0$ and the mutant frequency rapidly reaches a steady state value $\tilde{x}_i \approx Q_{id}$ (Assuming $x_i, Q_{id} \ll 1$ and $^i k_E$ or $^i k_A = 0$).

Under the described conditions, inferring mutation rates from mutant frequencies is possible. The experimental difficulties in applying this method are i) the preparation of a homogenous genotype population of sufficient size and ii) the need for a very sensitive method to determine small x_i-values. Using an approach of this kind, reliable values for mutation rates for DNA polymerases have been determined[33,34]. Unfortunately, the method of using genetic markers for detection, even though very sensitive, limits the application to a few special base exchanges. For determining mutation rates in populations of self-replicating RNA, a generally applicable and sufficiently sensitive chemical method must be developed. The method we are using now for screening a large number of mutants is able to measure x_i values down only to 0.01. It should be possible to improve the sensitivity.

Additional information can be obtained by computer simulations. Discrepancies with values calculated by the above formula are obtained. For neutral or nearly neutral mutants, the observed slope of mutant frequency was found in simulations to be nearly twice the expected value. This results from the fact that the rate-determining step of replication occurs after synthesis of the complementary RNA strand: When in a lethal mutant one of the steps before liberating the replica is blocked, the expected values are found; but when a step after the liberation of the replication is affected, the discrepancy appears. Thus, for measuring mutation rates, we have to take into account that replication of the free strands produced by mutation can start immediately. Since this part is due to replication of the mutant, the simulation calculates the normal value of μ_i, but adds to the autocatalytic replication part $W_{ii} - \bar{E}$ another contribution $f_i\mu_i$ due to the immediate replication of the free strands produced by mutation. f_i is the ratio of the composite initiation-elongation rate to the normal elongation rate, i.e.,

$$f_i = \{(^i k_A[\mathrm{E}])^{-1} + ^i k_E^{-1}\}^{-1}/\,^i k_E = \,^i k_A[\mathrm{E}]/(^i k_A[\mathrm{E}] + ^i k_E). \text{ For } ^i k_A[\mathrm{E}] \gg \,^i k_E$$
$$f_i \approx 1.$$

4.3.2. *Measurement of selection rate values.* Can the relative selection rate be derived if the mutation rate is known? One could try the Ansatz $\varsigma_i \approx Q_{ii}\,^i\kappa - \,^d\kappa$, assuming again $^i k_E[^i\mathrm{EI}] = \,^i\kappa[^i\mathrm{I}_o]$, but this Ansatz is not justified when age equilibration is not given. As can be seen in Fig. 4, deviations from the linear increase are observed when selection takes place: the straight line is modulated by a superimposed exponential function. Simulations show that the linear part of the curve is correctly described when the "youth bias" correction factor f_i is accounted for, while the exponential curve is indeed accurately described by $Q_{ii}\,^i\kappa - \,^d\kappa$. Thus as long as d is dominant the equation for the evolution rate of type i is

$$\xi_i \equiv \frac{\mathrm{d}x_i}{x_i\,\mathrm{d}t} = Q_{ii}\,^i\kappa - \,^d\kappa\left(1 - \frac{Q_{id}x_d}{x_i}(1 + f_i)\right)$$

If the mutation rates are fairly uniform over the sequence, the curve can be then exactly predicted for a self-replicating species with a chain length 100, since $Q_{ii} = 0.97$ and $Q_{ij} = 3 \times 10^{-4}$ for a one-error-mutant.

With increasing mutant frequencies the selection terms become more and more dominant and, if the replication rate $^i\kappa$ is not too small, a *selection phase* will be reached where the mutation term may be neglected (Fig. 5). In this period the age distribution of all viable mutants is equilibrated and $\xi_i \approx \varsigma_i \approx Q_{ii}\,^i\kappa - \bar{E}$, as described in Section 3.2. During this period, macroscopic selection takes place and the mutant frequency profiles the dominant mutant is displaced can be recorded readily with sequencing methods[21], as can the change of $\bar{E}$ by the increase of the incorporation rate.

4.4. *The quasispecies distribution*

Even after macroscopic selection has ceased the selection process continues to function in weeding out disadvantageous mutants. As their mutant frequencies x_i become smaller, the contributions of mutation to the evolution rate increase. With increasing contributions from mutation, however, equilibration of the age structure is again perturbed and the W_{ii}-values increase. Finally, after a large number of replication rounds a stable mutant distribution is obtained where mutation and selection balance each other with $\tilde{\varsigma}_i = \tilde{\mu}_i$. The evolution rates ξ_i of all mutants disappear and their relative populations reach steady state values (As $t \to \infty$, $x_i \to \tilde{x}_i$ and $\xi_i \to 0$). The stable mutant distribution $\{\tilde{x}_i\}$ is called a *quasispecies*. It is usually dominated by the master type o.

The quasispecies is of particular interest for theoretical treatment because the system of linear differential equations for it can be solved analytically[35,36]. Unfortunately, the analytical solutions are found only when the A_i, D_i and w_{ik} terms are constants, an assumption that is justified in the RNA replication system only under exponential growth conditions.

The mutation term $\mu_i = \sum_{j \neq i} w_{ij} x_j / x_i$ looks much more complicated than it is in real systems, because contributions from sparsely populated neighbours in the sequence space or from distant mutants are negligible. In a rugged fitness landscape, the mutation rate is usually dominated by a single term, e.g. by mutation of the master. However, the evolution rate of a subpopulation forming a broad hill in the fitness landscape may exceed the evolution rate of a sharp isolated peak with a higher maximum fitness value[19,29]. The calculation is particularly easy when the population is heavily dominated by the master ($\tilde{x}_o \approx 1$). Then mutation contributions from other mutants can be neglected and we obtain for neighbours of the master in the fitness landscape $\tilde{\mu}_i \tilde{x}_i \approx Q_{io}\,^o\kappa$ or

$$\tilde{x}_i \approx (1 + f_i) Q_{io} \tilde{x}_o \frac{^o\kappa}{^o\kappa - {}^i\kappa}$$

Since mutations are rare events and the probabilities Q_{ij} very low, small differences in the fitness landscape can cause drastic differences in the population landscape, i.e., the stringency for neutrality is very high.

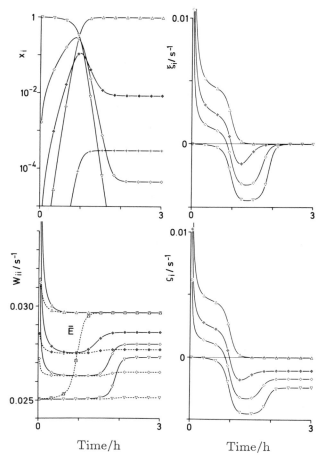

Fig. 5. Computer simulation, with simplified mechanism, of selection of advantageous mutants and establishment of a quasispecies distribution. At the beginning, mutant 1 (∇, standard rate values) is present as homogenous population. Serial mutations occur with the mutation rate 3×10^{-4} $I_1(\nabla) \rightleftharpoons I_2(\diamond) \rightleftharpoons I_3(\diamond) \rightleftharpoons I_4(\Delta) \rightleftharpoons I_5(+)$ where ${}^2k_E = 1.1\times$ standard, ${}^3k_E = 1.2\times$ standard, ${}^4k_E = 1.3\times$ standard and ${}^5k_A = 0$. Mutational gain values μ_i (not shown) contribute to the evolution rate values ξ_i only in the initial and the terminal phases. Selection is accompanied with a raise of $\bar{E}$. W_{ii}-values are influenced by the youth bias at periods where mutational gain is important. When the contributions of the youth bias are subtracted (dashed lines), nearly constant intrinisic selection rate values are found.

4.5. Mutation and selection in the linear growth phase

As can be expected from the discussion in Section 3.3, calculation of the evolution rates in the linear growth phase is more complicated than for the exponential growth phase. Nevertheless, computer simulations of the evolution process show essentially the same features as in the exponential growth phase when the population is kept constant by a regulated flux of nutrients (monomers and enzyme) and removal of excess RNA population. If, however, the population grows without control, the selective values change with time and a steady state quasispecies distribution is never obtained. Under these conditions, the evolution process is so complex that predictions are only possible when the reproduction mechanism as well as the interactions among the individuals of the population are fully understood and numerical integration becomes feasible. In the case of self-replicating RNA, we are close to this possibility.

4.6. *Stochastic effects*

Theoretically, a true quasispecies distribution is possible only after an infinite number of generations and for an infinitely large population. This assumption is not realistic, because in real systems the population size is limited and very small compared to the possible number of different genotypes, even for the system of self-replicating RNA with its large population sizes (10^{12} to 10^{15} for typical test tube experiments) and very small genome complexity (nucleotide chain lengths < 100). Deterministic mutant frequencies $\tilde{x}_i$ are reached only for the most frequent mutations, while rare mutations are usually not populated at all. The quasispecies distribution may be temporarily trapped in a local fitness optimum until a rare mutational jump reaches a more distant fitness mountain with a higher fitness optimum (punctuated equilibrium[37]). The effect of finite population size was taken into account in simulations by switching on replication of mutant i when the computed $[^iI_0]$ value was greater than the threshold corresponding to appearance of one strand per typical sample. This effected restriction of mutants to those present in at least one copy without otherwise interfering with the deterministic character of the simulation program.

The simulations show clearly the importance of maintaining high population sizes in evolution experiments. Fig. 6 simulates a classical evolution experiment of Spiegelman and coworkers[21]. An adapted mutant with three point mutations was obtained, and during the course of evolution single point mutations followed one another; the next point mutation occurred only after its predecessor was strongly enriched in the population. This evolution course was dictated, however, by the experimental conditions, which involved dilution after each transfer to a very small population. Most mutants, except the most frequent ones, were thus discarded after every transfer. Under conditions where the drop of the population to a small size was avoided, i.e., where larger F_{dil} values were assumed, a three-error mutant can be obtained directly. Since neutral mutants usually contain several, compensating mutations[28,30], it is likely that species can be selected that are more adapted to the conditions by chosing evolution experiments with large population sizes.

5. DISCUSSION

How general are the results obtained with self-replicating RNA molecules? Without doubt, RNA molecules are not living organisms. Nevertheless, the observed features of Darwinian evolution are the same as with eukaryotes and particularly with prokaryotes. The general theories for evolution of Mendelian populations as well as for vegetatively reproducing organisms[1,2,16,17,38] are confirmed by the results with self-replicating RNA. This confirmation is reassuring, but it has been expected; the previously used mathematical models for the replication process with overall reproduction rates had obviously described the real behaviour quite adquately.

However, this is only true for the limited case of truly exponential growth. In the linear growth phase and in the steady-state, which certainly mimic animal populations more realistically, unexpected features wer found. Of course, selection for effective competition and for loss reduction have always been considered as highly important factors in Darwinian evolution, but are usually treated only qualitatively.

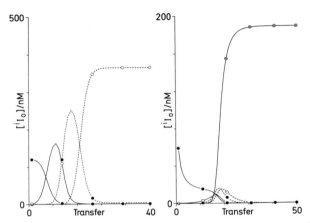

Fig. 6. Selection of adapted mutants. The experiments described by [21] are simulated a) (left) under the described conditions (transfers with $F_{dil} = 10^{-6}$ and transfer times of 15 min), b) (right) with transfer times of 5 min and a dilution factor of 10^{-2}. Standard values except for $^2k_D = 1.2\times$, $^3k_D = 1.5\times$, $^4k_D = 1.8\times$ and $^5k_D = 2.25\times$ standard rates. Mutants 2, 3 and 4 are formed by sequential one-error mutations (a ridge in the fitness sequence space); mutant 5 ($\oplus$) is an advantageous three-error mutant separated by a valley in the fitness sequence space. The outcome of the evolution experiment described in [21] is the consequence of the chosen experimental conditions.

In the self-replicating RNA system, these phenomena can be described quantitatively, even though the calculation is more difficult than for simple exponential growth. In the linear growth phase, there is no selection for maximum replication rate; experiments have shown indeed that the master in the quasispecies distribution does not have the highest replication rate. The mathematical forms of competition depend on the mechanism of interaction among the different mutants and can thus not be modelled with simple equations as is the case in Malthusian growth theory. Some examples where such interactions were considered can be found in the game theory of evolving evolutionary stable strategies[16]. However, the principle that these interactions are of fundamental importance in deriving fitness values must be broadly applied to theories of evolution.

6. GLOSSARY

6.1. *Statistical quantities*

$\mathcal{N}_i$ — Number of individuals of type i in population

x_i — Mutant frequency; Fraction of type i in the total population; $x_i = [{}^iI_o]/\sum_j[{}^jI_o]$

$\tilde{x}_i$ — Mutant frequency of type i in the quasispecies distribution

Q_{ij} — Probability of producing type i as replica per replication round of type j as template

Q_{ii} — Probability of producing a correct copy by one replication round of mutant i

6.2. *Concentrations*

$[{}^iI]$	Concentration of free single-stranded RNA of type i	$[mol/l]$	*
$[{}^iEI]$	Concentration of active replication complexes with template of type i	$[mol/l]$	
$[{}^iIE]$	Concentration of inactive replication complexes with template of type i	$[mol/l]$	
$[E]$	Concentration of free enzyme	$[mol/l]$	
$[{}^iE_c]$	Total concentration of template strands of type i complexed to enzyme; $[{}^iE_c] = [{}^iEI] + [{}^iIE]$	$[mol/l]$	*
$[E_o]$	Total concentration of enzyme; $[E_o] = [E] + \sum_i[{}^iE_c]$; standard value 2×10^{-7}	$[mol/l]$	**
$[{}^iI_o]$	Total concentration of template strands of type i; $[{}^iI_o] = [{}^iI] + [{}^iE_c]$	$[mol/l]$	*

6.3. *Elementary Rate Constants*

$^{ij}k_{ds}$ — Rate constant for double strand formation between one strand of type i and the complementary strand of type j; for $i = j$ homoduplex, for $i \neq j$ heteroduplex formation; standard value 5×10^4 — $[l/mol\ s^{-1}]$ *

333

$^i k_A$ Rate constant for replication complex formation between an $[l/\text{mol } s^{-1}]$ *
RNA strand of type i with replicase; standard value 10^7

$^i k_E$ Rate constant for synthesizing and releasing a replica from a $[s^{-1}]$
replication complex of type i, leaving behind an inactive complex; standard value 10^{-1}

$^i k_D$ Rate constant for dissociation of the inactive complex of type $[s^{-1}]$ *
i into free RNA and replicase; standard value 10^{-2}

6.4. Composite Rate Coefficients

$^i\kappa$ Experimentally measured overall replication rate constant for a $[s^{-1}]$ *
homogenous population of type i as template in the exponential
growth phase, miscopying included; $^i\kappa = \mathrm{d}[^i\mathrm{I_o}]/([^i\mathrm{I_o}]\,\mathrm{d}t)$

$^i\rho$ Experimentally measured relative rate of RNA synthesis per $[s^{-1}]$ *
template strand of type i bound to replicase, including miscopying; $^i\rho = \mathrm{d}[^i\mathrm{I_o}]/([^i\mathrm{E_c}]\,\mathrm{d}t)$

A_i Relative rate of RNA synthesis, miscopying included, for a homogenous population of type i as template; $A_i = \, ^i k_E[^i\mathrm{EI}]/[^i\mathrm{I_o}]$ $[s^{-1}]$

D_i Relative template loss rate; mainly governed by double strand $[s^{-1}]$
formation; $D_i = [^i\mathrm{I}]/[^i\mathrm{I_o}]\sum_j \, ^{ij}k_{ds}[^j\mathrm{I}]$

E_i Relative net excess RNA production rate with type i as template, including miscopying; $E_i = A_i - D_i$ $[s^{-1}]$ *

$\tilde{E}_i$ Relative net excess RNA production rate with type i as template in the quasispecies steady state, including miscopying; $[s^{-1}]$ *
$\tilde{E}_i = A_i - D_i$, usually $\tilde{E}_i = \, ^i\kappa$

$\bar{E}$ Relative net excess production rate for all RNA types; $\bar{E} = \sum_i E_i x_i$ $[s^{-1}]$ **

$\tilde{\bar{E}}$ Relative net excess production rate for all RNA types in the $[s^{-1}]$ **
quasispecies steady state; $\tilde{\bar{E}} = \sum_i \tilde{E}_i \tilde{x}_i$

W_{ii} Intrinsic selection rate value $W_{ii} = A_i Q_{ii} - D_i$ $[s^{-1}]$

w_{ij} Relative rate of producing type i by replication of type j; $w_{ij} = Q_{ij}\,^j k_E[^j\mathrm{EI}]/[^j\mathrm{I_o}]$ $[s^{-1}]$

ξ_i Evolution rate: Relative rate for relative increase of type i; $[s^{-1}]$
$\xi_i = \mathrm{d}x_i/(x_i\,\mathrm{d}t) = \varsigma_i + \mu_i$

μ_i Mutational gain rate: Relative rate for synthesis of type i by $[s^{-1}]$
miscopying other templates; $\mu_i = \sum_{j \neq i} w_{ij} x_j / x_i$

ς_i Relative selection rate value; $\varsigma_i = A_i Q_{ii} - D_i - \bar{E}$ $[s^{-1}]$

$\mathcal{F}_i$ Relative fitness $\mathcal{F}_i = W_{ii}\tau$

6.5. Other parameters

λ_i Population change factor per generation

f_i Correction factor for mutation term

τ Average generation time $[s]$

Quantities that can be measured are marked with an $*$; quanitities that are easy to measure are marked with $**$.

REFERENCES

1. T. Dobzhansky, F.J. Ayala, G.L. Stebbins and J.W. Valentine, *Evolution*, Freeman, San Francisco (1977).
2. M. Eigen, Selforganization of matter and the evolution of biological macromolecules, *Naturwissenschaften* 58:465 (1971).
3. D.R. Mills, R.L. Peterson and S. Spiegelman, An extracellular Darwinian experiment with a self-duplicating nucleic acid molecule, *Proc. Nat. Acad. Sci. USA* 58:217 (1967).
4. L.E. Orgel, Selection *in vitro*, *Proc. R. Soc. Lond.* B 205:435 (1979).
5. C.K. Biebricher, Darwinian selection of self-replicating RNA, in: *Evolutionary Biology*, M.K. Hecht, B. Wallace and G.T. Prance, eds., Vol. 16, Plenum Press, New York (1982).
6. C. Dobkin, D.R. Mills, F.R. Kramer, and S. Spiegelman, RNA replication: Required intermediates and the dissociation of template, product, and $Q\beta$ replicase, *Biochemistry* 18:2038 (1979).
7. C.K. Biebricher, M. Eigen and R. Luce, Product analysis of RNA generated *de novo* by $Q\beta$ replicase, *J. Mol. Biol.* 148:369 (1981).
8. C.K. Biebricher, M. Eigen and R. Luce, Kinetic analysis of template-instructed and *de novo* RNA synthesis by $Q\beta$ replicase, *J. Mol. Biol.* 148:391 (1981).
9. C.K. Biebricher, S. Diekmann and R. Luce, Structural analysis of self-replicating RNA synthesized by $Q\beta$ replicase, *J. Mol. Biol.* 154:629 (1982).
10. C.K. Biebricher, Darwinian evolution of self-replicating RNA, *Chemica scripta* 26B:51 (1986).
11. C.K. Biebricher and M. Eigen, Kinetics of RNA replication by $Q\beta$ replicase, in *RNA Genetics, Vol. I: RNA-directed Virus Replication*, E. Domingo, P. Ahlquist and J.J. Holland, eds., CRC Press, Boca Raton, FL (1987).

12. C.K. Biebricher, M. Eigen and W.C. Gardiner, Kinetics of RNA replication: Plus-minus asymmetry and double-strand formation, *Biochemistry* 23:3186 (1984).

13. C.K. Biebricher, M. Eigen and W.C. Gardiner, Kinetics of RNA replication: Competition and selection among self-replicating RNA species, *Biochemistry* 24:6550 (1985).

14. C.K. Biebricher, M. Eigen and W.C. Gardiner, Kinetics of RNA replication, *Biochemistry* 22: 2544 (1983).

15. H. Otten, *Ein Beitrag zur Durchführung von kontrollierten Evolutionsexperimenten mit biologischen Makromolekülen,* Dissertation, Technical University Braunschweig (1988).

16. J. Maynard Smith, *Evolutionary Genetics,* Oxford University Press, Oxford (1989).

17. M. Eigen and P. Schuster, The hypercycle—a principle of natural self-organization, Part A: Emergence of the hypercycle, *Naturwissenschaften* 64:541 (1977).

18. M. Eigen, The physics of evolution, *Chemica scripta* 26B:13 (1986).

19. P. Schuster, The physical basis of molecular evolution, *Chemica scripta* 26B:27 (1986).

20. D.E. Dykhuizen and D.L. Hartl, Selection in Chemostats, *Microbiol. Rev.* 47:150 (1983).

21. F.R. Kramer, D.R. Mills, P.E. Cole, T. Nishihara and S. Spiegelman, Evolution *in vitro,* Sequence and phenotype of a mutant RNA resistant to ethidium bromide, *J. Mol. Biol.* 89:719 (1974).

22. S.E. Luria and M. Delbrueck, Mutations of bacteria from virus sensitivity to virus resistance, *Genetics* 28:491 (1943).

23. J. Lederberg and M. Zinder, Concentration of biochemical mutants of bacteria with penicillin, *J. Am. Chem. Soc.* 70:4267 (1948).

24. D.E. Lea, and C.A. Coulson, The distribution of the number of mutants in bacterial populations, *J. Genet.* 49:264(1949).

25. A. Novick and L. Szilard, Experiments with the chemostat on spontaneous mutations of bacteria, *Proc. Nat. Acad. Sci. USA* 34:708 (1950).

26. H. Moser, Structure and dynamics of bacterial populations maintained in the chemostat, *Cold Spring Harbor Symp. Quant. Biol.* 22:121 (1957).

27. R. Levisohn and S. Spiegelman, The cloning of a self-replicating RNA molecule, *Proc. Nat. Acad. Sci. USA* 60:866 (1968).

28. C.K. Biebricher, Replication and evolution of short-chained RNA species replicated by Qβ replicase, *Cold Spring Harb. Symp. Quant. Biol.* 52:299 (1987).

29. M. Eigen and C.K. Biebricher, Sequence space and quasispecies distribution, in *RNA Genetics, Vol. III: Variability of RNA Genomes,* E. Domingo, P. Ahlquist and J.J. Holland, eds., CRC Press, Boca Raton FL (1987).

30. N. Hilliger, *Bestimmung der Selektionswerte der Mutanten einer Quasispeziesverteilung,* Masters thesis, Göttingen University (1990).

31. Batschelet, E. Domingo and C. Weissmann, The proportion of revertant and mutant phage in a growing population, as a function of mutation and growth rate, *Gene* 1:27 (1976).

32. E. Domingo, D. Sabo, T. Taniguchi and C. Weissmann, Nucleotide sequence heterogeneity of an RNA phage population, *Cell* 13:735 (1978).

33. A.R. Fersht, Fidelity of replication of phage $\phi X174$ by DNA polymerase III holoenzyme: Spontaneous mutation by misincorporation, *Proc. Natl. Acad. Sci. USA* 76:4946 (1979).

34. L.A. Loeb and T.A. Kunkel, Fidelity of DNA synthesis, *Annu. Rev. Biochem.* 52:429 (1982).

35. J.S. McCaskill, A localization threshold for macromolecular quasispecies from continuously distributed replication rates, *J. Chem. Phys.* 80(10):5194 (1984).

36. C.L. Thompson and J.L. McBride, On Eigen's theory of the self-organization of matter and the evolution of biological macromolecules, *Math. Biosci.* 21:127(1974).

37. S.J. Gould and N. Eldredge, Punctuated equilibria: the tempo and mode of evolution reconsidered, *Paleobiology* 3:115–151 (1977).

38. L. Demetrius, Growth rate, population entropy, and perturbation theory, *Math. Biosc.* 93:159 (1989).

DISORDERED SYSTEMS AND EVOLUTIONARY MODELS

L. Peliti

Dipartimento di Scienze Fisiche and Unità INFM
Mostra d'Oltremare, Pad. 19, 80125 Napoli (Italy)
Associato INFN, Sezione di Napoli

1. INTRODUCTION

Darwinian evolution is the outcome of the interplay of reproduction, mutation, and natural selection. One of the most important and difficult problems in understanding the emergence of life is to pinpoint the necessary conditions for the onset od Darwinian evolution. The beautiful experiments of Biebricher and collaborators[1] show that a form of it may arise in very simple systems, which cannot possibly be considered living.

We have investigated theoretically the evolutionary behavior of models at a comparable—or even smaller—degree of complexity. The original motivation was to understand the properties of a model of prebiotic evolution introduced by Anderson[2]. We have found it necessary, however, to introduce further simplifications, in order to obtain a model whose treatment could be made as explicit as possible.

We consider a population made up of a fixed number, M, of individuals, whose "genetic structure" is identified by the state of N binary variables. The individuals may be interpreted as self-replicating polynucleotide strands, if one is thinking of experiments analogous to those of Biebricher et al [1], or of hypothetical prebiotic evolution mechanisms. In this case, the N binary variables represent the nature of the nucleotides which may be found along the strand.

It may well be the case that the very mechanism of replication is more complex than the evolutionary effects we model, and that the real problem lies in understanding the emergence of the replication mechanism itself[3,4]. However, our analysis is not necessarily attached to a strict prebiotic interpretation: some of the features we have highlighted appy to more general situations.

The population evolves according to a selection-mutation mechanism. Selection is introduced via a fitness function, defined on "genome space", which determines the probability that a given genotype produces offspring. We have explicitly considered two simple limiting cases, which are amenable to detailed treatment. The simplest case is the one in which this function is a constant, meaning that all genotypes have equal *a priori* chances of reproducing: in Sewall Wright's useful terminology, this is called a flat fitness landscape[6]. In the other case[7], one distinguishes once and for all between "viable" genotypes, which are allowed to reproduce, and "unviable" ones, which leave behind no offsprings. Each genome is deemed unviable with probability x, independently of all the others. The model becomes a realization of Kimura's neutral evolution[8], in a fully uncorrelated (rugged) landscape[9]. We have found that the properties of these evolutionary models may be understood by means of conceptual instruments forged in statistical mecanics for the treatment of disordered systems. The relevance of the theory of disordered systems for biological problems has already been stressed in the context of neural network theory[10], immunology[11], and protein theory[12]. We see here a novel application of the same ideas, and expect that new ones will soon be found.

2. MODEL

The model we consider belongs to a class introduced by Fontana and Schuster[13] on the one hand, and by Amitrano et al [14] on the other. One considers a population Ω made up of a fixed number, M, of individuals, whose "genetic structure" is identified bu the state of N binary units

$$S_i^\alpha = \pm 1; \qquad \alpha = 1, 2, \ldots, M; \qquad i = 1, 2, \ldots, N. \tag{1}$$

The population evolves according to the combined effects of selection, replication, and mutation.

(i) *Selection:* a "death probability" $p(\mathbf{S})$ is assigned to each genotype $\mathbf{S}$, where $\mathbf{S} = (S_1, S_2, \ldots, S_N)$. At each generation, for each individual α, the death probability $p(\mathbf{S}^\alpha)$ is evaluated. The individual α is then removed from the population with this probability.

(ii) *Reproduction:* all surviving individuals of the "old" population Ω are removed, and a new one is formed by their offsprings. For each label α of the population, a parent $G(\alpha)$ is chosen independently at random among the individuals which have survived the previous step.

(iii) *Mutation:* the genotype $\mathbf{S}^\alpha$ of each individual α would be identical to that of its parent, were it not for the occurrence of mutations. We consider only point mutations, which occur with probability μdt during the time interval dt. We have therefore

$$S_i^\alpha(t) = \begin{cases} S_i^{G(\alpha)}(t-1), & \text{with probability } \frac{1}{2}\left(1 + e^{-2\mu}\right), \\ -S_i^{G(\alpha)}(t-1), & \text{with probability } \frac{1}{2}\left(1 - e^{-2\mu}\right). \end{cases} \tag{2}$$

The definition of the model is made complete by the choice of the death probability $p(\mathbf{S})$, which plays the role of a fitness function. We have considered the following two simple cases:

(i) *Flat fitness landscape[6]:* $p(\mathbf{S})$ =const., corresponding to the case in which all genotypes have the same *a priori* probability of reproducing;

(ii) *Neutral evolution in a rugged landscape[7]:* $p(\mathbf{S}) = 1$ with probability x, and 0 otherwise. The value of $p(\mathbf{S})$ is drawn at random, for each of the 2^N possible genotypes, independently for different genotypes. This model corresponds to *neutral evolution[8]*, since all viable genotypes are equivalent.

We now discuss the behavior of the model in the two cases.

3. FLAT FITNESS LANDSCAPE

When the death probability is a constant, the reproduction step decouples from the mutation one, and may be interpreted by means of a model of stochastic dynamics: the Annealed Random Map model, whose solution has been obtained by Derrida and Bessis some time ago[15].

Let us take the simple case in which the death probability p vanishes. One can then easily calculate the probability $w_n(t)$ that n individuals, picked at random in the population, had n different ancestors t generations ago. One may easily see that

$$w_n(t) = w_n(t-1) \left(1 - \frac{1}{M}\right) \left(1 - \frac{2}{M}\right) \cdots \left(1 - \frac{n-1}{M}\right). \tag{3}$$

By expanding to first order in $1/M$ and integrating one obtains

$$w_n(t) = \exp\left[-\frac{n(n-1)}{2} \frac{t}{M}\right]. \tag{4}$$

Thus, the relevant time scale (in number of generations) is proportional to M. We introduce therefore the reduced time variable

$$\tau = \frac{t}{M}. \tag{5}$$

In the more general case, where the death probability does not vanish, all the results given below apply, provided that the reduced time variable τ is defined by

$$\tau = \kappa \frac{t}{M}, \tag{6}$$

where the factor κ is equal to the variance $[m^2]_{av} - 1$ of the number of offsprings of a given individual, which is itself a function of the death probability.

From this result, it is straightforward to obtain the probability $p(\tau)$ that the last common ancestor of two individuals picked at random existed τM generations ago:

$$p(\tau) = \frac{d}{dt}[1 - w_2(\tau)] = e^{-\tau}. \tag{7}$$

The relatedness of two individuals—the smaller or larger antiquity of their last common ancestor—is reflected in the larger or smaller similarity of their genotypes $\mathbf{S}$. Let us assume that the last common ancestor of the individuals α and β existed τM generations ago. Then their genomes performed a random walk in genome space, originating from the genotype of the ancestor, and making on average μN steps per generation. The probability $\Phi_\nu(t)$ that in such a random walk ν mutations are accumulated in t geenrations, is given by

$$\Phi_\nu(t) = \frac{N!}{2^N \, \nu! \, (N-\nu)!} \left(1 - e^{-2\mu t}\right)^\nu \left(1 + e^{-2\mu t}\right)^{N-\nu}. \tag{8}$$

If the last common ancestor of the the two individuals existed τM generations ago, the probability that their genotypes are different by ν mutations is given by this formula, with $t = 2\tau M$. We can thus obtain the probability P_ν that two individuals, picked at random in the population, have genotypes different by ν mutations:

$$P_\nu = \int_0^\infty d\tau \, e^{-\tau} \, \Phi_\nu(2\tau M) = \frac{\lambda N!}{2^N \, (N-\nu)!} \sum_{\ell=0}^{N-\nu} \binom{N-\nu}{\ell} \frac{\Gamma(\lambda+\ell)}{\Gamma(\lambda+\ell+\nu+1)}, \quad (9)$$

where the parameter λ is defined by $\lambda = (4\mu M)^{-1}$ and Γ is Euler's gamma function. This probability may be converted into the probability distribution of the overlap $q^{\alpha\beta}$, defined by

$$q^{\alpha\beta} = \frac{1}{N} \sum_{i=1}^{N} S_i \alpha S_i^\beta, \quad (10)$$

by means of the relation

$$q = 1 - \frac{2\nu}{N}. \quad (11)$$

Therefore the genetic structure of the population may be described by the probability distribution $P(q)$ that two individuals, picked at random in Ω, have an overlap $q^{\alpha\beta}$ equal to q. This quantity is analogous to the Parisi order parameter in spin glasses[10]. In fact, the quantity $P(q)$ is obtained as a population average, as the ratio of the number of pairs (α, β) whose overlap $q^{\alpha\beta}$ is equal to q to the total number of pairs of individuals in the population:

$$\langle P(q) \rangle = \left[\binom{M}{2} \right]^{-1} \sum_{(\alpha,\beta)} \delta(q^{\alpha\beta} - q). \quad (12)$$

The quantity we have calculated above is the average of this expression over all possible histories of the population, defined by the collection of all the applications $\alpha \to G(\alpha)$ of individuals on their parents. One may show that $\langle P(q) \rangle$ is in fact a random quantity. In particular, the average Q of q, which is a measure of the genetic variability of the population, fluctuates rather strongly as time goes on. It may be shown that, if $M \to \infty$ for fixed λ (admittedly a rather academic limit), the average $\overline{Q}$ of Q over all possible population histories is given by

$$\overline{Q} = \frac{\lambda}{\lambda+1}. \quad (13)$$

In the same limit, the expected variance of Q is given by

$$\Delta Q^2 = \overline{Q^2} - \overline{Q}^2 = \frac{2\lambda^3}{(\lambda+1)^2(\lambda+2)(3\lambda+1)(3\lambda+2)}. \quad (14)$$

Therefore, this simple model of an evolving population exhibits the lack of self-averaging typical of spin glasses[10].

The evolution of the average "genetic" structure of the population can be described by the correlation function

$$\chi(t) = \frac{1}{N} \sum_{i=1}^{N} \overline{\langle S_i(t) \rangle \langle S_i(0) \rangle}, \quad (15)$$

where the angular brackets, as above, denote the population average, and the bar denotes the average over histories. One has

$$\chi(t) = \overline{Q}e^{-2\mu^*|t|}, \tag{16}$$

which defines the effective mutation rate μ^*. It turns out that the effective mutation rate μ^* is equal to the bare one μ:

$$\mu^* = \mu, \tag{17}$$

in agreement with a famous result of Kimura[8]. All these predictions are successfully compared with computer simulations of the model[6].

4. RUGGED FITNESS LANDSCAPE

The model (ii) corresponds to neutral evolution in a rugged fitness landscape: *neutral* because all viable genotypes are equivalent; *rugged* fitness landscape since the viability of each genotype is independent of that of its neighbors. It turns out that the behavior of this model is most simply understood in the infinite genome limit[8]

$$N \to \infty, \qquad \sigma = \mu N = \text{const.} \tag{18}$$

Let us consider an individual α, possessing a viable genome at generation t. The probability δ that its immediate offspring possesses an unviable genome—and is hence removed from the game—is given by the product of the probability of having one mutation, leading to a brand new genotype, times the probability x that the new genotype is unviable:

$$\delta = \sigma x. \tag{19}$$

From the point of view of genealogy statistics, therefore, the present model is equivalent to the flat fitness landscape one, with a fixed death probability given by eq. (19). In particular, the distribution of antiquities of the last common ancestors is given by $e^{-\tau}$, where $\tau = \kappa t/M$, and

$$\kappa = \frac{1}{1-\delta}. \tag{20}$$

This prediction may be easily checked by computer simulations. One has to remark, however, that eq. (19) holds only in the $N \to \infty$ limit, while finite N corrections yield a systematically *smaller* value for the death probability. The genealogy results agree with the predictions, provided that the effective death probability is introduced in eq. (20) instead of the estimated one.

Genotype distributions may be inferred from genealogy statistics in a way analogous to the one sketched in the previous section. One should keep in mind that the walk in genome space connecting each individual to its ancestor is constrained to step only in viable genotypes. This is not quite the problem of random walks on hypercubes with randomly placed holes, considered in the context of spin glass dynamics[16], since, in our case, attempt to step on an unviable genotype (a "hole") leads to the removal of the walker, whereas it leaves the position of the walker unchanged in the

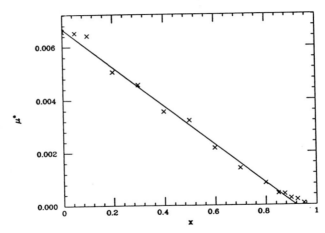

Fig. 1. Effective mutation rate μ^* as a function of the fraction of unviable genotypes x, for a population of 200 individuals with genome size N equal to 15. The "bare" mutation rate μ is identified by the diamond on the μ^* axis. The percolation threshold x_c is identified by the diamond on the x axis. From ref. 7.

other case. In fact simulations show[7] that the effective mutation rate μ^* is consistently higher than the effective mobility of the walkers calculated in ref. 16.

The behavior of μ^* as a function of the fraction x of unviable genotypes is more or less linear, interpolating between $\mu^*(x{=}0) = \mu$, in agreement with the result of the previous section, and $\mu^*(x{=}x_c) = 0$, when the percolation threshold x_c is reached. For $x > x_c$, the connected clusters of viable genotypes fail to percolate, and the average genotype of the population keeps forever memory of its ancestral genotype, as anticipated in the first of ref. 14.

5. DISCUSSION

The point I wish to make in expounding these results is that concepts arising in the theory of disordered systems, like the existence of two kinds of averages, the lack of self-averaging properties, etc., are called to play a growing role in evolutionary theory. They provide indeed a way to define sharply concepts like variability, its fluctuations, and the genetic structure of the populations, in a way amenable to explicit mathematical treatment. The main thrust of the present research is to go at a more elementary level, by understanding the way in which self-replication mechanisms may arise in the context of a network of chemical reactions, represented as a disordered dynamical system.

ACKNOWLEDGMENTS

I am grateful to my collaborators, C. Amitrano, B. Derrida, M. Saber, for having embarked with me on such an odd enterprise. I thank M. Mézard and M. Serva for valuable suggestions, and R. Adler for encouragement.

REFERENCES

1. C. K. Biebricher, M. Eigen, W. C. Gardiner, Jr., in these Proceedings.
2. P. W. Anderson, *Proc. Natl. Acad. Sci. USA*, 80, 3386 (1983).
3. L. F. Abbott, *J. Mol. Evol.*, 27, 114 (1988).
4. G. Wächtershäuser, *Microbiological Reviews*, 52, 452 (1988).
5. S. Wright, *Proc. Natl. Acad. Sci. USA*, 23, 307 (1937).
6. B. Derrida, L. Peliti, *Bull. Math. Biol.*, in press.
7. U. Bastolla, L. Peliti, in preparation.
8. M. Kimura, *The neutral theory of molecular evolution*, Cambridge: Cambridge University Press (1983).
9. S. A. Kauffman, S. Levin, *J. theor. Biol.*, 128, 11 (1987).
10. M. Mézard, G. Parisi, M. A. Virasoro, *Spin glass theory and beyond*, Singapore: World Scientific (1987).
11. G. Weisbuch, in these Proceedings.
12. P. Wolynes, in these Proceedings. G. Parisi, in these Proceedings.
13. W. Fontana, P. Schuster, *Biophys. Chem.* 26, 123 (1987); W. Fontana, W. Schnabl, P. Schuster, *Phys. Rev.*, A40, 3301 (1989).
14. C. Amitrano, L. Peliti, M. Saber, *C. R. Acad. Sci. Paris*, III 307, 803 (1988); *J. Mol. Evol.*, 29, 513 (1989).
15. B. Derrida, D. Bessis, *J. Phys. A: Math. Gen.*, 21, L509 (1988).
16. J.-M. Flesselles, R. Botet, *J. Phys. A: Math. Gen.*, 22, 903 (1989).

WHAT CONTROLS THE EXOCYTOSIS OF NEUROTRANSMITTER?

Lee A. Segel[1] and Hanna Parnas[2]

1. Department of Applied Mathematics and Computer Science
   Weizmann Institute of Science, Rehovot, Israel
2. Department of Neurobiology and Otto Loewi Center
   for Cellular and Molecular Neurobiology
   The Hebrew University, Jerusalem, Israel

1. INTRODUCTION

Before commencing our answer to the question posed in the title, we present a sketch of some required terms and concepts for those unfamiliar with basic neurobiology. The most evident features of a *neuron* (nerve cell) are a *cell body* together with a long *axon*. See Figure 1. Extending from the cell body are a number of finger-like processes called *dendrites* that usually transmit the input into the cell. The axon branches into many fibres at the end of which are typically located small button-like structures called *nerve endings*. If the input to a cell is sufficiently excitatory (i.e. if the depolarizing voltage change exceeds a certain threshold) an electrical wave called an *action potential* is generated, which travels down the axon and passes into each nerve ending. The resulting depolarization brings about the *exocytosis* (secretion) of special chemicals called *neurotransmitters* that are believed to be located in spheroidal *vesicles* (diameter 100–500Å) close to the membrane of the nerve ending at specific *release sites*. Opposite these sites on the membrane of the adjacent *postsynaptic cell* is a special *active zone* with receptors that reversibly bind the released neurotransmitter after it traverses an *intercellular gap* of width approximately 500Å. The two special release and receptor regions together with the gap constitute a *synapse*. The average number of vesicles released by an action potential (or by a similar artificial stimulus) is the *quantal content*, called here *total release*. *Spontaneous release* of individual vesicles also occurs, at a low rate.

The bound postsynaptic receptors induce a signal that is excitatory or inhibitory depending on various parameters, among them the type of transmitter. If the time and space integrated electrical signals reach the required threshold at the region associated with the response generation an action potential is formed. Depending on the nature of the "downstream" cell, the new action potential can act toward the formation of another release of transmitter, or toward the induction of muscle contraction or gland secretion.

From an input/output perspective, the action potential results in exocytosis of a number of neurotransmitter-containing vesicles. The key biophysical events of exocytosis are the fusion of the vesicular membrane to the cell membrane and the appearance of a small aperture in the fused membrane. The entire release process, starting from the arrival of the action potential, takes less than 2 msec at room temperature.

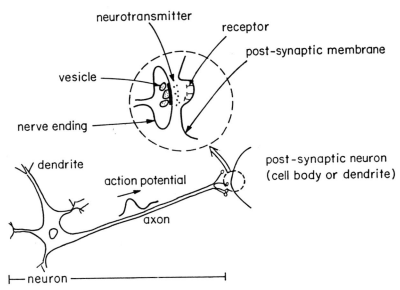

neurotransmitter

receptor

post-synaptic membrane

vesicle

nerve ending

dendrite

action potential

post-synaptic neuron
(cell body or dendrite)

axon

⊢— neuron

Fig. 1. Schematic diagram of a neuron. Magnified circle: a nerve-nerve synapse.

The major part of this paper will be concerned with a review of experimental and theoretical work on the control of neurotransmitter release. In the concluding portion we point out the linkage between this work and recent findings concerning the biophysics of release.

The first major contribution to our understanding of the control of neurotransmitter release is usually dated from work of Katz and Miledi[7], that showed that the presence of calcium ions (Ca^{2+}) in the extracellular medium is required for release. This conclusion of Katz and Miledi together with numerous other authors was echoed in experiments with artificial membranes. These demonstrated that the introduction of calcium causes the membranes to fuse. (See the review of Prestegard and O'Brien[20].) The parallel between artificial and natural systems is not complete, however, for calcium in the millimolar range was required to induce fusion in artificial systems. By contrast, it is believed that the concentrations of calcium that are required to induce fusion in neurotransmitter release are in the range of perhaps one to at most 50 micromolar. One way around this difficulty is to postulate the existence of some proteins that cause a local high concentration of calcium. Be that as it may, it is a major message of this communication that although calcium is necessary for neurotransmitter release it is insufficient, and that a second factor is also required.

2. A BASIC FRAMEWORK FOR UNDERSTANDING THE ROLE OF CALCIUM IN CONTROLLING TOTAL RELEASE

The many experiments demonstrating that release requires Ca^{2+} led to the *calcium hypothesis*, that the presence of sufficient (intracellular) Ca^{2+} near release sites is necessary and sufficient to induce release. This hypothesis spurred further experimental investigations into the role of Ca^{2+}, some of which we now mention.

Figure 2 shows data on total release L_T for several different values of the extracellular calcium concentration C_e. It is seen that release is an increasing function of C_e, which tends toward saturation at higher values of C_e. For small values of C_e, a power law dependence of L_T on C_e can be inferred. A second type of experiment is explained in Figure 3. Two impulses, separated by a time interval Δt, are applied to the terminal. It is found that the total release induced by the second impulse, $L_T^{(2)}$, is larger than that induced by the first impulse, $L_T^{(1)}$. The ratio of these two releases is called the *facilitation F*:

$$F = L_T^{(2)}/L_T^{(1)}. \tag{1}$$

Katz and Miledi[7] suggested the basic reason for facilitation: residual calcium that remains after the first impulse supplements the calcium that enters the terminal after the second impulse, thereby increasing release.

Figure 4 schematically depicts results concerning the dependence of facilitation F on the extracellular calcium concentration C_e at two different time intervals Δt between impulses, one small (say 5 ms) and the other large (say 30 ms). It is seen that F decreases with C_e when Δt is small but increases when Δt is relatively large.

Particularly valuable in studying release are measurements of release kinetics, i.e. the distribution over time of vesicular release. An example of such a measurement is reported in Figure 5.[3] It is seen that the shape of the kinetics is virtually independent of the extracellular calcium concentration even though the amplitude of release is greatly augmented when the extracellular calcium is increased. In Figure 5 the time is measured from the negative peak of a presynaptic action potential, which has effectively terminated by $t = 1.5$ msec. Thus the figure also exemplifies the (often repeated) observation that the maximum release occurs after the conclusion of the triggering pulse. This point is decisively illustrated in Fig. 7 of Ref. 3.

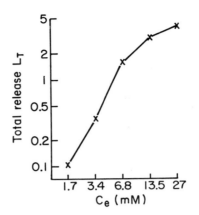

Fig. 2. A log-log plot of total release L_T as a function of extracellular calcium C_e. The maximum slope is about 3.5, suggesting that $L_T \sim C_e^4$ for small C_e. From distal synapses in crayfish opener muscle fibers. Releases calibrated as fraction of release from proximal synapses at physiological C_e of 13.5 mM. Redrawn from Figure 3 of Ref. 21.

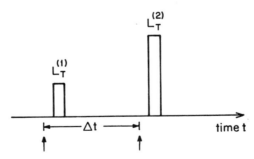

Fig. 3. Diagram of total release L_T induced by two successive impulses (arrows).

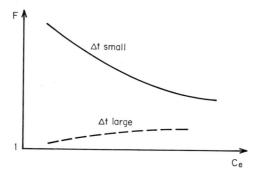

Fig. 4. Schematic depiction of the dependence of facilitation F on the extracellular calcium concentration C_e, as measured by Rahamimoff[22].

We turn to theory. Release kinetics will be considered later. We now present a theory[15] for the dependence of the total release L_T on calcium, emphasizing the explanation of the facilitation result of Figure 4. (Here and elsewhere, we have tried to select experiments with well marked qualitative behaviours in order to distinguish possible theories.)

Three major processes govern the relationship between calcium and the total neurotransmitter release. These are the (total) *entry* E_T as a function of the extracellular calcium concentration C_e, the (total) *release* L_T as a function of the intracellular calcium concentration C, and the *removal* of intracellular calcium once it has entered. Following entry (at $t = 0$) $C = C_r + E_T$ where C_r is the (constant) resting level of intracellular calcium. With this in mind we constructed our first model as follows.

$$\text{Entry: } E_T = E_T(C_e). \qquad \text{Release: } L_T(C) = \lambda[C/(K_\lambda + C)]^n. \qquad (2a, b)$$

$$\text{Removal: } dC/dt = -kC; \quad \text{at } t = 0, \quad C = C_r + E_T(C_e). \qquad (3)$$

For present purposes, we do not have to specify anything about the entry E_T except that it is an increasing function of C_e. The saturation and *cooperativity* (power law dependence), demonstrated in Figure 2 as a function of extracellular calcium, formed part of the reasons, among several others[15,11] to choose a saturating cooperative dependence of release on intracellular calcium. The particular function taken is not important. What is important is that L_T is proportional to C^n when C is small and approaches a constant when C greatly exceeds K_λ. Finally, removal is taken to be a linear process, resulting in an exponential decay of the intracellular calcium once it has entered.

From (1), (2) and (3) it follows that

$$\frac{dF}{dC_e} = \frac{dF}{dE_T}\frac{dE_T}{dC_e}, \qquad (4a)$$

$$\frac{dF}{dE_T} = -(1 + e^{-kt})e^{-kt}K_\lambda[K_\lambda + (1 + e^{-kt})E_T]^{-2}. \qquad (4b)$$

Since E_T is by assumption an increasing function of extracellular calcium we see that according to (4) the facilitation F is always a decreasing function of C_e. As shown in Figure 4, this is in accord with the experiments when Δt is small but is not in accord with experiments when Δt is large. Thus we must alter our model. Consideration of the alternatives resulted in the conclusion that neglect of saturation in removal is the flaw in

351

our earlier model[15]. We thus modified that model to make removal a saturating process:

$$\frac{dC}{dt} = -\frac{\mu C}{K_\mu + C}.$$
(5)

With this assumption, the two different qualitative results of Figure 4 are indeed recovered.

Considerable further work was done developing and testing the model of (2a,b) and (5). In particular, it was concluded that entry is a saturating function (without cooperativity):

$$E_T(C_e) = \eta C_e / (K_\eta + C_e).$$
(6)

The presence of saturation is in contrast with the Nernst diffusion law that has often been assumed[6]. An important topic for current research is to understand how the mechanisms of channel conductance differ from simple diffusion[4].

As demonstrated in the sample calculation (4), facilitation is an important probe for understanding aspects of neurotransmitter release. For a survey of such uses of facilitation see Ref. 17.

3. THE SECOND FACTOR

It has long been recognized that the presence of facilitation threatened a naive view that Ca^{2+} is the sole controller of release. One must assume that release ceases because Ca^{2+} has been reduced to a low level but yet sufficient residual Ca^{2+} remains to cause facilitation. It was thought that these conflicting demands could be harmonized by the presence of cooperativity[7] but later investigations showed that difficulties persist if results at several different levels of extracellular Ca^{2+} are considered simultaneously[19].

If calcium were the sole controller of release, one would expect strong changes of the release kinetics when extracellular calcium concentrations are changed. In fact, as illustrated in Figure 5, the shape of the release kinetics is remarkably well preserved as the extracellular calcium concentration is changed over quite a large range: only the amplitude is affected. Also see Figures 6 and 8 below. A priori it is quite reasonable to expect the presence of a second factor, in addition to Ca^{2+}, to control release. The biological functions of calcium are numerous. Because of the incessant traffic of calcium molecules attending to their various tasks, we would expect significant fluctuations in local calcium concentrations. If only a small fluctuation of calcium is both necessary and sufficient to induce release, it would be hard to explain the highly controlled nature of the observed release process.

If a second factor is needed, what can it be? Much evidence points to depolarization.

It is now understood that depolarization causes the opening of transmembrane channels, which permits the entry of calcium into the nerve ending. This is a necessary step, for as we have seen it is intracellular calcium that is required for the triggering of the release process. In sketching evidence that levels of Ca^{2+} are insufficient to control release we first argue that Ca^{2+} is *insufficient to evoke* release and then that removal of Ca^{2+} from near the release sites is *insufficient to terminate* release.

Concerning release onset we call attention to experiments in which Ca^{2+} was present at the right place and yet evoked release did not commence until the presynaptic cell was depolarized. Figure 6 shows that when a Ca^{2+} ionophore is introduced (a method of increasing membrane permeability to Ca^{2+}) the level of intracellular Ca^{2+} rises significantly, as indicated by the large increase in the amount of release in the presence of the ionophore. Yet, in spite of the elevated intracellular Ca^{2+} prior to stimulation, evoked release only commences, after a latent period, following membrane depolarization. The latent period in the presence of the ionophore is the same as in the control experiment.

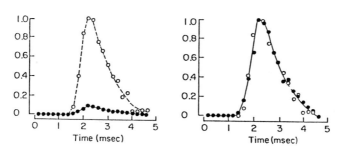

Fig. 5. The time course of release in mouse phrenic nerve terminal when extracellular Ca^{2+} is 0.5 mM (filled circles) and 1 mM (empty circles). Temperature 14.5°C. A: Curves normalized to maximum of larger release. B: Each curve normalized to its own maximum. From Figure 8 of Ref. 3.

Figure 7 shows results from an experiment wherein intracellular Ca^{2+} was elevated by means of illuminating a caged Ca^{2+} compound, nitr-5, which was injected into the cell. (The illumination "opens the cage" and releases calcium intracellularly.) The elevation of intracellular Ca^{2+} indeed increased the level of the randomly secreted spontaneous release. However, evoked *synchronous* release took place only following a brief depolarizing pulse, in this case an action potential. Virtually no Ca^{2+} was present in the bathing solution, so that calcium entry was negligible.

We now briefly discuss experiments in which the removal of Ca^{2+} was modulated, yet the time course of release, in particular its termination were not at all affected.

It is believed that a major mechanism for the removal of excess intracellular Ca^{2+} is by means of a molecular "machine" that exchanges intracellular calcium and extracellular sodium. Figure 8 shows that reduction in extracellular Na^+ indeed inhibited the extrusion of Ca^{2+}, as seen by the marked prolongation of facilitation. Yet the time course of release remained unaltered. In addition Hochner, H. Parnas and I. Parnas (unpublished) have demonstrated that injection of a very efficient Ca^{2+} chelator (i.e., neutralizer; nitr-5, in the dark) reduced the amount of release, shortened and reduced facilitation, but did not affect the time course of release.

The experiments in Figures 6–8 clearly suggest that the second factor, in addition to Ca^{2+}, becomes available upon depolarization. Under normal conditions, depolarization thus serves two roles. Firstly, it opens Ca^{2+} channels, hence causes an influx of Ca^{2+} that increases its intracellular level. The second role of depolarization involves activation of an as yet unknown entity, making it ready to participate in the release process.

Figures 6 and 7 show that a depolarization-activated factor, in the presence of Ca^{2+} initiates release. Figure 8 shows that facilitation, and hence intracellular Ca^{2+}, remains elevated for hundreds of msec at 50% Na^+. Yet release terminates normally. It must be that when voltage returns to its resting value the "second factor" also takes a resting form and release ceases.

4. THE CALCIUM-VOLTAGE HYPOTHESIS FOR RELEASE KINETICS

A simple kinetic theory that includes the effects of depolarization is symbolized by the scheme

$$S + C \underset{k_{-2}}{\overset{k_2}{\rightleftharpoons}} Q, \qquad nQ + V \overset{k_3}{\rightarrow} L, \qquad (7a,b)$$

where at least one of the rate constants depends on depolarization. For example k_3 might take on a relatively large value during depolarization, compared to its relatively small value under rest conditions.

In (7), S represents an entity whose union with intracellular calcium C results in a complex Q, n of which induce release L. The kinetic scheme (7) corresponds to the following differential equations (with suitable initial conditions):

$$dS/dt = -k_2 SC + k_{-2}Q, \quad dQ/dt = k_2 SC - k_{-2}Q - k_3 Q^n V, \qquad (8a,b)$$

$$dC/dt = E(t, C_e, \phi) - k_2 SC - R(t), \quad dL/dt = k_3 Q^n V. \qquad (8c,d)$$

Here E and R are suitable descriptions of the instantaneous rates of entry and removal. In particular the entry E depends on the depolarization ϕ. The vesicle density V is taken to be constant. Total release is computed by

$$L_T = \int_0^\tau (dL/dt)dt \qquad (8e)$$

where τ is the duration of the release process.

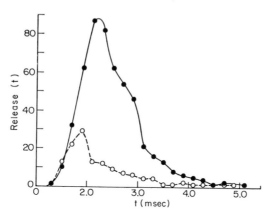

Fig. 6. The time course of release in the absence (empty circles) and presence (filled circles) of a calcium ionophore—which increases total release five fold. An action potential (duration less than 1 msec; temperature 16°C) was applied at $t = 0$, in crayfish. Redrawn from Figure 2 of Ref. 13. (Here and in Figures 8 and 10, the curve connecting the experimental points is solely to aid visualization of the data.)

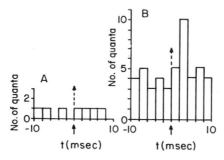

Fig. 7. Vesicle releases, spontaneous and induced by a brief depolarizing action potential (arrow). The extracellular medium was devoid of Ca^{2+}. A: No illumination. B: Following illumination, which released intracellular Ca^{2+} "caged" by nitr-5. Note the induced release 2 msec after the stimulus. Redrawn from Figure 3 of Ref. 5.

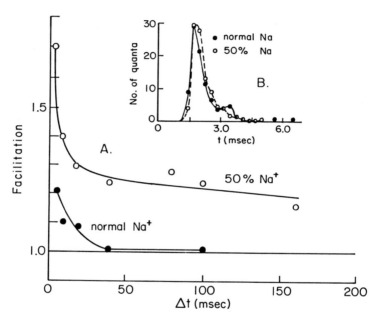

Fig. 8. A: Facilitation in normal extracellular Na^+ (220 mM) and in reduced Na^+ (110 mM) as a function of the time Δt between pulses. Compare Figure 3 and Eq. (1). B: Time course of induced release. Crayfish neuromuscular; notion, 18°C. Redrawn from Ref. 1.

Solution of these equations does not lead to satisfactory results. It is predicted that the peak of release never occurs after the termination of the impulse[13], contrary to experiment (see Figure 5).

The calcium voltage hypothesis[11] introduces the required lag by postulating that the key entity S normally is in an inactive state T. Depolarization brings about a $T \rightarrow S$ transition. The scheme (7) is consequently supplemented by

$$T \; \underset{k_{-1}^{\phi}}{\overset{k_1^{\phi}}{\rightleftharpoons}} \; S. \tag{9}$$

where k_1^{ϕ} increases with depolarization ϕ while k_{-1}^{ϕ} decreases. Thus depolarization causes a shift of T to its active conformation S. Because the peak of the release histogram always follows the termination of the pulse, parameters must be such that there is a slow process late in the chain of events that control release. In this simple version of the theory, the slow step is (7b).

Figure 9 shows simulations employing the calcium voltage hypothesis. Note that the peak of release indeed occurs after the conclusion of the 0.5 msec pulse, and that the shapes of the kinetics are the same at the two rather different levels of extracellular Ca^{2+} used in the simulations.

Throughout the development of the calcium voltage hypothesis, efforts were made to rule out various alternative mechanisms. For example an alternative to (7) (as an explanation to the cooperativity shown for example in Figure 2) is that n calcium molecules bind to a single molecule S in order to constitute the complex Q:

$$S + nC \; \underset{k_{-2}}{\overset{k_2}{\rightleftharpoons}} \; Q, \qquad Q + V \overset{k_3}{\rightarrow} L. \tag{10}$$

This last hypothesis is attractive a priori, for it is known that calcium regulation often involves the molecule calmodulin, which binds four calcium molecules. Nonetheless hypothesis (10) can be ruled out with the aid of a demonstration[12] that the initial slope in a graph of release kinetics provides a good estimate for the cooperativity parameter n. As is seen in Figure 10, experiments show that this slope is approximately 4. Theory shows, however, that if hypothesis (10) is adopted, then the slope will be 1.

The calcium voltage model has been applied to a wide variety of experiments[14]. In doing so certain modifications to the basic kinetic scheme of (7) and (9) were mandated. The first change was required in order to take account of experiments wherein hyperpolarization reduces the quantal content. Hence a more inactive state, T_1, of the control molecule was hypothesized. This is symbolized by

$$T_1 \; \underset{k_{-0}^{\phi}}{\overset{k_0^{\phi}}{\rightleftharpoons}} \; T. \tag{11}$$

In order that correct quantitative behavior both as a function of temperature and of the extracellular calcium concentration could be obtained, the simplified version of release that was adopted in (7) had to be made somewhat more detailed. Thus the final stages of the model are now symbolized as follows:

$$nQ + V \xrightleftharpoons[k_{-3}]{k_3} V^*, \qquad V^* \xrightarrow{k_4} L. \tag{12}$$

A final modification was required when the spontaneous release phenomenon was investigated. Here so called miniature endplate potentials (mepp) are observed to occur sporadically in a resting preparation, without stimulation. To analyze such a phenomenon, a recovery step was postulated:

$$L \xrightarrow{k_5} T. \tag{13}$$

Note that the recovery step is properly neglected in theories of induced release, for these events occur so rapidly that recovery is quite negligible.

Although conceptually simple, the full calcium voltage model requires a rather large number of equations. Computer simulations indicate that the model agrees well with a wide variety of experiments, keeping the same basic values for parameters. For example results concerning L_T that we originally obtained from overall models such as (2b), (5) and (6) can be recovered from the full kinetic model via (8e). Analytic approximations, wherein accuracy is often sacrificed in favor of simplicity, give clear indications of the qualitative behavior predicted by the model. Such an approximation is one for the spontaneous release frequency Fr at relatively low values of the intracellular calcium concentration[9]:

$$Fr \approx \frac{k_3 k_4 V_T T_0^{n-1}}{k_4 + k_{-3}} \left(\frac{C}{\frac{k_{-2}}{k_2}(1 + \frac{k_{-1}^\phi}{k_1^\phi}) + C} \right)^n. \tag{14}$$

Both a cooperative dependence on C, for small C, and saturation are exhibited. Compare the analogous formula (2b) for evoked release. The level of saturation is not well estimated by (14), however. Instead one should employ the alternative formula, for larger C

$$Fr \approx \frac{1}{n} \left(\frac{C}{\frac{1}{k_2}(1 + \frac{k_{-1}^\phi}{k_1^\phi}) + (\frac{1}{k_1^\phi} + \frac{1}{k_4} + \frac{1}{k_5})C} \right). \tag{15}$$

Figure 11 shows a log log plot of simulations of spontaneous release as a function of depolarization. Experimental results[2] are provided for comparison. The same qualitative behavior is observed.

The review by H. Parnas, I. Parnas and Segel[14] describes a number of experiments that are in agreement with the calcium-voltage hypothesis. Later work includes not only further results for total release L_T, release kinetics $L(t)$, but also stochastic release phenomena[8,9]. Also reviewed[14] is the competing calcium hypothesis, which to explain various experiments relies solely on Ca^{2+} to control release, with (in the present refined version of the calcium hypothesis) three-dimensional diffusion as the dominant removal process from the neighborhood of the release sites.

5. CONCLUSIONS

We have presented a comprehensive theory for the control of neurotransmitter release. We have argued that release is unlikely to be controlled by a single factor. This point may be a general one — key biological processes perhaps seldom rely on a single controller.

An outline has been provided of some of the experimental evidence that the "second factor" is depolarization dependent, similarly to the influx of Ca^{2+}. For clarity in exposition we first described this evidence and then developed the corresponding the-

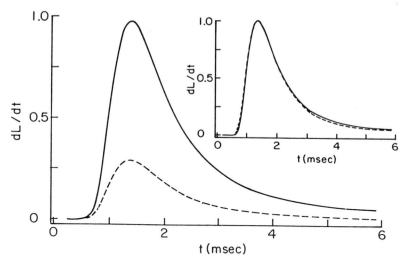

Fig. 9. Simulation of release kinetics at $C_e = 0.5$ mM (dashed line) and $C_e = 1$ mM (solid line). Full calcium-voltage model. 0.5 msec pulse. Main graph: normalization to maximum release when $C_e = 1$ mM. Inset: each curve normalized to its own maximum[9].

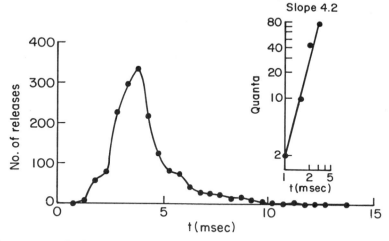

Fig. 10. Time course of release, obtained by recording individual vesicle releases, in crayfish, following each of 3650 1 msec stimulations, commencing at $t = 0$. Temperature 9° C. Inset: Log-log plot of the initial phase of release. Redrawn from Figure 1 of Ref. 14.

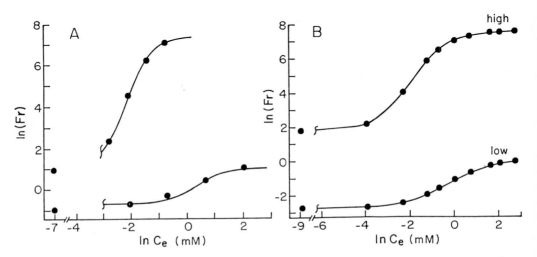

Fig. 11. A: Measurements[2] of spontaneous release at high (upper curve) and low depolarizations, at different concentrations of extracellular calcium C_e. B: Corresponding simulation results[9].

ory. Historically, however, the importance of depolarization was gradually manifested in a theory-experiment dialog. This began with explicit suggestions of H. Parnas and Segel[16], based on the experiments of Cooke and his associates[2], that depolarization played an important direct role in the control of release.

We have shown that in its present state — symbolized by (11),(9),(7a), (12) and (13) — the theory agrees with a wide variety of experiments and that various alternatives can be ruled out. This is not to say that there do not exist variants of the theory that may in the end be correct. For example the activated control molecule S may communicate with the vesicle via Ca^{2+} with the aid of intermediate molecules. The essential point is that release is controlled by a molecule T that is reversibly activated by depolarization (to S) and that communicates with the vesicle in a calcium-mediated fashion.

The agreement of theory with experiment is encouraging, but of course not decisive. What would be decisive is identification of the $S - T$ molecule or entity. There begins to be some reason for optimism that this might be accomplished. By membrane capacitance measurements, single vesicle opening events can be observed[10]. Many experiments have been carried out on non-neural cells, especially mast cells of the immune system, whose vesicles are much larger than those of neurons. For these larger cells the release process lasts for seconds, not milliseconds, and presumably is much easier to monitor. Yet it seems that the initial event of fusion lasts less than a millisecond and may be universal.

A recent experimental paper in the line pioneered by Neher and Marty[10] "speculated that fusion begins when a preassembled, ion channel-like structure opens in response to a cytosolic messenger, forming a fusion pore"[23]. It is not yet known whether evidence for this pore also will be found in neurons, as well as in mast cells. If it indeed appears relevant to neurons, one way to harmonize the above speculation[23] with the calcium-voltage hypothesis is to assume that the channel is a tetramer composed of four T subunits. Depolarization activates these subunits but the channel remains closed. Only when Ca^{2+} binds to all four activated subunits does the channel open.

Investigation of this and other concrete implementations of the calcium-voltage hypothesis is a subject of our current research.

ACKNOWLEDGEMENTS

HP acknowledges support from the US-Israel Binational Foundation (grant no. 88–100) and from the National Institute of Health (grant NS 24940).

REFERENCES

H. Arechiga, A. Cannone, H. Parnas and I. Parnas, Blockage of synaptic release by brief hyperpolarizing pulses, *J. Physiol. (Lond.)* (1990) in press.

J. D. Cooke, K. Okamoto and D. M. J. Quastel, The role of calcium in depolarization-secretion coupling at the motor nerve terminal, *J. Physiol. (Lond.)* 228:459 (1973).

N. B. Datyner and P. W. Gage, Phasic secretion of acetylcholine at the mammalian neuromuscular junction, *J. Physiol. (Lond.)* 303:299 (1980).

B. Hille, *Ionic channels in excitable membranes*, Sinauer Associates, Sunderland, MA (1984).

5. B. Hochner, H. Parnas and I. Parnas, Membrane depolarization evokes neurotransmitter release in the absence of calcium entry, *Nature* 342:433 (1989).

6. B. Katz, *Nerve, Muscle and Synapse,* McGraw-Hill, NY, (1966).

7. B. Katz and R. Miledi, The role of calcium in neuromuscular facilitation, *J. Physiol. (Lond.)* 195:481 (1968).

8. C. Lustig, Neural transmitter release: models and mechanisms, Ph.D. Thesis, Weizmann Institute of Science (1989).
   C. Lustig, H. Parnas and L. Segel, Analysis of spontaneous neurotransmitter release, (1991) unpublished.

9. E. Neher and A. Marty, Discrete changes of cell membrane capacitance observed under conditions of enhanced secretion in bovine adrenal chromaffin cells, *Proc. Natl. Acad. Sci. USA* 79:6712 (1982).

10. H. Parnas, J. Dudel and I. Parnas, Neurotransmitter release and its facilitation in crayfish. VII. Another voltage dependent process besides Ca entry controls the time course of phasic release, *Pflügers Arch.* 406:121 (1986a).

11. H. Parnas, I. Parnas and L. Segel, A new method for determining cooperativity in neurotransmitter release, *J. Theor. Biol.* 119:481 (1986b).

12. H. Parnas and I. Parnas, The Ca-voltage hypothesis for neurotransmitter release, in: *Intracellular Communication, Theory and Experiment*, A. Goldbeter, ed., Academic Press, NY (1989).

13. H. Parnas and I. Parnas and L. Segel, On the contribution of mathematical models to the understanding of neurotransmitter release, in: *International Review of Neurobiology,* J.R. Smythies and R.J. Bradley, eds, Acadamic Press: Orlando (1990).

14. H. Parnas and L. Segel, A theoretical explanation for some effects of calcium on the facilitation of neurotranmitter release, *J. Theor. Biol.* 91:125 (1980).

15. H. Parnas and L. Segel, A theoretical study of calcium entry in nerve terminals, with applications to neurotranmitter release, *J. Theor. Biol.* 91:125 (1981).

16. H. Parnas and L. Segel, Facilitation as a tool to study the entry of calcium and the mechanisms of neurotransmitter release, *Progress in Neurobiology* 32:1 (1988).

17. I. Parnas, J. Dudel and H. Parnas, Neurotransmitter release and its facilitation in crayfish. II. Duration of facilitation and removal processes of calcium from the terminal, *Pflügers Arch.* 393:232 (1982a).

18. I. Parnas and H. Parnas, Calcium is essential but insufficient for neurotransmitter release: the calcium-voltage hypothesis, *J. Physiol. (Paris)* 81:289 (1986).

19. G. H. Prestegard and M. P. O'Brien, Membrane and vesicle fusion, *Ann. Rev. Phys. Chem.* 38:383 (1987).

20. I. Parnas, H. Parnas and Dudel, Neurotransmitter release and its facilitation in crayfish muscle. V. Basis for synapse differentiation of the fast and slow type in one axon, *Pflügers Arch.* 395:261 (1982).

21. R. Rahamimoff, A dual effect of calcium ions on neuromuscular facilitation, *J. Physiol. (Lond.)* 195:471 (1968).

22. A. E. Spruce, L. J. Breckenridge, A. K. Lee and W. Almers, Properties of the fusion pore that forms during exocytosis of a mast cell secretory vesicle, *Neuron* 4:643 (1990).

STRUCTURE AND FUNCTION OF THE PHOTOSYNTHETIC APPARATUS

Pierre Joliot

Institut de Biologie Physico-Chimique, Université de Paris
75005 Paris, France

In photosynthesis, the process of conversion of light energy into chemical energy occurs in specialized membrane apparatus: external membranes of photosynthetic bacteria or intrachloroplastic membranes in the case of the oxygenic photosynthesis in green algae and higher plants. Most of the pigments which collect light energy of electron carriers are associated with protein complexes including several polypeptides non-covalentely bounded. Due to purification of these protein complexes are now possible, without a loss of their functional activity. On the other hand, two types of bacterial photocenters have been recently crystallized and their structure is now known with a resolution of a few angstroms.

The photosynthetic energy conversion basically implies two processes - excitation transfer and electron transfer - which both represent a privileged subject of studies for the physicists. Excitation transfer between pigments allows the concentration of light energy on the photocenter, where the primary process of charge separation occurs. These charges are then transferred along the photosynthetic electron transfer chain, which includes a large number of electron carriers. In the case of isolated photocenters, the kinetics of electron transfer have been characterized with a time resolution of about hundred of femtoseconds. The theoretical analysis of the mechanism of electron tunneling in bacterial photocenters and more specifically, coupling between electron and atomic movments is now developed by several groups of solid state physicists.

It is important to stress that excitation and electron transfer also occur between complexes embedded in the same membrane. Then, photosynthetic process should also be studied on non-perturbed membrane such as isolated cloroplast or living cells. The understanding of the mechanism of the intercomplex energy transfer requires a precise knowledge of the relative localization of the complexes in the membrane and of the domains in which the few mobile carriers which establish a link between the complexes can diffuse. Studies on the lateral organization of the photosynthetic membrane which is somewhere between order and disorder should involve a collaboration betweeen biophysicists and specialists of statistical physics.

RELAXATIONAL PROCESSES IN FRUSTRATED RANDOM SYSTEMS:
A SPHERICAL-SPIN MODEL

A. Jagannathan

Centre de Physique Théorique de l'Ecole Polytechnique
91128 Palaiseau Cedex, France

1. INTRODUCTION

A large variety of problems can be entered under the classification of random frustrated systems. The dynamical response of such systems appears to possess some universal characteristics, features common to systems very different in nature. Such features may be considered as being consequences of these two properties: of randomness, which implies that the system is characterised by a number of parameters that may be chosen from a set of values all of which are physically permitted, and of frustration, which implies that there are a set of constraints, not mutually compatible, that the system must satisfy in an optimal way. The prototypical examples of this kind of system are the much-studied spin glasses. Here the randomness and frustration enter in the magnetic interactions between spins. The sign and strength of the interaction between pair of spins can take on a number of different possible values, in a real spin glass. In consequence there arises a mutually incompatible set of rules that say, allow local but not global minimisation of the interaction energy. In a nonmagnetic context, alternatively, one may envisage polymers on a random substrate (Parisi, in this Workshop) which must satisfy connectivity constraints while trying to minimise local potential energy. The result is that, broadly speaking, the system finds a number of relatively satisfactory solutions to its problem of optimization, and the number of solutions grows extremely rapidly with the size of the system. The existence of multiple states of the system, associated with a spectrum of excitations of different energies, is what then gives rise to some characteristic time dependent properties.

Here I present a spin model that is related to the Ising model for spin glasses, but is simpler and exactly soluble in the thermodynamic limit. This is the spherical-spin model introduced long ago as a soluble version of the Ising ferromagnet and subsequently used[1] to describe a particularly simple form of spin glass. An extension of a modified version of this model to study dynamics is now presented and shown to lead in a natural way to the nonexponential relaxation behavior observed in many contexts, including the spin glasses and polymers.

Biologically Inspired Physics, Edited by L. Peliti
Plenum Press, New York, 1991

2. THE MODEL

As in the Ising case, the Hamiltonian for the spherical model spin glass is written

$$H = -\frac{1}{2} \sum J_{ij} S_i S_j \qquad (1)$$

where the bonds J_{ij} linking two sites $\{i,j\}$ are randomly chosen from a gaussian distribution and, as in the Ising model considered by Sherrington and Kirkpatrick[2] are of infinite range (i.e. all pairs of sites interact). Unlike the Ising model, the spins are allowed to be continuous-valued, with the average value per site being equal to unity. This is the spherical constraint

$$\frac{1}{N} \sum_i <S_i^2> = 1 \qquad (2)$$

where the angular brackets denote thermal averaging (N being the number of sites). The presence of the brackets significantly affects the fluctuation properties of the model, and makes it very different from the version originally studied in Ref.1, in which the spin length normalization is rigorously enforced. The relaxed form of the constraint, as expressed in Eq.2, allows for all of the interesting fluctuation properties that lead to the dynamical behavior discussed in this article. In the limit that N is taken to infinity, the system undergoes a phase transition at $T_c=J$ (the variance of the gaussian distribution for J_{ij}). The low temperature phase is characterized by an order parameter, q, which is the spin amplitude in the highest eigenstate of the matrix of J_{ij}

$$q = \frac{1}{N} <S_{\lambda max}^2> = \frac{T_c - T}{T_c} \qquad (3)$$

As a consequence of the mean spherical constraint, Eq.2, and the known properties of the spectrum of the random J_{ij} matrix, the quantity q undergoes large fluctuations in the low temperature phase[3]. Correspondingly, also the free energy undergoes large fluctuations. This property is now carried over in the dynamical formulation of the spherical model for the spin glass. The dynamical evolution is assumed to be given by the simplest possible relaxational form that gives the correct long time stationary properties. Assuming a microscopic spin flip time of $\tau_0 = \Gamma_0^{-1}$ in the Langevin spin equation of motion[4]

$$dS_i(t)/dt = -\Gamma_0 \delta H_{eff}/\delta S_i + \eta_i(t) \qquad (4)$$

where the effective Hamiltonian includes a term with a "chemical potential" z put in to ensure that Eq.2 is satisfied, $H_{eff} = -1/2 \Sigma(2z - \beta J_{ij})S_i S_j$, where β denotes the inverse temperature in units of k_B. It is now simplest to transform to the basis that diagonalizes the J-matrix, and rewrite Eq.4 in terms of the $\{S_\lambda\}$ variables. The correlation functions in the new variables (which are obtained from the site variables by an orthogonal transformation) are defined by taking averages over the noise in the usual manner. Thus the on-site correlation function averaged over sites is

$$\bar{C}(t) = 1/N \ \Sigma_i <S_i(t') \ S_i(t'+t)>$$
$$= 1/N \ \Sigma <S_\lambda(t') \ S_\lambda(t'+t)> = \ \Sigma \ C_\lambda(t) \qquad (5)$$

(Henceforth the thermal average is understood to mean averaging over η using the properties $<\eta(\tau)> = 0$ and $<\eta(\tau)\,\eta(\tau')> = 2\Gamma_o\,\delta(\tau - \tau')$.) Implicit in the relations above is the assumption that the system has evolved away from dependence on initial conditions and is in the stationary state described by the effective Hamiltonian. Thus, correlation functions only depend on time differences, and in particular, equal time correlations are equal to the corresponding thermodynamic expectation values. This allows to reexpress Eq.2 as

$$\Sigma\, C_\lambda(0) = 1/(2z - \beta\lambda) = 1 \tag{6}$$

using the equation of motion (4) to calculate the intermediate step of the equation. The last equation, along with the well-known formula for the distribution of eigenvalues of the random J-matrix[5]

$$\rho(\lambda) = (2\pi J^2)^{-1}\sqrt{4J^2 - \lambda^2}\;\theta(4J^2 - \lambda^2) \tag{7}$$

allows to solve for the temperature dependent chemical potential z.

3. CORRELATIONS AT HIGH TEMPERATURE

In the high temperature phase, one obtains that $z = (1 + \beta^2 J^2)/2$, and using the equation of motion, one obtains the average onsite time correlation function to decay exponentially with a characteristic time given by $\tau_o = \Gamma_o^{-1}$. The solution at high temperature is

$$\bar{C}(t) = \frac{1}{\sqrt{4\pi t}}\,(\frac{\Gamma_o J}{T})^{-3/2}\,e^{-\Gamma_o t} \tag{8}$$

Close to the transition, one may expand in $\theta = (T_c - T)/T_c$ and one obtains the critical correlation function:

$$\bar{C}(t) = \frac{1}{\sqrt{2\pi\theta^2}}(2\pi\Gamma_o t)^{-3/2}\,e^{-\theta^2\Gamma_o t}\;, \tag{9}$$

which indicates that the characteristic time of relaxation diverges as T_c is approached, with an exponent of two, just as in the Ising model[6], in which correlations decay with the same law given by Eq. 8.

4. DYNAMICS IN THE LOW TEMPERATURE PHASE

The dynamics in the paramagnetic phase is seen to be conventionally behaved in the spherical spin glass model. The low temperature phase, being the region of interest, is however slightly more complicated. In the condensed phase, the static model has been shown[3] to admit macroscopically large fluctuations in thermodynamic properties (a property it shares with the mathematically equivalent problem of Bose condensation in of an ideal Bose gas in dimensions greater than two). There is a spectrum of allowed values of the order parameter q defined earlier, and correspondingly a spectrum of possible free energies. Both distributions can be deduced from a knowledge of a basic probability density function called the Kac density and denoted by $\rho(x;T)$. This function gives the probability density for the system to have a spin normalization of $N^{-1}\Sigma S_i^2 = x$, given that on average the spin normalization is unity (as expressed by Eqs.2 and 6).

In most statistical mechanical systems, the Kac density is extremely sharply peaked - essentially a delta function in the thermodynamic limit - at the value $x=1$ (to use our choice of normalization). This means that thermodynamic properties are well-defined and have the property of self-averaging.

In the case of the mean spherical model, however, the function $\pi(x;T)$ can be exactly evaluated, and everywhere below T_c it is found to be a broad function, with nonvanishing width in the thermodynamic limit. It is given by

$$\rho(x;T) = \frac{1}{\sqrt{2\pi q(x-x_o)}}\, e^{-(x-x_o)/2q} \tag{10}$$

with an upper cutoff at the value $x_o=\beta J$. (The parameter q is the average value of the order parameter and is simply given by $q=(T_c-T)/T_c$). Thus now in the dynamical evolution, we must take into account the existence of the large fluctuations embodied in Eq.10 above. The different states labelled by the value of x are separated by large (macroscopic) energy differences, and hence do not, we assume, communicate with each other at finite times. The spin correlations must therefore for all finite times of observation be averaged over the response in different states that are not in communication. This is done by evaluating the spin correlation in each state x, and only then taking the average with respect to the distribution ρ. Doing this (see ref.7 for details) results in the following time dependence for correlations in the mode corresponding to λ_{max}, the condensed eigenmode and the one with the macroscopic fluctuations:

$$C_{\lambda max}(t) = q\left[1 + \sqrt{\frac{t}{\tau_{av}}}\right] e^{-\sqrt{t/\tau_{av}}} \tag{11}$$

valid for finite times $t \ll N\tau_0$. A new time scale of relaxation appears here, $\tau_{av} = q/2\Gamma_0$. The spin correlations thus have stretched exponential time dependence at all finite times, arising in a natural fashion out of the distribution $\rho(x;t)$. There is a second piece in the total spin correlations arising from the noncritical modes which in fact have a very slow decay (at finite time scales). This part, obtained by summing over all of the noncritical modes gives

$$C_{rest}(t) = \frac{T}{J}\sqrt{\frac{T}{\pi\Gamma_o J}}\, t^{-1/2} \tag{12}$$

The total spin correlation is then $\bar{C}(t) = C_{\lambda max}(t) + C_{rest}(t)$. At macroscopically large times, of the order of $N\tau_0$, the dependence crosses over to simple exponential decay, as the system then averages its dynamical properties over all the possible states. This yields decay of spin correlations which decay with time as $e^{-\Gamma_o t/Nq}$. The relaxation time in the asymptotic regime is seen thus to be proportional to the system size.

5. DISCUSSION

This study of the simplest correlations shows how the large fluctuations and nonselfaveraging properties of a model as simple as the mean spherical spin glass model can lead in a natural way to the nonexponential behavior ubiquitous in random systems. In addition to the time dependences calculated, the model provides for several different time scales of relaxation. The critical correlations are as in the Ising model. The temperature dependence of the time scale τ_{av} in the condensed phase is of interest, as it predicts this time scale to increase, in proportion with the mean value of the order parameter, with decreasing temperature. The power of the exponent in the stretched exponential in this model on the other hand is always one-half, independent of temperature. This is in contrast with the result obtained for a random free energy model by de Dominicis et al.[8], where the exponent is temperature dependent. With respect to some other models that predict stretched exponentials we also point out a difference in interpretation of the time scales that lead to the stretched exponential form. In this calculation the time scales arise from the different intrinsic local harmonic oscillator frequencies in the different states. This is in contrast with models that assume time scales arising due to the existence of multiple barrier heights[9], where the time scales may follow a temperature law characteristic of activated processes. The situation may be considered in terms of the picture of protein dynamics[10] that is emerging from numerous experimental studies. The glassy behavior of proteins is linked with the existence of a very large number of conformational substates. It has been proposed [Frauenfelder in this Workshop] that the energy landscape may be hierarchical, with large valleys containing smaller ones. In a two-step version of this picture one may conceive of a short time dynamics where independent substates (a working definition of which must be provided) fluctuate independently. The calculation described earlier corresponds in this scenario to taking an average over the sub-valley fluctuations. The second step is the crossover to activated dynamics connecting different subvalleys in a single bigger valley (or state). The present calculation does not consider these activated processes. (The situation in proteins is not as simple as the two step process as there are activated processes occurring on many time scales.)

Work is currently in progress to extend these calculations to describe the relaxation processes in flexible polymers.

ACKNOWLEDGMENTS

I wish to acknowledge the hospitality of the Centre de Physique Théorique of Ecole Polytechnique where this work was performed.

REFERENCES

1. J.M.Kosterlitz, D.J.Thouless and R.C.Jones, Phys.Rev.Lett. **36**,1217 (1976).
2. D.Sherrington and S.Kirkpatrick, Phys.Rev.Lett. **35**,1972 (1975); S.Kirkpatrick and D.Sherrington, Phys.Rev.B **17**, 4384 (1978).
3. A.Jagannathan, J.Rudnick and S.Eva, to be published.
4. C.Itzykson, Quantum Field Theory and Critical Phenomena, Clarendon Press, Oxford (1989).
5. M.L.Mehta, Random Matrices (Academic Press, New York, 1967).
6. K.Binder and A.P.Young, Rev.Mod.Phy. **58**, 801(1986).
7. A.Jagannathan, preprint.
8. C.de Dominicis, H.Orland, F.Lainée, J. Physique **46**, L463 (1985).
9. R.G.Palmer, D.L.Stein, E.Abrahams and P.W.Anderson, Phys.Rev.Lett. **53**, 958 (1984).

10. I.E.T.Iben, D.Braunstein, W.Doster, H.Frauenfelder, M.K.Hong, J.B.Johnson, S.Luck, P.Ormos, A.Schulte, P.J.Steinbach, A.H.Xie, R.D.Young, Phys.Rev.Lett. **62**, 1916 (1989).

CRYSTALLOGRAPHY OF COMPOSITE FLOWERS:

MODE LOCKING AND DYNAMICAL MAPS

N. Rivier*, A.J. Koch and F. Rothen

Institut de Physique Expérimentale
Université de Lausanne
CH-1015 Dorigny-Lausanne, Switzerland
*Blackett Laboratory,Imperial College
London SW7 2BZ, UK

1. INTRODUCTION: PHYLLOTAXIS AS CRYSTALLOGRAPHY

Compositae (daisies, pinecones, asters, sunflowers) have a structure shown in Fig.1. Understanding this structure —one aspect of the field of phyllotaxis (leaf or floret arrangement) [1-6]— is a problem of crystallography: A surface is tiled with florets (the "atoms") of roughly the same size, the majority of which are hexagonal (daisy) or rhombus-shaped (sunflower). The

Fig. 1. The Daisy. All cells are hexagonal except o=pentagon, +=heptagon, dipole o+=dislocation. Grain boundaries are quasicrystalline. (From [6,9]).

unusual feature as far as crystallography is concerned, is that the pattern has cylindrical symmetry. Florets sprout from the central stem (strictly, from a circle, the meristem, surrounding the geometrical center), one after the other, the younger ones pushing out their older siblings. We are dealing therefore with close-packing of deformable florets in cylindrical symmetry.

In Fig.1, one also recognizes grains, regions entirely made of hexagons, limited by circular grain boundaries, which are quasicrystalline (aperiodic, inflatable sequences)[7] arrays of dislocations (dipoles pentagon(o)-hexagon(+))[8] and isolated hexagons.

The reticular planes in crystals are here spirals, called *parastichies*, immediately noticeable in Fig.1. Three spirals traverse a hexagonal floret (two, through a rhombus), thus 3 (resp. 2) families of parastichies characterize a grain. The number of parastichies of a given family in a grain (like the number of spokes in a bicycle wheel) is a finite integer B_m (Fig.1), and it is this number whose universality will concern us here. At a grain boundary, the family of flattest spirals is replaced by the steepest ones, thereby increasing the number of reticular "planes", while the other two families go through the boundary unchanged. This is indeed the role of edge dislocations in crystallography. The switch at a grain boundary to more reticular spirals is necessary to accommodate more florets of the same size on an longer circular perimeter, as the flower grows.

The remarkable, universal fact is that almost all *compositae* have Fibonacci numbers of spirals $B_m=1,1,2,3,5,8,13,21,34...$ (the remaining few percent follow the similar Lucas sequence, $B_m=1,3,4,7,11,18,29...$)[*]. (Outer grains have higher numbers than inner ones, running through the sequence in an automatic and self-similar fashion.[6,4,10] 95% of Norwegian pinecones have Fibonacci numbers[1]). Moreover, a given grain has three (resp. two) successive Fibonacci or Lucas numbers numbering its three (two) families of parastichies.

This universality occurs in a given flower, and can easily be checked in Fig.1.[**] But it also covers all *compositae*, everywhere.[***] The numerology is thus either the (unlikely) product of an infinitely accurate biochemical protractor (whose evolution should have left a fossil record), or the result of a structurally stable partition of space. We shall show that it can be explained as the latter, by using crystallography and physics (anisotropic deformation of florets due to packing

---

[*] Fibonacci and Lucas series are such that a number is the sum of the two preceding ones in the series (as in eq.(3) below). They differ by their initial numbers only.
[**] At a grain boundary, the B_{m-2} flattest spirals give way to B_{m+1} steepest, through a grain boundary consisting of B_{m-2} isolated hexagons, B_{m-1} dislocations, i.e. B_m objects or B_{m+1} florets (since each dislocation has two florets, one pentagon, one heptagon), sources of the new parastichies. All these numbers satisfy relation (3).
[***] By contrast, as many daisies have their $B_m=13$, say, parastichies left- as right-handed.

costs energy). The ground state of the florets, soft bodies repelling each other, must be structurally stable, ie. invariant under small fluctuations of its parameters.[6]

We have therefore a cylindrical cellular structure which is universal. This universality, already noticed by Leonardo, Kepler, Goethe and others, remains the major outstanding puzzle of phyllotaxis.[2,4,11,19]

2. CONSTRUCTION OF THE DAISY

A crystallographic structure is *labelled* and described by a *local reference frame*. In conventional, translationally invariant systems, the frame is the fundamental domain or unit cell, and each atom is labelled by the position of its unit cell (ie. by D integers in D dimension), with, if necessary, an index giving its position within the cell. In cylindrical symmetry, we are limited to one integer $s=1,2,...$, labelling each cell from the center outwards (from younger to older florets). The center of floret s is given in cylindrical coordinates by

$$\theta(s) = 2\pi\lambda s \tag{1}$$

$r(s) = a\sqrt{s}$, or any monotonic increasing function of s.

Cells are then drawn by democratic partition of space between centers (Voronoi construction:[8] interfaces are perpendicular bisectors between neighbouring centers, and every point inside a cell is closer to its center than to any other's). This produces the Daisy of Fig.1. The structure is parametrized essentially by one quantity, the divergence angle $2\pi\lambda$, $\lambda \in [0,1]$, separating two successive florets.* (a is the average cell radius). The cells lie on a generative spiral $r=r(\theta/2\pi\lambda)$, not visible in Fig.1 because successive florets are not neighbours (the pitch of the generative spiral is much smaller than the cell radius). Only visible are the reticular spirals (parastichies), which are ordered as $\{\lambda s\}$, as we shall see below. ($\{x\}$ is the fractional part of x, $x=\{x\}$ mod.1. The ordering follows from the fact that eq.(1) with r=cst is the circle map). The function r(s) can be chosen as proportional to $\sqrt{s}$ in order to generate equal-sized florets, but any monotonically increasing function of s would retain the same ordering.

Let us label two neighbours by s and $s+B_m$. They have nearly the same azimuth $\theta(s+B_m) \approx \theta(s)$, with $\lambda \approx A_m/B_m$. If these were equalities (λ rational), their centers would lie on the

---

* Conventionally, one writes $r(s)=az^s$ or $\ln r(s)=2\pi\beta s$, with $\beta=\ln z/2\pi$, so that the two equations (1) appear similar. z(s) is called plastochronic ratio. This similarity is only illusory since z(s) must decrease with s to produce florets of equal sizes. If $r=a\sqrt{s}$, $\beta=\ln s/(4\pi s)$. Only in the case of a logarithmic spiral[12] is z constant. Then, there are no dislocations and the number of parastichies remains the same regardless of the size of the flower. The size of the florets grows outwards.[12] The fact that $\beta(s)$ decreases with s forces grain boundaries and transitions in the Farey tree, as we shall see.

same spoke, and the Voronoi cellular structure would resemble a spider web.[6] λ is therefore irrational, and representable uniquely as a continued fraction,

$$\lambda = 1/\{c_1+1/[c_2+1/(c_3+1/...)]\} \approx 1/\{c_1+1/[...+1/c_m]\} = A_m/B_m \qquad (2)$$

$(0<c_i, A_m=c_mA_{m-1}+A_{m-2}, B_m=c_mB_{m-1}+B_{m-2}, s\in Z^+$ are natural numbers). A_m/B_m is called the m'th convergent to the irrational λ. The neighbour label B_m is also the number of cells in a circular shell (one cell thick), or the number of reticular spirals introduced in §1. (A_m is the number of turns in the generative spiral necessary to fill the shell).

An hexagonal cell in a grain has six neighbours, and belongs to three families of reticular spirals. Its neighbours are labelled by three B_m's, denominators of successive convergents to λ. Thus, the *local reference frame* is the triangle of labelled neighbours $(s+B_m, s+B_{m-1}, s+B_{m-2})$ to a given cell s (Fig.2). One goes from s to $s+B_m$ either directly, or by a jog through the other two parastichies and the final label must be independent of the path and of the origin cell s (homogeneity), $s+B_m=(s+B_{m-2})+B_{m-1}$, thus,

$$B_m = B_{m-1} + B_{m-2} \qquad (3)$$

which is indeed the relation generating Fibonacci, Lucas, etc., sequences. The triangular relation (3) is crystallographically central and overriding.

Consider a given grain $\{B_m, B_{m-1}, B_{m-2}\}$, ie. a set of three successive rational convergents to λ, with A_m/B_m bounded by the other two. One can show* that **any** $\lambda \in I_m =]A_{m-1}/B_{m-1}, A_{m-2}/B_{m-2}[$

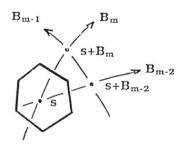

Fig. 2. Reference frame demonstrating relation (3).

---

* Setting r(s)=cst in eq.(1), one obtains the circle map $\theta(s)=2\pi\lambda s$, with ordering $\{\lambda s\}$, and the result follows either by inspection of the ordering produced at the three rational convergents of λ, or by using the fact (noted above) that neighbours to floret s in a circular shell of B_m florets are $s+B_{m-1}$ and $s+B_{m-2}$, for any λ which has A_{m-2}/B_{m-2}, A_{m-1}/B_{m-1} and A_m/B_m as successive rational approximants. Note that A_m/B_m is inside the interval I_m. The argument is also valid (mutatis mutandis) for intermediate convergents and non-noble λ.

generates the same ordering of (up to) B_m reticular spirals: $s, s+B_{m-2}, s+2B_{m-2},...,$ or $s, s+B_{m-1},...$ (mod. B_m), clockwise or anticlockwise, ie. **the same structure** $\{\lambda s\}$. This is topological mode-locking: The structure is locked on that given by the rational convergent A_m/B_m for any λ within the interval I_m. If λ lies within a different interval, the locked-in structure is different, with neighbours ordered by the relevant B's. Possible structures (A_m/B_m) follow an incomplete devil's staircase as a function of λ, exactly as in conventional mode-locking of two coupled oscillators,[13] where the abscissa of the staircase represents the ratio of their natural frequencies, and the ordinate, the observed ratio, is locked to rationals.

So far, so general. The result that ordering $\{\lambda i\}$ (i=1,..., n) of a finite number $n \leq B_m$ of florets is structurally stable for any λ within the interval I_m holds equally well for any model of phyllotaxis based on a divergence angle between successive florets and thus reducible to a circle map: For the crystallographic packing of florets described here, but also for models based on inhibitor/activator-induced growth on the meristem of the flower (a circle surrounding the stem where generation of new florets is supposed to happen)[14,15], for which it constitutes a topological framework.

3. BIFURCATION AT GRAIN BOUNDARY: FAREY TREE OF BIFURCATIONS

As one goes outwards, the number B_m of a given family of parastichies must increase if the size of the florets is to remain approximatively constant. One goes from one set of approximants $\{B_m, B_{m-1}, B_{m-2}\}$ to the next, from one grain to the next across a grain boundary, circular array of dislocations (which play indeed their crystallographic part of intercalating an additional half reticular plane). In other words, the ordinate of the Farey construction (Fig.3) $\beta = \ln z/2\pi$ must decrease as one goes outwards, and forces a bifurcation out of a given grain (triangle in Fig.3). The bifurcation abscissa is the approximant A_m/B_m. See also Fig.5 of [10], or Fig.3.1 of [5].

It might seem that eq.(3) is all that is needed to prove universality of phyllotaxis. This is not so, as there are two arrangements compatible with the triangular rule (3), corresponding to regular (true phyllotaxis) and singular transitions at the grain boundary. At a grain boundary, two families of reticular spirals go through, and the third is replaced by a larger B (better convergent), through dislocations which do indeed add material. There are only two possibilities, regular (RT) and singular (ST) transitions,[6,10]

$$\{B_m, B_{m-1}, \mathbf{B_{m-2}}\} \quad \rightarrow \quad \{\mathbf{B_{m+1}}, B_m, B_{m-1}\} \qquad \text{(RT)} \qquad (4)$$
$$\{B_m{}^{(p)}, \mathbf{B_m{}^{(p-1)}}, B_{m-1}\} \rightarrow \{\mathbf{B_m{}^{(p+1)}}, B_m{}^{(p)}, B_{m-1}\} \qquad \text{(ST)}$$

(discontinued families indicated in bold), consistent with the homogeneity relation (3). We have therefore an apparent bifurcation as $\beta=\ln z/2\pi$ decreases. Only the RT alternative gives the observed Fibonacci or Lucas phyllotaxis. λ is then *noble* ($\{c_i\}=1$, $i\geq i_o$). Noblest is the golden mean $1/\tau=(1-\sqrt{5})/2$ ($i_o=1$) generating the Fibonacci sequence. Singular transitions never occur in nature[6,10] (λ is then non-noble and has $c_i\neq 1$, but homogeneity relation (3) is still preserved by introducing intermediate convergents[11] $B_m^{(p)}\equiv pB_{m-1}+B_{m-2}=B_m^{(p-1)}+B_{m-1}$). So, universality of phyllotaxis is due to the fact that singular transitions never occur. Why?

The bifurcation scheme can be set on a Farey tree.* At a grain boundary, it is necessary for the structure to lock on rationals with larger denominators $B_m\rightarrow B_{m+1}$ (4) (Recall that B_m is the number of cells filling a circular shell. This number must increase as one goes outwards, in order to keep the cell size constant). In ordinary mode-locking, the coupled oscillators achieve a finer tuning if instead of locking on p_1/q_1 (one oscillator making p_1 oscillations in q_1 periods of the other, before the two are back in phase) or on p_2/q_2, they lock on the new rational frequency

$$(p_1+p_2)/(q_1+q_2) \equiv p_1/q_1 \oplus p_2/q_2 \qquad (5)$$

(Farey sum), exactly like successive rational convergents to λ. Note that the denominators $q_1+q_2=q_1\oplus q_2$ satisfy triangular rule (3). Eq.(5) generates and orders all the rationals $\in [0,1]$ uniquely as in Fig.3, the Farey construction.[2]

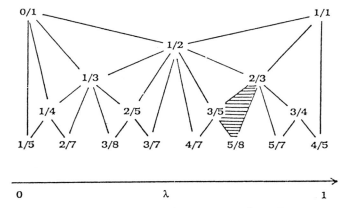

Fig. 3. Farey construction of rationals. A tri-angle defines a (structurally locked-in) grain. In phyllotaxis, the abscissa is the divergence λ, the ordinate, the logarithmic plastochronic ratio β.

---

* The tree already appears in [5,10] (Figs.3.1 and 5, resp.), but the remark that it was isomorphic to Farey's is due to NR. It was also made independently by Levitov.[18]

Figure 3 can be seen as a tiling by triangles, each triangle representing a grain $\{B_m, B_{m-1}, B_{m-2}\}$. A grain boundary is a bifurcation in the Farey tree. The triangle is replaced by either one of the two Farey offsprings, coinciding with the regular or singular transition of eq.(4) (Fig.4).

4. UNIVERSALITY AND STRUCTURAL STABILITY OF PHYLLOTAXIS

Universality and eq.(1) seem to imply that all *compositae* own the same, infinitely accurate protractor: Successive florets sprout as if separated by the same angle $2\pi\lambda$ exactly, with $\lambda=1/\tau$ (Fibonacci) or $[1+\tau]/[2+\tau]$ (Lucas), at all times and places. Such accuracy is unbelievable, especially since λ and β must fluctuate. The correct explanation is that sprouting and fluctuations of the structure are described by a dynamical map; universality and stability of the structure are then consequences of universality and stability of dynamical maps.

To prove universality, we must understand why the regular transition option is always preferred at a bifurcation (giving rise to Fibonacci numerology), except possibly for the first bifurcation in ~5% of cases (Lucas numerology). Singular transition never occurs in nature, despite satisfying triangular relation (3). This is again a crystallographic problem, involving the strain energy required to pack soft florets with a divergence angle λ within the interval $(I_m=]A_{m-1}/B_{m-1}, A_{m-2}/B_{m-2}[)$ yielding the required, mode-locked structure.

Structure-invariant fluctuations of λ at level i in the Farey tree form a group, and can be represented generically by a dynamical map c_i acting on λ (sending any interval I_m into itself, with A_m/B_m as fixed point and the interval boundaries A_{m-1}/B_{m-1} and A_{m-2}/B_{m-2} mapped into each other). This map is a homographic function (or a 2x2 matrix) with integer coefficients. Homographic functions with *rational* fixed points have

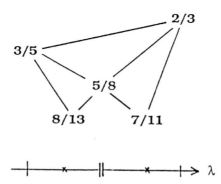

Fig. 4. Bifurcation at a grain boundary. Here, the regular transition is on the left (8/13), the singular transition on the right (7/11).

remarkable properties (their fixed points have marginal stability and they are involutions (if det $c_i = -1$)). Such a map c_i
can be constructed for all grains, ie. for all triangles in the
Farey construction. Similarly, one can construct homographic
functions for the bifurcation, c'_{i+1} and c''_{i+1}, fixing the
smaller intervals $I'_{m+1} =]A_{m-1}/B_{m-1}, A_m/B_m[$ (through RT, say) and
$I''_{m+1} =]A_m/B_m, A_{m-2}/B_{m-2}[$ (through ST) respectively. Finally,
since c_i sends I'_{m+1} onto I''_{m+1} and vice-versa, the homographic
map $c'_{i+1} \cdot c_i$ shifts any λ from the "wrong" interval I''_{m+1} into
I'_{m+1}, where it undergoes a regular transition (funnel map).

Each alternative at a bifurcation costs some energy, a
measure of the elastic strain needed to confine the structure
crystallographically, observable in the anisotropy (elongation)
of florets, especially near the grain boundaries.* The energy
of the structure is a functional of the *fluctuation map* $c_i - 1$
(a 2x2 matrix representing generically structure-conserving deviations from an exact value of λ throughout the grain).** It
must be a non-negative quantity (an energy cost), which vanishes only in the absence of any fluctuation. The simplest functional satisfying both requirements is

$$E_i = \mathrm{Tr}\{(c_i - 1)^+ (c_i - 1)\}. \tag{6}$$

With c'_{i+1} or c''_{i+1} instead of c_i, eq. (6) gives energies for
both alternatives at a bifurcation E_{i+1}^{RT} and E_{i+1}^{ST}. Moreover,
since the funnel operation (forcing a divergence on the ST side
I''_{m+1} of the bifurcation to undergo a RT) is a shift of divergence (effected by c_i) followed by a regular transition c'_{i+1},
its energy is the sum of the energies associated with each successive operation, $E_{i+1}^F = E_i + E_{i+1}^{RT}$.

One finds readily that $E^{RT} < E^F < E^{ST}$, so that funnel is preferred to singular transition at a bifurcation, as indeed happens in a growing plant. This completes the crystallographic
explanation of the universality and structural stability of
phyllotaxis of *compositae*.*

Each (forced) bifurcation in the Farey construction (Fig.3)
narrows the range of divergences: Either λ is on the right side
of the bifurcation point A_m/B_m and the structure undergoes a regular transition, or it is on the wrong side, and shift of λ to
the right side followed by a regular transition is energetically more favorable than a singular transition. Eventually, the

---

* Compare the shape of the florets in true and false phyllotaxis, generated by RT or ST, with their respective energies E^{RT} and E^{ST}. Notice the
anisotropy of the florets in the case of false phyllotaxis.[4,6]
** Exactly as a Ginzburg-Landau free energy is a functional of the gradient
$\nabla \Phi$ of some spatially non-uniform order parameter Φ.
* "...not the least curious feature of the case is the limited, even the
small number of possible arrangements which we observe and recognise."
([1], p.912).

range of possible divergences λ is *funnelled* by successive forced bifurcations to the golden angle $1/\tau$ (or to Lucas's).

5. CONCLUSIONS

The remarkable universality and stability of the structure of composite flowers, which have always successive Fibonacci or Lucas numbers of visible spirals, has been explained by crystallographic means, but a crystallography of soft, deformable florets, in cylindrical symmetry. Soft materials are known to exist (colloidal crystals, bubble rafts, magnetic bubbles, Bénard convection cells[6,9,16] or flux lines in superconductors[17] - the structure in the latter two cases has been analyzed as a phyllotaxis). The ordering of florets is summarized as a Farey tree of successive bifurcations. A given structure is the unique representative of a finite range of the growth parameters (structural mode-locking). Structural bifurcations are imposed by growth. Universality of phyllotaxis reduces to the fact that one alternative is selected at each bifurcation. This is explained by introducing an energy cost (elastic energy necessary to deform a floret to fit in the structure) for each branch at a bifurcation. Thus, diverse and fluctuating growth conditions yield a unique structure, or to its closest numerical relative.

Phyllotaxis is a very simple example of a structural archetype produced by a sequential process. However, this sequential "evolution" is automatic and physical - biologically-inspired physics indeed; it is not a Darwinian evolution, and no fossil record of false (non-noble) phyllotaxis should be found.

We acknowledge helpful suggestions by J. Guerreiro, M. Jorand, G. Bernasconi, M. Kunz and H. Meinhardt. N.R. would like to thank M. Feigelman for showing him Levitov's preprint[17] after his talk. After completion of this work, we received a second preprint by Levitov,[18] discussing the same question of universality. N.R's work in Lausanne was supported by the Herbette Foundation.

REFERENCES

[1] d'A. W. Thompson, "On Growth and Form," Cambridge Univ. Press (1942).

[2] H. S. M. Coxeter, "Introduction to Geometry," Wiley, New York (1961).

[3] R. V. Jean, "Phytomathématique," Presses Univ. du Québec, Montréal (1978); "Mathematical Approach to Pattern and Form in Plant Growth," Wiley, NY (1984); Nomothetical modeling of spiral symmetry in biology, Symmetry 1:81 (1990).

[4] N. Rivier, Crystallography of spiral lattices, Mod. Phys. Lett. B 2:953 (1988).

[5] A. J. Koch, "Des Cristaux Colloidaux à la Phyllotaxie, deux Exemples de Réseaux Cristallins en Géométrie Cylindrique," Thèse, Université de Lausanne (1989).

[6] N. Rivier, R. Occelli, J. Pantaloni and A. Lissowski, Structure of Bénard convection cells, phyllotaxis and crystallography in cylindrical symmetry, J. de Physique 45:49 (1984).

7 N. Rivier, A botanical quasicrystal, <u>J. de Physique (Coll.)</u> **47**:C3-299 (1986).

8 D. Weaire and N. Rivier, Soap, cells and statistics - Random patterns in two dimensions, <u>Contemp. Phys.</u> **25**:59 (1984).

9 R. Occelli, "Transitions Ordre-Désordre dans les Structures Convectives Bidimensionelles," Thèse, Université de Provence, Marseille (1985).

10 F. Rothen and A. J. Koch, Phyllotaxis, or the properties of spiral lattices. II. Packing of circles along logarithmic spirals, <u>J. de Physique</u> **50**:1603 (1989).

11 H. S. M. Coxeter, The role of intermediate convergents in Tait's explanation for phyllotaxis, <u>J. of Algebra</u> **20**:167 (1972).

12 F. Rothen and A. J. Koch, Phyllotaxis, or the properties of spiral lattices. I. Shape invariance under compression, <u>J. de Physique</u> **50**:633 (1989).

13 A.B. Pippard, "The Physics of Vibration," vol.II, pp. 391-413 .(Cambridge Univ. Press, 1978, 1989); D. Baums, W. Elsässer and E.O. Göbel, Farey tree and devil's staircase of a modulated external-cavity semiconductor laser, <u>Phys. Rev. Letters</u> **63**:155 (1989).

14 C. Mazrec and J. Kappraff, Properties of maximal spacing on a circle related to phyllotaxis and to the golden mean, <u>J. Theor. Biol.</u> **103**:201 (1983).

15 J. Guerreiro, A.J. Koch and F. Rothen, in preparation.

16 N. Rivier, The structure and dynamics of patterns of Bénard convection cells, IUTAM Symp. on Fluid Mechanics of Stirring and Mixing, UCSD (1990), to appear.

17 L. S. Levitov, Phyllotaxis of flux lattices in layered superconductors, Landau Institute preprint (1990).

18 L. S. Levitov, Hamiltonian approach to phyllotaxis, Landau Inst. preprint (1990).

PARTICIPANTS

Amalia Anagnostopoulou-Konsta
Dept. of Physics
Nat.Tech.Univ.Athens
15773 Zografou (Greece)

David Andelman
Dept. of Physics and Astronomy
Tel Aviv University
Ramat-Aviv (Israel)
ANDELMAN@TAUNIVM.TAU.AC.IL

Robert Austin
Dept. of Physics
Princeton University
Jadwin Hall
Princeton, NJ 08544-0708 (USA)
RHA@PUCC

Richard Bagley
Center for Nonlinear Studies
Los Alamos Nation.Lab.
B258
Los Alamos, NM 87545 (USA)

Artur Baumgärtner
Inst.f. Festkorperforschung
Forschungszentrun
5170 Jülich (RFA)
IFF07@DJUKFA11

David Bensimon
Lab.Phys.Statist.
Ecole Normale Sup.
24, rue Lhomond
75231 Paris Cedex 05 (France)
DAVID@FRULM62

Antonio Bianconi
Dpt. Medicina Sperim.
Univ. dell'Aquila
Via San Sisto 20
67100 L'Aquila (Italy)
BIANCONI@IRMLNF

Christof Biebricher
Biophys. Chemie
Max Planck Inst.
Postfach 2841
400 Göttingen (RFA)
CBIEBRI@DGOGWDG1

Meyer Bloom
Dept. of Physics
Univ. of British Columbia
6224 Agriculture Road
V6T2A6 Vancouver BC (Canada)

Antonio Borsellino
SISSA
Strada Costiera 10
34100 Trieste (Italy)

Søren Brunak
Dept. Str.Prop.Mat.
Tech.Univ. Denmark
Bldg 307
2800 Lyngby (Denmark)
BRUNAK@NBUVAX.NBI.DK

M.-France Carlier
Lab. d'Enzymologie
CNRS
91198 Gif-sur-Yvette Cedex (France)

Antonio Cattaneo
Inst.Neurobiol. CNR
Via C.Marx 15
00137 Rome (Italy)

Shirley S.Chan
Chemistry Dept.
Rutgers Univ.
New Brunswick, NJ 08903 (USA)
75231 Paris Cedex 05 (France)

Didier Chatenay
Section Phys.Chim.
Inst.Curie
11 rue P.M. Curie
75005 Paris (France)
FISMOL@FRCURI61

Germinal Cocho-Gil
Inst. de Física
Univ.Nac.Auton.Mexico
Ap.Post.20-364
01000 Mexico DF (Mexico)
COCHO@UNAMVM1

Paolo Del Giudice
Lab.Fisica - Ist.Sup.Sanità
Viale Regina Elena 299
00161 Rome (Italy)
VAXSAN::GIUDICE

Winfried Denk
IBM Research Lab.
Zürich (Suisse)
DEN@ZURLVM1

Tom Duke
Laboratoire de Physico-Chimie Théorique
ESPCI
10, rue Vauquelin
75231 Paris Cedex 05 (France)

Mikhail Feigelman
Theoretical Physics
Landau Institute
Kosygina 2
Moscow (URSS)

M.Frank-Kamenetskii
Inst.Medic.Genetics
URSS Academy of Sciences
123182 Moscow (URSS)
CDPLIMG%LABRIA@STANFORD

Hans Frauenfelder
Physics Dept.
Univ. of Illinois
Urbana, IL 61801 (USA)

Francesco Gaeta
Int.Inst.Genetics Biophys.
CNR
Via G.Marconi 10
80125 Naples (Italy)

Françoise Gaill
Cent.Biologie Cellulaire
CNRS
67, rue M.Gunsbourg
94205 Ivry Cedex (France)

Andrea Giansanti
Dpt. di Fisica
Università "La Sapienza"
P.le A.Moro, 2
00185 Rome (Italy)
GIANSANTI@ROMA1.INFN.IT

Gerhard Gompper
Sektion Physik
Univ.München
Theresienstr 37
8000 München 2 (RFA)
UH3D101@DM0LRZ01

Hans Gruler
Abteilung Biophysik
Univ.Ulm
Oberer Eselsberg P.4060
7900 Ulm (Donau) (RFA)
GRULER@DULRULL51

Sol Gruner
Physics Dept.
Princeton Univ.
PO Box 70 8
Princeton, NJ 08544 (USA)
GRUNER@PUCC

Wolfgang Hebel
Directorate-General for Science
Research and Development
Commission of the European Communities
Brussels (Belgium)

Benno Hess
Ernahungsphysiologie
Max Planck Institute
Rheilanddaumm 20
4600 Dortmund 1 (RFA)

Anu Jagannathan
Dept. of Physics
UCLA
Los Angeles, CA 90024 (USA)
INB5JAG@UCLAVMXA

Fritz Jähnig
Biologie
Max Planck Institute
Corrensstrasse 38
7400 Tübingen 2 (RFA)
JAEHNIG@DTUMPI51

Pierre Joliot
Inst.Biol.Phys.Chim.
13, rue P. et M. Curie
75005 Paris (France)

Josef Käs
Physics Dept. (E22)
Tech.Univ.München
James Franck Str.
8046 Garching (RFA)
E22@DGATUM5P

Stephen Langer
James Franck Inst.
Univ.Chicago
5640 S.Ellis Avenue
Chicago, IL 60637 (USA)
LANGER@CONTROL.UCHICAGO.ED

Stanislas Leibler
Serv.Phys.Théor.
CEN Saclay
91191 Gif-sur-Yvette Cedex (France)
LEIBLER@FRSAC11

Anthony Maggs
Laboratoire de Physico-Chimie Théorique
ESPCI
10, rue Vauquelin
75231 Paris Cedex 05 (France)
MAGGS@FRSAC11

Mirella Matzeu
Istituto Superiore di Sanità
Viale Regina Elena, 299
00161 Roma (Italy)

Hans Meinhardt
Entwicklungsbiologie
Max Planck Inst.
Spemannstr. 35
7400 Tübingen (RFA)

Marc Mezard
Labor.Phys.Statistique
Ecole Normale Sup.
24, rue Lhomond
75231 Paris Cedex 05 (France)
MEZARD@FRULM62

K.Mortensen
Physics Dept.
Risø National Lab.
4000 Roskilde (Denmark)
MORTENSEN@RISOE.DK

Jean Pierre Nadal
Labor.Phys.Statistique
Ecole Normale Sup.
24, rue Lhomond
75231 Paris Cedex 05 (France)
NADAL@FRULM62

George Oster
Dept. of Molecular and Cellular Biology
Univ. of California
Berkeley, CA 94720 (USA)
GOSTER@CAVEBEAR.BERKELEY.EDU

Giorgio Parisi
Dipartimento di Fisica
Univ. Roma 'Tor Vergata'
Via del Fontanile di Carcaricola
00173 Roma (Italy)
PARISI@VAXTOV.INFN.IT

Pierre Pelce
Labor.Rech.Combustion
Univ. Provence St Jérome
13397 Marseille (France)
PELCE@FRMRS11

Luca Peliti
Dipartimento Scienze Fisiche
Univ. Napoli
Mostra d'Oltremare - Pad.19
80125 Naples (Italy)
PELITI@ROMA1.INFN.IT

Alain Pocheau
Labor.Rech.Combustion
Univ. Provence St Jérome
13397 Marseille (France)
PELCE@FRMRS11

Nicolas Rivier
Blackett Laboratory
Imperial College
London, SW7 2BZ (GB)
NICK@SST.PH.IL.AC.UK

Tim Ryan
Dept.Molecul.Cellul.Physiol.
Stanford Univ.
Stanford, CA 94305 (USA)
MA.TIM@STANFORD

Erich Sackmann
Dept.Physics
Techn.Univ.Munich
8046 Garching-Bei-München (RFA)

Enrico Scalas
Dipartimento di Fisica
Univ. di Genova
Via Dodecaneso
16146 Genova (Italy)
SCALAS@GENOVA.INFN.IT

Lee Segel
Dept.Applied Mathem.
Weizmann Institute
Rehovoth (Israel)
MASEGEL@WEIZMANN

David Sherrington
Dept. of Physics
1 Keble Road
Oxford OX1 3NP (UK)
SHERRNGTNV1.PH.OXFORD.AC.UK

Adam Simon
J.Frank-E. Fermi Inst.
5640 S.Ellis Avenue
Chicago, IL 60637 (USA)
SIMON@CONTROL.UCHICAGO.EDU

Eugenio Tabet
Laboratorio di Fisica
Istituto Superiore di Sanità
Viale Regina Elena, 299
00161 Rome (Italy)

Gerard Toulouse
Labor.Phys.Statistique
Ecole Normale Sup.
24, rue Lhomond
75231 Paris Cedex 05 (France)
TOULOUSE@FRULM11

G.Venturi
Scientific Affairs Division
NATO Headquarters
1110 Brussels (Belgium)

Mikhail Volkenstein
Inst.Molecular Biology
USSR Academy of Sciences
Moscow (URSS)

Watt W.Webb
Applied Physics
Cornell University
Clark Hall
Ithaca, NY 14853 (USA)

Peter Weichman
Caltech 114-36
Pasadena, CA 91125
PBW@CALTECH.BITNET

Gérard Weisbuch
Labor.Phys.Statistique
Ecole Normale Sup.
24, rue Lhomond
75231 Paris Cedex 05 (France)
WEISBUCH@FRULM62

Bostjan Zeks
Inst. of Biophysics
Medical Faculty
Lipiceva 2
61105 Ljubljana (Yugoslavia)

Peter G.Wolynes
Dept. of Chemistry
Univ. of Illinois
505 South Mathews Ave.
Urbana, IL 61801 (USA)
WOLYNES@UIUCSCS

AUTHORS

Amalia Anagnostopoulou-Konsta
Dept. of Physics
Nat.Tech.Univ.Athens
15773 Zografou (Greece)

L.Apekis
Dept. of Physics
Nat.Tech.Univ.Athens
15773 Zografou (Greece)

E.Ascolese
Int.Inst.Genetics Biophys.
CNR
Via G.Marconi 10
80125 Naples (Italy)

David Bensimon
Lab.Phys.Statist.
Ecole Normale Sup.
24, rue Lhomond
75231 Paris Cedex 05 (France)
DAVID@FRULM62

Karin Berndl
Sektion Physik
Universität München
Theresienstr. 37
8000 München 2 (RFA)

Antonio Bianconi
Dpt. Medicina Sperim.
Univ. dell'Aquila
Via San Sisto 20
67100 L'Aquila (Italy)
BIANCONI@IRMLNF

Christof Biebricher
Biophys. Chemie
Max Planck Inst.
Postfach 2841
3400 Göttingen (RFA)
CBIEBRI@DGOGWDG1

Meyer Bloom
Dept. of Physics
Univ. of British Columbia
6224 Agriculture Road
V6T2A6 Vancouver BC (Canada)

Giorgio Careri
Dpt. di Fisica
Università "La Sapienza"
P.le A.Moro, 2
00185 Rome (Italy)

C.Christodoulides
Dept. of Physics
Nat.Tech.Univ.Athens
15773 Zografou (Greece)

Kelvin Chu
Physics Dept.
Univ. of Illinois
1110 W.Green St.
Urbana, IL 61801 (USA)

Germinal Cocho-Gil
Inst. de Física
Univ.Nac.Auton.Mexico
Ap.Post.20-364
01000 Mexico DF (Mexico)
COCHO@UNAMVM1

Nicolas J.Cordova
Dept. of Applied Mathematics
Weizmann Institute of Science
Rehovot (Israel)

A.Congiu Castellano
Biostructure Research unit of GNCB
CNR - Dipartimento di Fisica
Univ. di Roma 'La Sapienza'
P.le A.Moro, 2
00185 Roma (Italy)

D.Daoukaki
Dept. of Physics
Nat.Tech.Univ.Athens
15773 Zografou (Greece)

Paolo Del Giudice
Lab.Fisica - Ist.Sup.Sanità
Viale Regina Elena 299
00161 Rome (Italy)
VAXSAN::GIUDICE

S.Della Longa
Dpt. Medicina Sperim.
Univ. dell'Aquila
Via San Sisto 20
67100 L'Aquila (Italy)

Klaus Dornmair
Biologie
Max Planck Institute
Corrensstrasse 38
7400 Tübingen 2 (RFA)

Tom Duke
Laboratoire de Physico-Chimie Théorique
ESPCI
10, rue Vauquelin
75231 Paris Cedex 05 (France)

M.Eigen
Biophys. Chemie
Max Planck Inst.
Postfach 2841
3400 Göttingen (RFA)

Evan Evans
Dept. of Physics
Univ. of British Columbia
6224 Agriculture Road
V6T2A6 Vancouver BC (Canada)

M.Frank-Kamenetskii
Inst.Medic.Genetics
URSS Academy of Sciences
123182 Moscow (URSS)
CDPLIMG%LABRIA@STANFORD

Hans Frauenfelder
Physics Dept.
Univ. of Illinois
1110 W.Green St.
Urbana, IL 61801 (USA)

Francesco Gaeta
Int.Inst.Genetics Biophys.
CNR
Via G.Marconi 10
80125 Naples (Italy)

Françoise Gaill
Cent.Biologie Cellulaire
CNRS
67, rue M.Gunsbourg
94205 Ivry Cedex (France)

William C.Gardiner, Jr.
Biophys. Chemie
Max Planck Inst.
Postfach 2841
3400 Göttingen (RFA)

Andrea Giansanti
Dpt. di Fisica
Università "La Sapienza"
P.le A.Moro, 2
00185 Rome (Italy)
GIANSANTI@ROMA1.INFN.IT

Gerhard Gompper
Sektion Physik
Univ.München
Theresienstr 37
8000 München 2 (RFA)
UH3D101@DM0LRZ01

Sol M.Gruner
Physics Dept.
Princeton Univ.
PO Box 70 8
Princeton, NJ 08544 (USA)
GRUNER@PUCC

Hans Gruler
Abteilung Biophysik
Univ.Ulm
Oberer Eselsberg P.4060
7900 Ulm (Donau) (RFA)
GRULER@DULRULL51

Anu Jagannathan
Dept. of Physics
UCLA
Los Angeles, CA 90024 (USA)
INB5JAG@UCLAVMXA

Fritz Jähnig
Biologie
Max Planck Institute
Corrensstrasse 38
7400 Tübingen 2 (RFA)
JAEHNIG@DTUMPI51

Pierre Joliot
Inst.Biol.Phys.Chim.
13, rue P. et M. Curie
75005 Paris (France)

Josef Käs
Physics Dept. (E22)
Tech.Univ.München
James Franck Str.
8046 Garching (RFA)
E22@DGATUM5P

A.J.Koch
Inst. de Phys. Expérimentale
Université de Lausanne
1015 Dorigny-Lausanne (Switzerland)

Stanislas Leibler
Serv.Phys.Théor.
CEN Saclay
91191 Gif-sur-Yvette Cedex (France)
LEIBLER@FRSAC11

Andrea C.Levi
SISSA
Strada Costiera, 11
34014 Trieste (Italy)

Reinhard Lipowsky
Sektion Physik
Universität München
Theresienstr. 37
8000 München 2 (RFA)

L.Medrano
Facultad de Ciencias
Univ.Nac.Auton.Mexico
Ap.Post.20-364
01000 Mexico DF (Mexico)

Hans Meinhardt
Entwicklungsbiologie
Max Planck Inst.
Spemannstr. 35
7400 Tübingen (RFA)

Emilio Merlo Pich
Istituto di Fisiologia Umana
Università
Modena (Italy)

P.Miramontes
Facultad de Ciencias
Univ.Nac.Auton.Mexico
Ap.Post.20-364
01000 Mexico DF (Mexico)

D.G.Mita
Int.Inst.Genetics Biophys.
CNR
Via G.Marconi 10
80125 Naples (Italy)

K.Mortensen
Physics Dept.
Risø National Lab.
4000 Roskilde (Denmark)
MORTENSEN@RISOE.DK

M.Mutz
Lab.Phys.Statist.
Ecole Normale Sup.
24, rue Lhomond
75231 Paris Cedex 05 (France)

George Oster
Dept. of Molecular and Cellular Biology
Univ. of California
Berkeley, CA 94720 (USA)
GOSTER@CAVEBEAR.BERKELEY.EDU

Giorgio Parisi
Dipartimento di Fisica
Univ. Roma 'Tor Vergata'
Via del Fontanile di Carcaricola
00173 Roma (Italy)
PARISI@VAXTOV.INFN.IT

Hanna Parnas
Dept. of Neurobiology and
Otto Loewi Center for Cellular and
Molecular Neurobiology
The Hebrew University
Jerusalem (Israel)

M.A.Pecorella
Int.Inst.Genetics Biophys.
CNR
Via G.Marconi 10
80125 Naples (Italy)

Luca Peliti
Dipartimento Scienze Fisiche
Univ. Napoli
Mostra d'Oltremare - Pad.19
80125 Naples (Italy)
PELITI@ROMA1.INFN.IT

Robert Philipp
Physics Dept.
Univ. of Illinois
1110 W.Green St.
Urbana, IL 61801 (USA)

P.Pissis
Dept. of Physics
Nat.Tech.Univ.Athens
15773 Zografou (Greece)

J.L.Rius
Inst. de Física
Univ.Nac.Auton.Mexico
Ap.Post.20-364
01000 Mexico DF (Mexico)

Nicolas Rivier
Blackett Laboratory
Imperial College
London, SW7 2BZ (GB)
NICK@SST.PH.IL.AC.UK

Giacomo Rizzolatti
Istituto di Fisiologia Umana
Università
Parma (Italy)

F.Rothen
Inst. de Phys. Expérimentale
Université de Lausanne
1015 Dorigny-Lausanne (Switzerland)

P.Russo
Int.Inst.Genetics Biophys.
CNR
Via G.Marconi 10
80125 Naples (Italy)

Erich Sackmann
Dept.Physics
Techn.Univ.Munich
8046 Garching-Bei-München (RFA)

Enrico Scalas
Dipartimento di Fisica
Univ. di Genova
Via Dodecaneso, 33
16146 Genova (Italy)
SCALAS@GENOVA.INFN.IT

Lee Segel
Dept.Applied Mathem.
Weizmann Institute
Rehovoth (Israel)
MASEGEL@WEIZMANN

Udo Seifert
Sektion Physik
Universität München
Theresienstr. 37
8000 München 2 (RFA)

Bostjan Zeks
Inst. of Biophysics
Medical Faculty
Lipiceva 2
61105 Ljubljana (Yugoslavia)

David Sherrington
Dept. of Physics
1 Keble Road
Oxford OX1 3NP (UK)
SHERRNGTNV1.PH.OXFORD.AC.UK

Sasa Svetina
Inst. of Biophysics
Medical Faculty
Lipiceva 2
61105 Ljubljana (Yugoslavia)

Ronald D.Vale
Dept. of Pharmacology
University of California
School of Medicine
San Francisco, CA 94143 (USA)

Mikhail Volkenstein
Inst.Molecular Biology
USSR Academy of Sciences
Moscow (URSS)

Gérard Weisbuch
Labor.Phys.Statistique
Ecole Normale Sup.
24, rue Lhomond
75231 Paris Cedex 05 (France)
WEISBUCH@FRULM62

Peter G.Wolynes
Dept. of Chemistry
Univ. of Illinois
505 South Mathews Ave.
Urbana, IL 61801 (USA)
WOLYNES@UIUCSCS

K.Y.M.Wong
Dept. of Physics
1 Keble Road
Oxford OX1 3NP (UK)

INDEX